BIOLOGICALLY ACTIVE SMALL MOLECULES

Modern Applications and Therapeutic Perspectives

BIOLOGICALLY ACTIVE SMALL MOLECULES

Modern Applications and Therapeutic Perspectives

Edited by

Debarshi Kar Mahapatra, PhD

Sanjay Kumar Bharti, PhD

AAP | APPLE ACADEMIC PRESS

First edition published 2023

Apple Academic Press Inc.
1265 Goldenrod Circle, NE,
Palm Bay, FL 32905 USA

760 Laurentian Drive, Unit 19,
Burlington, ON L7N 0A4, CANADA

CRC Press
6000 Broken Sound Parkway NW,
Suite 300, Boca Raton, FL 33487-2742 USA

4 Park Square, Milton Park,
Abingdon, Oxon, OX14 4RN UK

Library and Archives Canada Cataloguing in Publication

Title: Biologically active small molecules : modern applications and therapeutic perspectives / edited by Debarshi Kar Mahapatra, PhD, Sanjay Kumar Bharti, PhD.
Names: Mahapatra, Debarshi Kar, editor. | Bharti, Sanjay Kumar, editor.
Description: First edition. | Includes bibliographical references and index.
Identifiers: Canadiana (print) 20220275319 | Canadiana (ebook) 20220275335 | ISBN 9781774910689 (hardcover) | ISBN 9781774910696 (softcover) | ISBN 9781003283119 (ebook)
Subjects: LCSH: Biopharmaceutics. | LCSH: Biomolecules.
Classification: LCC RM301.4 .B56 2023 | DDC 615.7—dc23

Library of Congress Cataloging-in-Publication Data

CIP data on file with US Library of Congress

ISBN: 978-1-77491-068-9 (hbk)
ISBN: 978-1-77491-069-6 (pbk)
ISBN: 978-1-00328-311-9 (ebk)

About the Editors

Debarshi Kar Mahapatra, PhD
Assistant Professor, Department of Pharmaceutical Chemistry,
Dadasaheb Balpande College of Pharmacy,
Rashtrasant Tukadoji Maharaj Nagpur University, Nagpur, Maharashtra, India

Debarshi Kar Mahapatra is currently an assistant professor at the Department of Pharmaceutical Chemistry, Dadasaheb Balpande College of Pharmacy, Rashtrasant Tukadoji Maharaj Nagpur University, Nagpur, India. He taught medicinal and computational chemistry at both undergraduate and post-graduate levels and has mentored students in various research projects. His area of interest includes computer-assisted rational designing, synthesis of low molecular weight ligands against druggable targets, development of drug delivery systems, and optimization of unconventional formulations. He has published several research papers, review articles, imperative case studies in various reputed international journals and presented his original contributions at several international platforms, for which he received several awards by a number of scientific and professional bodies. He has contributed a number of edited books, textbooks, lab manuals, book chapters, and guide books on pharmaceutical chemistry, natural products, and pharmaceutics. Presently, he is serving as a reviewer and editorial board member for several journals of international repute. He is a member of a number of professional and international scientific societies.

Sanjay Kumar Bharti, PhD
Institute of Pharmaceutical Sciences, Guru Ghasidas Vishwavidyalaya
(A Central University), Bilaspur, India

Sanjay Kumar Bharti is working as assistant professor in the Institute of Pharmaceutical Sciences, Guru Ghasidas Vishwavidyalaya (A Central University), Bilaspur, India. He has been working in reputed organizations like Win-Medicare Pvt. Ltd, Meerut, India, as a chemist, and the National

Institute of Pharmaceutical Education and Research (NIPER), Hajipur, as a lecturer. He has published several research papers, books, book chapters, and review articles in various reputed journals. His research interest includes synthesis of Schiff's base, heterocyclic compounds, and metallopharmaceuticals for therapeutics. He is an active member of various scientific and pharmaceutical organizations like IPA, IPGA, ISCA, etc. He has completed his BPharm at IT-BHU, Varanasi, India; MPharm from RGPV, Bhopal; and PhD from IIT-BHU, Varanasi, India.

Contents

Contributors

Sevim Akyuz
Department of Physics, Faculty of Science and Letters, Istanbul Kultur University, Bakirkoy-Istanbul 34156, Turkey

Ahmad Ali
Department of Life Sciences, University of Mumbai, Vidyanagari, Santacruz (East), Mumbai 400098, Maharashtra, India

Vivek Asati
Department of Pharmaceutical Chemistry, NRI Institute of Pharmacy, Bhopal 462021, Madhya Pradesh, India

Virupaksha A. Bastikar
Amity Institute of Biotechnology, Amity University, Mumbai, India

Sanjay Kumar Bharti
Institute of Pharmaceutical Sciences, Guru Ghasidas Vishwavidyalaya (A Central University), Bilaspur 495009, Chhattisgarh, India

Harpal S. Buttar
Department of Pathology & Laboratory Medicine, Faculty of Medicine, University of Ottawa, Ontario, Canada

Sefa Celik
Department of Physics, Faculty of Science, Istanbul University, Vezneciler, Istanbul 34134, Turkey

Shital Chaskar
Department of Pharmaceutical Chemistry, MET's Institute of Pharmacy, Adgaon, Nashik, Maharashtra, India

Harshala Chaudhari
Department of Quality Assurance Techniques, MET's Institute of Pharmacy, Adgaon, Nashik, Maharashtra, India

Vandana Chaudhary
B. S. Anangpuria Institute of Pharmacy, Faridabad, Haryana, India

Santosh S. Chhajed
Department of Pharmaceutical Chemistry, MET's Institute of Pharmacy, Adgaon, Nashik, Maharashtra, India

Subrata Deb
Department of Pharmaceutical Sciences, College of Pharmacy, Larkin University, Miami, FL 33169, USA

Aliye Demet Demirag
Institute of Graduate Studies in Sciences, Istanbul University, Istanbul 34452, Turkey

Pramodkumar P. Gupta
School of Biotechnology & Bioinformatics, D Y Patil University, Plot no 50, Sector 15, CBD Belapur, Navi-Mumbai 400614, Maharashtra, India

Animeshchandra G. M. Haldar
Department of Applied Chemistry, Priydarshini Bhagwati College of Engineering, Nagpur, Maharashtra, India

Ginpreet Kaur
Department of Pharmacology, Shobhaben Pratapbhai Patel School of Pharmacy and Technology Management, SVKM'S NMIMS, Mumbai 400056, Maharashtra, India

T. N. V. Ganesh Kumar
Chebrolu Hanumaiah Institute of Pharmaceutical Sciences, Chandramoulipuram, Chowdavaram, Guntur 522019, Andhra Pradesh, India

Debarshi Kar Mahapatra
Department of Pharmaceutical Chemistry, Dadasaheb Balpande College of Pharmacy, Nagpur 440037, Maharashtra, India

Gyanshankar P. Mishra
Department of Pulmonary Medicine, Government Medical College, Nagpur, Maharashtra, India

Upendra Nagaich
Amity Institute of Pharmacy, Amity University, Noida 201301, Uttar Pradesh, India

Aysen E. Ozel
Department of Physics, Faculty of Science, Istanbul University, Vezneciler, Istanbul 34134, Turkey

Jaimin Patel
Department of Pharmacology, Shobhaben Pratapbhai Patel School of Pharmacy and Technology Management, SVKM'S NMIMS, Mumbai 400056, Maharashtra, India

Parag A. Pathade
Department of Pharmaceutical Chemistry, MET's Institute of Pharmacy, Adgaon, Nashik, Maharashtra, India

Vaibhav P. Patil
Department of Pharmaceutical Chemistry, MET's Institute of Pharmacy, Adgaon, Nashik, Maharashtra, India

Buchi Reddy Purra
Department of Chemistry, REVA University, Bangalore, Karnataka, India

Madhusudana Reddy
Department of Chemistry, REVA University, Bangalore, Karnataka, India

Erica Sequeira
Department of Pharmacology, Shobhaben Pratapbhai Patel School of Pharmacy and Technology Management, SVKM'S NMIMS, Mumbai 400056, Maharashtra, India

Aarti Sonar
Department of Pharmaceutical Chemistry, MET's Institute of Pharmacy, Adgaon, Nashik, Maharashtra, India

Sandeep S. Sonawane
Department of Pharmaceutical Chemistry, MET's Institute of Pharmacy, Adgaon, Nashik, Maharashtra, India

Jayakumar Surendradoss
The Centre for Drug Research and Development, Vancouver, Canada

Mobashshera Tariq
Department of Life Sciences, University of Mumbai, Vidyanagari, Santacruz (East), Mumbai 400098, Maharashtra, India

Alice Varghese
Apotex Inc., Toronto, Canada

Abbreviations

ABAP	2,2'-azo-bis(2-amidino-propane)hydrochloride
ACE	angiotensin converting enzyme
Ach	acetylcholine
AGE	advanced glycation end
AhR	aryl hydrocarbon receptor
AP	alkaline phosphatase
ASA	acetyl salicylic acid
AUC	area under curve
BAC	boronate affinity chromatography
BP	blood pressure
CAR	constitutive androstane receptor
Cmax	peak plasma level
COPD	chronic obstructive pulmonary disease
COX	cyclooxgenase
CRE	carbonyl reducing enzymes
CYP	cytochrome P450
DHEA	dihydro epiandrosterone
DMPD	*N*, *N*-dimethyl-p-phenylenediamine dihydrochloride
DMSO	dimethyl sulphoxide
DNA	deoxyribonucleic acid
DPPH	1,1-diphenyl-2-picryl hydrazyl
*d*Tc	*d*-tubocurarine
ECMDB	E. coli metabolome database
EGF	epidermal growth factor
FRAP	ferric reducing ability of plasma
FRET	fluorescence resonance energy transfer
FXR	farnesoid X receptor
GA	general anesthetics
GDT	global distance test
GDTTS	global distance test total score
GI	gastrointestinal
GSH	glutathione
GST	glutathione transferases
HMDB	human metabolome database

HO-2-dG	β-hydroxy-2-deoxyguanosine
HPETE	hydroperoxy eicosatetraenoic
HTRF	homogeneous time-resolved fluorescence
HTS	high-throughput screening
IBD	inflammatory bowel disease
Iκκ	IκBα kinase
IL-1	interleukin-1
LOX	lipoxygenase
LPS	lipopolysaccharide
LXR	liver X receptor
LT	leukotriene
MAC	minimum alveolar concentration
MAO	monoamine oxidase
MAOI	monoamine oxidase inhibitor
NAR	nucleic acid research
NAT	*N*-acetyltransferase
NBT	nitroblue tetrazolium
NCBI	national center for biotechnology information
NF-κB	nuclear factor kappa B
NLM	national library of medicine
NMR	nuclear magnetic resonance
NrF2	nuclear factor erythroid 2-related factor 2
NSAIA	non-steroidal anti-inflammatory agent
NSAIC	non-steroidal anti-inflammatory candidate
NSAID	non-steroidal anti-inflammatory drug
ORAC	oxygen radical absorbance capacity
PAF	platelet activating factor
PC	prostacyclin
PDB	protein data bank
PG	prostaglandin
P-gp	P-glycoprotein
PPAR	peroxisome proliferator-activated receptor
PXR	pregnane X receptor
RA	rheumatoid arthritis
RMSD	root mean square deviation
RNA	ribonucleic acid
RNS	reactive nitrogen species
ROS	reactive oxygen species
SAR	structure-activity relationship

SPA	scintillation proximity assays
SULT	sulfotransferase
TDI	time-dependent inhibition
TG	thymidine glycol
TNF	tumor necrosis factor
TPTZ	2,4,6-tri-(pyridyl)-s-triazine
TRAP	total radical-trapping
TX	thromboxane
UGT	uridine diphosphate-glucuronosyltransferase
UHTS	ultra high-throughput screening
YMDB	yeast metabolome database

Foreword

It is a moment of pleasure for me to write the foreword for this amazing book, *Biologically Active Small Molecules: Modern Applications and Therapeutic Perspectives,* edited by Debarshi Kar Mahapatra, PhD, and Sanjay Kumar Bharti, PhD. With the unprecedented progress in drug discovery techniques, the emergence of low molecular weight ligands remains a burning topic. In the modern day, when the world is moving toward "nano," the chemical entities are becoming smart. The structure-based drug design (SBDD)-based approach has revolutionized the screening of low molecular weight ligands in the rational development through the knowledge of biological targets.

There are 15 interdisciplinary and multidisciplinary chapters included in the book that highlight the latest developments, modern concepts, innovations in synthetic routes, natural products, biological targets, pharmacological advances, drug-design techniques, screening methods, metabolic transformations, etc., in a very concise, complete manner. The chapters are thoroughly edited and written in an easy-to-understand language, along with figures, diagrams, and illustrations for a better understanding.

I undoubtedly recommend this edited book to undergraduate students, postgraduate students and research scholars, faculty members at colleges, institutes, and universities, researchers and scientists of research laboratories, and industry research and development.

—Neelam Singla, PhD
Managing Director,
Accura Research Chemicals Private Ltd.
Hyderabad, Telangana, India

Ex-Principal and Ex-Professor,
Department of Pharmaceutical Chemistry,
Advanced Institute of Pharmacy,
Palwal, Haryana, India

Preface

Biologically Active Small Molecules: Modern Applications and Therapeutic Perspectives is a new book that concentrates only on a hot subject in pharmacology. With advances in drug discovery and design, creating therapeutically advantageous compounds based on the understanding of biomolecular targets (receptors, enzymes, channels, and others) is the most important field and the future of contemporary pharmacotherapeutics.

Low-mass compounds have an important role in a wide range of pharmacological actions. They have become universally popular due to their simple chemistry, fatigue-free separation techniques, versatile acceptance for computational studies, a large number of places for the substitution of active chemical moieties by well-established synthetic routes with less effort, better quality attributes, and ability to demonstrate a wide range of biological activities.

There was a period, many decades ago, when large molecules were thought to be superior to the low-mass components. Using morphine as an example, the bond disconnection method produced many notable centrally acting analgesics that are much more powerful and have fewer adverse effects than morphine. Natural goods and their semisynthetic derivatives, which often exhibit low-mass features, have received considerable attention in our contemporary age as a result of advances in rational design. This book is intended to provide the most up-to-date information and multidimensional advancements of certain recently discovered therapeutically active low molecular weight compounds, following the research route.

This edited book contains essential information for postgraduate and research students, college and university faculty, scientists at various levels, research laboratories, and industry personnel, and is written in an easy-to-understand language by eminent authors from reputable institutions around the world. The book is divided into five sections comprising 15 chapters that are up-to-date, concise, and comprehensive. The material includes categorization, structures, chemical synthesis, medicinal chemistry, pharmacology, biochemical pathways, mechanism of action, side effects, and adverse effects, among other things.

While editing the contents, the World Health Organization (WHO), the United States Food and Drug Administration (USFDA), and the United States Pharmacopeia (USP) used the most recent therapeutic regulations. The book provides a comprehensive overview of the subject by emphasizing advancements in medicine, chemical sciences, and pharmaceuticals from a multidisciplinary and interdisciplinary perspective. The flowcharts, figures, drawings, and diagrams will offer sufficient information while also piquing the readers' attention.

A multidimensional approach has been taken in the selection of the chapters. Chapter 1 focuses on the advances of bedaquiline, a low molecular weight ligand known to exhibit tremendous antitubercular activity. Chapter 2 highlights the role of volatile hydrocarbon gases used for operative surgical procedures. Chapter 3 stressed on the role of nuclear factor-kappa B (NF-κβ), a new molecular target involved in inflammation and their inhibition by very small molecules of natural and semisynthetic origin called "chalcone." Chapters 4 and 5 present in-depth details with classification, structures, biochemical pathways, pharmacology, mechanism of actions, side effects, adverse effects of low molecular weight adrenergic agonists, and antagonists. Chapter 6 corresponds to the role of small-sized skeletal muscle relaxants in neuromusculoskeletal pharmacotherapeutics. Chapter 7 represents the necessary concepts, developments, and latest innovations in the area of nonsteroidal anti-inflammatory agents.

Chapter 8 underlines the low molecular mass natural products of varied origin having noteworthy antihyperglycemic activity. Chapter 9 is a concise chapter briefing the methods and models for screening the antioxidant activity of small size ligands of both natural and synthetic origin. Chapter 10 is a very interesting chapter highlighting metabolism and enzymatic transformation of the ligands after administration. Chapter 11 is a generalized content on the possible interactions of a biologically privileged small compound with another drug, food, herb, and other components, which will be an appealing matter for the prescribers. Chapter 12 provides essential information of the numerous online databases which are of vital interest for the researchers in selecting and duly screening the low-mass ligands and the possible biological targets. Chapter 13 throws light on the high-throughput screening technique, an emerging technique for rapid virtual screening and speeding up drug discovery process of therapeutically active ligands of low mass. Chapter

14 provides the basic role of homology modeling technique for the rational design of low molecular weight ligand.

Chapter 15 is based on the current coronavirus pandemic situation (COVID-19). The molecular modeling and docking studies of potential drug candidates against various drug targets of the virus are comprehensively highlighted.

—Debarshi Kar Mahapatra, PhD
Sanjay Kumar Bharti, PhD

PART I

Small Molecules as Active Pharmacological Agents

CHAPTER 1

Bedaquiline: The New Antituberculosis Drug on the Horizon

GYANSHANKAR P. MISHRA

Department of Pulmonary Medicine, Government Medical College, Nagpur, Maharashtra, 440009, India

E-mail: gpmishra81@gmail.com

ABSTRACT

Multidrug-resistant tuberculosis (MDR-TB) is tuberculosis that is resistant to at least rifampicin and isoniazid, which threatens to erode the current progress in the global control of TB. Drug resistance may further worsen with the appearance of drug resistance to at least one of the injectable agents (amikacin, capreomycin, and kanamycin) or to fluoroquinolones (pre-extensively drug-resistant (pre-XDR)-TB) or both (extensively drug-resistant (XDR)-TB). Bedaquiline, a diarylquinoline compound with a novel mechanism of anti-TB action, has been granted accelerated or conditional approval in the US (2012) and Europe (2014) for use in MDR-TB, with interim guidance provided for its use by the World Health Organization (WHO). Recently, it was included in the revised WHO classification of second-line antituberculosis drugs in the management for drug resistant TB. It has a potential to reduce the current long duration of treatment of MDR-TB and XDR-TB, which is, as of now, adversely affecting the observance of these patients. The accessibility of this relatively new drug can revolutionize the therapeutics of TB today. This chapter deals with the development, pharmacology, clinical trials, and other essential aspects of bedaquiline.

Biologically Active Small Molecules: Modern Applications and Therapeutic Perspectives.
Debarshi Kar Mahapatra, & Sanjay Kumar Bharti (Eds.)

1.1 INTRODUCTION

Multidrug-resistant tuberculosis (MDR-TB) is tuberculosis that is resistant to at least isoniazid and rifampicin, which threatens to erode the current progress in the global control of TB.[1] Drug resistance may worsen further with the emergence of resistance to at least one of the injectable agents (amikacin, kanamycin, and capreomycin) or to fluoroquinolones (pre-extensively drug-resistant (pre-XDR)-TB) or both (extensively drug-resistant (XDR)-TB),[2,3] Diarylquinoline, bedaquiline (trade name Sirturo, code names TMC207 and R207910), is a new TB drug discovered by Koen Andries and colleagues. This drug has been granted accelerated or conditional approval in the US (2012) and Europe (2014) for use in MDR-TB, with interim guidance provided for its use by the World Health Organization (WHO).[4] In a span of almost 40 years, bedaquiline is the first new drug with a novel mechanism of action to appear on the horizon of antituberculosis treatment.

Adding bedaquiline to a favored environment schedule for 24 weeks results in a faster culture alteration and appreciably higher frequency of culture conversions at 120 weeks in comparison to the placebo. Udwadia et al. reported striking improvement in the first case series of five patients from India on the use of bedaquiline with the individualized optimal background regimen (MDR/XDR-TB) based on their drug sensitivity testing (DST) results.[5] In a recent study by Skrahina et al. from Belarus, the use of bedaquiline in XDR, pre-XDR, and MDR- pulmonary TB cases has been reported to achieve sputum culture conversion in 94% of such patients in 6 months.[6]

Bedaquiline probably kills both dormant and semidormant bacilli also along with the actively replicating mycobacteria by inhibiting mycobacterial ATP synthase, an essential membrane-bound enzyme by interfering with energy production and disrupting intracellular metabolism. This potential action of bedaquiline on dormant and semidormant bacilli has the potential to decrease both the duration of therapy as well as the risk of relapse. Both these aspects are much needed in the current management strategies of drug-resistant TB. This is an area of focus for future research with bedaquiline-based regimes.[7]

Recently, off-label use of bedaquiline has shown to have a potential role in the management of nontuberculous mycobacterial (NTM) lung disease with 50% patients showing culture conversion at 6 months. However, at present, bedaquiline is not recommended for use in NTM lung disease.[8]

1.2 PHARMACOLOGY

Bedaquiline is a diarylquinoline antimycobacterial with a novel mechanism of action. The drug inhibits adenosine 5'-triphosphate synthase present in mycobacteria. It is an enzyme that is necessary for the energy production in *M. tuberculosis*. It specifically binds to the c subunit and probably also to the ε subunit of the mycobacterial F-ATP synthase.[9–12]

Bedaquiline (Sirturo) for oral administration is available as 100 mg strength tablets. Each tablet contains 120.89 mg of bedaquiline fumarate drug substance, which is equivalent to 100 mg of bedaquiline.[13] The recommended dosage of bedaquiline is:

- Weeks 1–2: 400 mg (four tablets of 100 mg) once daily with food.
- Weeks 3–24: 200 mg (two tablets of 100 mg) 3 times per week with food (with at least 48 h between doses) for a total dose of 600 mg per week.

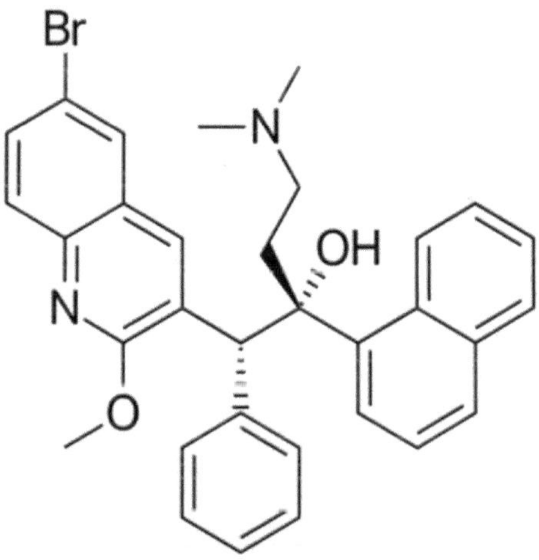

This novel anti-TB drug contains a central quinolinic heterocyclic nucleus with amine side chains and alcohol functional groups which are principally accountable for anti-infective response. It is a new class where the molecule directly targets the mycobacterial ATP synthase thereby crippling mycobacteria energy metabolism. Bedaquiline does not act exclusively against drug-resistant *M. tuberculosis* isolates, but also against drug-susceptible strains. Its effective half-life is >24 h (Table 1.1).[9–12]

TABLE 1.1 Pharmacodynamic and Pharmacokinetic Profile of Bedaquiline.

Parameters	Information
Group	Diarylquinoline antimycobacterial
Available as	Bedaquiline fumarate
IUPAC name	(1*R*,2*S*)-1-(6-bromo-2-methoxyquinolin-3-yl)-4-(dimethylamino)-2-naphthalen-1-yl-1-phenylbutan-2-ol
Molecular formula	$C_{32}H_{31}BrN_2O_2$
Molecular weight	555.516 g/mol
Metabolism	Primarily hepatic by CYP3A4, metabolized to M2 4–6 times less active in terms of antimycobacterial potency compared with parent drug) and M3
Elimination	Primarily excreted in the feces; renal elimination (<0.001%)
Half-life	5.5 months
Protein binding	>99%
T_{max}	~5 h

1.3 DRUG–DRUG INTERACTION

CYP3A4 inducers coadministration with bedaquiline reduces bedaquiline exposure whereas CYP3A4 inhibitors coadministration with bedaquiline increases the bedaquiline level.[14–16]

1.3.1 RIFAMPIN (STRONG CYP3A4 INDUCER)

Administration of bedaquiline (300 mg once daily) and rifampin (600 mg once daily) for 3 weeks leads to the reduction in the exposure by 52%. Decrease in the systemic exposure occurs when rifamycin is coadministered with bedaquiline, or systemic application of strong CYP3A4 inducers results. The application of bedaquiline as a first-line drug is a major concern where rifampicin remains a main-line component of the therapeutic schedule.

1.3.2 KETOCONAZOLE (STRONG CYP3A4 INHIBITOR)

Bedaquiline (400 mg once daily) administration for 14 days and ketoconazole (400 mg once daily) for 4 days results in an enhancement of exposure by 22%. As a result of adverse reactions with bedaquiline such as prolonged systemic exposure, coadministration of bedaquiline with systemically strong

CYP3A4 inhibitors for >14 days must be avoided, until the ratio of benefit/risk is greater than 1. When nevirapine was coadministered with bedaquiline, no dose adjustment was required.

1.4 SAFETY

At present, bedaquiline carries a black box warning alerting healthcare professionals to an augmentation in all-cause mortality and to a prolongation of the QT interval (QTcF) in patients receiving bedaquiline in contrast to controlled or placebo group. Currently, the only concern with bedaquiline is its safety. However, in a recently published study by Skrahina et al., it has been shown that the safety profile of this drug should not be a limitation to its efficacy potential of yielding a good culture conversion in advanced drug-resistant TB cases also.[6] Most of the side effects of this drug in their study were mild–to-moderate severity and could be managed. Overall, the following side effects may be observed during bedaquiline therapy.

1.4.1 CARDIAC RHYTHM DISTURBANCES

An electrocardiogram (ECG) should be obtained as a part of pretreatment evaluation, then daily for the first 2 weeks, subsequently every 2 weeks for 3 months, and then after every 30 days. QTcF is calculated by Fridericia's formula as QT interval divided by the cube-root of RR interval, where RR is the interval from the onset of one QRS complex (the graphical deflections seen on an ECG that correspond to the depolarization of the right and left ventricle with each heartbeat) to the onset of the next QRS complex, measured in milliseconds.[17] Coadministration of other QT interval prolonging drugs in the regime warrant at least weekly ECG monitoring throughout the bedaquiline therapy. QT prolongation can produce ventricular arrhythmias (Torsades de Pointes) and may cause unexpected deaths. Reports of major studies on the efficacy and safety of bedaquiline are summarized in Table 1.2.

1.4.2 HOW DOES ONE APPROACH PROLONGED QT INTERVAL DURING BEDAQUILINE THERAPY?[18]

- A QTc of >440 ms is deemed to be prolonged but no specific action is required awaiting >450 ms.

TABLE 1.2 Clinical Studies on Bedaquiline.

Group, Year	Type of study	Subjects	Efficacy findings	Tolerability findings	QTc effect
DIACON, 2009	Phase IIb, randomized, multicenter, double-blind, placebo-controlled study	23	Bedaquiline reduced the time to conversion to a negative sputum culture. Bedaquiline increased the proportion of patients with the conversion by sputum culture.	Most AEs were mild-to-moderate; only nausea occurred significantly more frequently in the bedaquiline group.	QT interval prolongation was observed in the bedaquiline and the placebo group (more pronounced in the bedaquiline group), with intergroup differences ranging from 1.0 to 10.8 ms ($p>0.05$). None of the absolute values for the corrected QT interval exceeding 500 ms, and no adverse events were associated with ECG changes.
DIACON, 2014	Phase IIb, randomized, multicenter, double-blind, placebo-controlled study	79	Bedaquiline reduced the median time to culture conversion from 125 days to 83 days	Bedaquiline had similar rates of AEs, treatment-related AEs, and AEs leading to study discontinuation than placebo The most frequent AEs were nausea, arthralgia, and vomiting. The severity of most adverse events was grade 1 or 2.	At study week 24, the mean change from baseline in the QTcF was an increase of 15.4 ms in the bedaquiline group and an increase of 3.3 ms in the placebo group ($p<0.001$). After bedaquiline treatment ended, the QTcF gradually decreased, and the mean value was similar to that in the placebo group by study week 60.
GUGLIELMETTI, 2015	Retrospective cohort study	35	Culture conversion rate was 97% after 6 months of therapy	Mild liver enzyme elevation (≥two-fold from baseline) was reported in 14% of patients, and a ≥five-fold increase occurred in two additional patients (6%). The confounding effect of concomitant drugs was mentioned.	QTc prolongation was greater in individuals exposed to bedaquiline and fluoroquinolones, or clofazimine

TABLE 1.2 *(Continued)*

Group, Year	Type of study	Subjects	Efficacy findings	Tolerability findings	QTc effect
NDJEKA, 2015	Interim cohort analysis	91 (54 HIV+)	In total, 48 (76%) out of 63 patients with 6 months of follow-up either achieved culture conversion or remained culture-negative 6 months after initiation of bedaquiline	Good profile	Clofazimine use and not HIV infection were associated with QTc increase. (ART based on either lopinavir/ ritonavir or nevirapine)
PYM, 2015	Phase 2, multicenter, multinational, open-label, noncomparative, single-arm trial	233	Culture conversion was 72.2% at 120 weeks, and 73.1%, 70.5% and 62.2% in MDR-TB, pre-XDR-TB and XDR-TB patients, respectively	The commonest AEs were similar to those generally reported in MDR-TB treatment cohorts; most were grade 1 or 2. Hyperuricemia and increased aspartate aminotransferase was the most frequent grade ≥3 AEs. The incidence of AEs was highest during the first 12 weeks. Serious AEs were reported in 47 (20.2%) patients: respiratory infections/ disorders were the most common	Prolongation of the QTcF interval was reported infrequently. Two patients had an increase in QTcF interval >500 ms; they were both taking clofazimine, and one had concurrent hypokalemia. No episode of clinically significant dysrhythmia was reported. Two deaths were considered doubtfully related to bedaquiline.
SKRAHINA, 2016	Interim cohort analysis	197	96% patients showed culture conversion at 6 months.	The most adverse events were mild or moderate in severity and reversible. One death was possibly related to MDR-TB therapy	41% patients had cardiac disorders (abnormal electrocardiogram and arrhythmia being the most common)

- A QTc value between 450 and 480 ms requires other causes of prolonged QTc interval to be ruled out before deciding to withhold bedaquiline.
- If the QTc value of >480 ms reaches, then following steps must be taken immediately:
  - Confirm the QTc prolongation by repeating the ECG.
  - Check the body electrolytes level. If found abnormal, therapy by bedaquiline should be terminated.
  - If the patient is quite stable and the electrolyte levels are normalized enough, however, the QTc interval is ranged between 480 and 500 ms, then repeat the ECG weekly to ascertain that the interval is stable.
  - If the ECG displays QTc interval of >500 ms, subsequently all the QTc prolonging medicines must be immediately stopped.
  - Bedaquiline and all QTc prolonging medicines must be stopped immediately, if ventricular arrhythmia is clinically diagnosed.
  - Because of the long half-life of bedaquiline, if the ECG has QTc prolongation, weekly monitoring of ECG should take place until the QTC interval normalizes even if the bedaquiline therapy is no longer being given.

1.4.3 MISCELLANEOUS ISSUES

1.4.3.1 AST AND/OR ALT ELEVATION

If the elevation is >3 times the upper limit of the normal, then the hepatotoxic drugs (e.g., pyrazinamide, ethionamide, PAS) of the regime may be withheld, and if levels do not return to normal, then bedaquiline needs to be withdrawn till the levels are normalized.[19]

1.4.3.2 AMYLASE AND OR LIPASE ELEVATION

Acute pancreatitis may rarely occur. If lipase is >2 times and/or amylase is >5 times the upper limit of normal, then bedaquiline should be immediately stopped.

1.4.3.3 MUSCULOSKELETAL SYSTEM–MYALGIA

Muscle tenderness at the site other than injection site, marked impairment of activity due to muscle involvement, or frank myonecrosis require bedaquiline to be stopped.

1.4.3.4 GASTROINTESTINAL SYSTEM DISORDERS

Nausea, Vomiting.

1.5 ISSUES OF DRUG RESISTANCE

The mutation in the target gene atpE of mycobacteria leads to bedaquiline resistance. Not all isolates with increased minimum inhibitory concentrations (MICs) have atpE mutations, suggesting the existence of at least one other resistance mechanisms.[20]

1.5.1 CROSS-RESISTANCE

The mutation in the gene transcriptional repressor, Rv0678, which encodes for efflux pump system (MmpS5-MmpL5) remains the prime nontarget-based bedaquiline resistance. The cross-resistance of bedaquiline with clofazimine is also reported. Provided the diversity of resistance-associated variants (RAVs) in the gene transcriptional repressor and their inconsistent consequences on the MIC, only the drug-susceptibility phenotypic methods can presently be promoted to evaluate the bedaquiline vulnerability.[21]

1.5.2 PRINCIPLES FOR FORMULATING A REGIME FOR DRUG-RESISTANT TB[22-23]

- At least a minimum of five different drugs (pyrazinamide and four s-line TB medicines must be chosen from category A – 1 drug / B – 1 drug / C – 2 drugs / D2 or D3 if needed) are strictly recommended for patients having multidrug-resistant TB or have developed resistance to rifampicin.
- Patients suffering from MDR-TB or rifampicin-resistant TB are recommended to continue high-dose ethambutol or high-dose isoniazid.
- Bedaquiline may classically be obligatory in the these conditions as stated by the WHO:
  - When an effectual therapeutic schedule involves 4 second-line anti-TB medicines in addition to pyrazinamide, it cannot be intended.
  - If resistance or multidrug resistance is reported for any fluoroquinolone class of drug.

- The most effective treatment regimen for multidrug-resistant TB as recommended by WHO includes pyrazinamide along with four second-line drugs. The drug resistance surveillance data and drug-susceptibility testing has indicated toward the use of a late-generation fluoroquinolone, a second-line injectable drug, and bacteriostatic drugs (ethionamide or prothionamide with cycloserine).
- Bedaquiline is primarily recommended if the antitubercular therapy is not practicable due to:
  - drug resistance
  - poor tolerance, adverse drug reactions, or contraindication to some specific constituent of the multidrug therapy, or
  - nonavailability of the drug from the supplier.

1.5.3 IMPORTANT ASPECTS OF THE BEDAQUILINE THERAPY OF THE WHO RECOMMENDATIONS[24]

- It needs a close observation.
- The application to HIV-affected patient, patients with comorbidities (especially cardiac, diabetes, and liver disease), and people addicted to habit-forming components must be in a state of adequate caution.
- A maximum dose of 400 mg daily for the first 2 weeks is recommended which is followed by 400 mg daily for the first 2 weeks of 200 mg dose for the remaining duration with a maximum duration of 6 months.
- It should not be appended to an unresponsive state as a sole agent.
- Continuous monitoring of cardiac parameters is essential to detect any kind of arrhythmia (specifically QT prolongation).
- Active pharmacovigilance systems must be kept in place among the patient groups before the commencement of therapy to report the spontaneously arising adverse drug reactions.
- The drug resistance should be determined by assessing the MICs, in those conditions when specific drug-susceptibility tests are absent.

Considering all these factors an effective background regime is very important in designing a bedaquiline-based regime as bedaquiline if used with a potentially weak/ineffective regime will be exposed to the risk of developing drug resistance.

1.6 THE CURRENT CHALLENGES AND WAY FORWARD WITH BEDAQUILINE

There are a few areas regarding bedaquiline which need to be focused on future avenues for potential research on this novel drug. At present, the WHO classification of medicines have been into application for managing the evidence-based drug-resistant tuberculosis at present. With the emergence of more evidence regarding the efficacy and safety of these medicaments, additional changes are expected to happen in this classification. One such possible future evolution of the current WHO classification has been predicted by Tiberi S et al. considering the current trend of recent emerging evidence (Table 1.3).[25] As seen from the proposed table of the possible future evolution of classification of anti-TB drugs, bedaquiline has the potential to be elevated to a higher group from its current status in the classification (i.e., elevated to future group B from current group D2). The possible inclusion of bedaquiline as a core drug of a regime to tackle drug-resistant TB cases will surely increase the overall efficacy of the regime.

TABLE 1.3 Possible Future Evolution of Classification of Anti-TB Drugs.

GROUP A (fuoroquinolones)	Levofloxacin, moxifloxacin, gatifloxacin
GROUP B (other core) Second-line agents	**Bedaquiline,** delamanid, ethionamide/ Prothionamide, cycloserine/terizidone, linezolid clofazimine
GROUP C (second-line injectable agents)	Amikacin, capreomycin, kanamycin, meropenem/ clavulanate
GROUP D (add-on agents) (not core MDR-TB regimen components)	Pyrazinamide, ethambutol, high-dose isoniazid, p-amino salicylic acid, amoxicillin-clavulanate, rifabutin

The introduction of bedaquiline in the pharmacotherapeutics of tuberculosis has validated adenosine triphosphate (ATP) synthase as an attractive target to kill *Mycobacterium tuberculosis*. Future research on this novel target may be the way forward.[26] Recently, the discovery of imidazo[1,2-a]pyridine ethers and squaramides as selective and potent inhibitors of mycobacterial ATP production has been reported by Tantry et al. signifying that research on these lines has already begun.[27] Bedaquiline has paved the way for opening up a whole new dimension to the future research in TB.

1.7 CONCLUSION

To conclude, bedaquiline is a novel molecule that has arrived in the field of tuberculosis. It needs to be utilized to its full potential in the most efficient possible way. A rationale and judicious use of this molecule are advisable to prevent resistance to this new and potentially lethal weapon in our armory against *Mycobacterium tuberculosis*.

KEYWORDS

- **bedaquiline**
- ***Mycobacterium tuberculosis***
- **antitubercular**
- **clinical studies**
- **drug resistance**
- **drug interactions**

REFERENCES

1. World Health Organization. Global Tuberculosis Report 2014. http://apps.who.int/iris/bitstream/10665/137094/1/9789241564809_eng.pdf?ua=1 (accessed May 5, 2018).
2. Shah, N. S.; Wright, A.; Bai, G-H. et al. Worldwide Emergence of Extensively Drug-Resistant Tuberculosis. *Emerg. Infect. Dis.* **2007,** *13*, 380–387.
3. Banerjee, R.; Allen, J.; Westenhouse, J. et al. Extensively Drug-Resistant Tuberculosis in California, 1993–2006. *Clin. Infect. Dis.* **2008,** *47*, 450–457.
4. Andries, K.; Verhasselt, P.; Guillemont, J.; Göhlmann, H. W.; Neefs, J. M.; Winkler, H.; Van Gestel, J.; Timmerman, P.; Zhu, M.; Lee, E.; Williams, P.; de Chaffoy, D.; Huitric, E.; Hoffner, S.; Cambau, E.; Truffot-Pernot, C.; Lounis, N.; Jarlier, V. A Diarylquinoline Drug Active on the ATP Synthase of *Mycobacterium tuberculosis*. *Science* **2005,** *307*, 223–227.
5. Udwadia, Z. F.; Amale, R. A.; Mullerpattan, J. B. Initial Experience of Bedaquiline Use in a Series of Drug-Resistant Tuberculosis Patients from India. *Int. J. Tuberc. Lung Dis.* **2014,** *18*, 1315–1318.
6. Skrahina, A.; Hurevich, H.; Falzon, D.; Zhilevich, L.; Rusovich, V.; Dara, M. et al. Bedaquiline in the Multidrug-Resistant Tuberculosis Treatment: Belarus Experience. *Int. J. Mycobacteriol.* **2016,** *5* (Suppl 1), S62–S63.
7. Andries, K.; Villellas, C.; Coeck, N.; Thys, K.; Gevers, T. et al. Acquired Resistance of *Mycobacterium tuberculosis* to Bedaquiline. *PLoS ONE* **2014,** *9* (7), e102135.

8. Villellas, C.; Coeck, N.; Meehan, C. J.; Lounis, N.; de Jong, B.; Rigouts, L. et al. Unexpected High Prevalence of Resistance-Associated Rv0678 Variants in MDR-TB Patients Without Documented Prior Use of Clofazimine or Bedaquiline. *J. Antimicrob. Chemother*. **2017,** *72* (3), 684–690.
9. Pontali, E.; Sotgiu, G.; D'Ambrosio, L.; Centis, R.; Migliori, G. B. Bedaquiline and Multidrug-Resistant Tuberculosis: A Systematic and Critical Analysis of the Evidence. *Eur. Resp. J*. **2016,** *47* (2), 394–402.
10. Sirturo (Bedaquiline) Full prescribing Information. http://www.accessdata.fda.gov/drugsatfda_docs/label/2015/204384s005lbl.pdf (accessed Jan 1, 2017).
11. Kundu, S.; Biukovic, G.; Grüber, G.; Dick T. Bedaquiline Targets the ε Subunit of Mycobacterial F-ATP Synthase. *Antimicrob. Agents Chemother*. **2016,** *60* (11), 6977–6979. doi:10.1128/AAC.01291-16.
12. Mase, S. et al. Provisional CDC Guidelines for the Use and Safety Monitoring of Bedaquiline Fumarate (Sirturo) for the Treatment of Multidrug-Resistant Tuberculosis. *MMWR Recomm. Rep*. **2013,** *62* (RR-09), 1–12.
13. Centers for Disease Control and Prevention. Provisional CDC Guidelines for the Use and Safety Monitoring of Bedaquiline Fumarate (Sirturo) for the Treatment of Multi Drug-Resistant Tuberculosis. https://www.cdc.gov/mmwr/preview/mmwrhtml/rr6209a1.htm (accessed May 1, 2017).
14. RNTCP, Central TB Division, MOHFW. Guidelines for Use of Bedaquiline in RNTCP through Conditional Access under Programmatic Management of Drug Resistant Tuberculosis in India. WHO. 2016. https://drive.google.com/file/d/0B8_9PlDVeBP6eEg1TGFjVHVGTFU/view (accessed May 3, 2018).
15. World Health Organization. *The Use of Bedaquiline in the Treatment of Multidrug-Resistant Tuberculosis: Interim Policy Guidance*. Document WHO/HTM/TB/2013.6; World Health Organization: Geneva, 2013.
16. Companion Handbook to the WHO Guidelines for the Programmatic Management of Drug-Resistant Tuberculosis. World Health Organization: Geneva, 2014. ANNEX 4, "How-to" Guide on the Use of Bedaquiline for MDR-TB Treatment. https://www.ncbi.nlm.nih.gov/books/NBK247434/
17. National Center for Biotechnology Information. PubChem Compound Database; CID–5388906. https://pubchem.ncbi.nlm.nih.gov/compound/5388906 (accessed May 1, 2018).
18. World Health Organization. *WHO Treatment Guidelines for Drug-resistant Tuberculosis*; World Health Organization, 2018 (accessed May 5, 2018).
19. Field, S. K. Bedaquiline for the Treatment of Multidrug-Resistant Tuberculosis: Great Promise or Disappointment? *Ther. Adv. Chronic Dis*. **2015,** *6* (4), 170–184.
20. Philley, J. V.; Wallace, R. J., Jr.; Benwill, J. L.; Taskar, V.; Brown-Elliott, B. A.; Thakkar, F. et al. Preliminary Results of Bedaquiline as Salvage Therapy for Patients with Nontuberculous Mycobacterial Lung Disease. *Chest* **2015,** *148* (2), 499–506.
21. Diacon, A. H.; Pym, A.; Grobusch, M. et al. The Diarylquinoline TMC207 for Multidrug-Resistant Tuberculosis. *N. Engl. J. Med*. **2009,** *360*, 2397–2405.
22. Guglielmetti, L.; Le Dû, D.; Jachym, M. et al. Compassionate Use of Bedaquiline for the Treatment of Multidrug-Resistant and Extensively Drug-Resistant Tuberculosis: Interim Analysis of a French Cohort. *Clin. Infect. Dis*. **2015,** *60*, 188–194.
23. Ndjeka, N.; Conradie, F.; Schnippel, K.; et al. Treatment of Drug-Resistant Tuberculosis with Bedaquiline in a High HIV Prevalence Setting: An Interim Cohort Analysis. *Int. J. Tuberc. Lung Dis*. **2015,** *19*, 979–985.

24. Pym, A. S.; Diacon, A. H.; Tang S-J. et al. Bedaquiline in the Treatment of Multidrug- and Extensively Drug-Resistant Tuberculosis. *Eur. Respir. J.* **2016,** *47*, 564–574.
25. Tiberi, S.; Scardigli, A.; Centis, R.; D'Ambrosio, L.; Munoz-Torrico, M.; Salazar-Lezama, M. A. et al. Classifying New Anti-TB Drugs: Rationale and Future Perspectives. *Int. J. Infect. Dis.* **2017,** *56*, 181–184.
26. Diacon, A. H.; Pym, A.; Grobusch, M. P.; de los Rios, J. M.; Gotuzzo, E.; Vasilyeva, I. et al. Multidrug-Resistant Tuberculosis and Culture Conversion with Bedaquiline. *N. Engl. J. Med.* **2014,** *371* (8), 723–732.
27. Tantry, S. J., Markad, S. D., Shinde, V., Bhat, J., Balakrishnan, G., Gupta, A. K. et al. Discovery of Imidazo[1,2-a]pyridine ethers and Squaramides as Selective and Potent Inhibitors of Mycobacterial Adenosine Triphosphate (ATP) Synthesis. *J. Med Chem.* **2017,** *60* (4), 1379–1399.

CHAPTER 2

Plant-Derived Natural Products as Antiglycating Agents

MOBASHSHERA TARIQ and AHMAD ALI*

Department of Life Sciences, University of Mumbai, Vidyanagari, Santacruz (East), Mumbai 400098, Maharashtra, India

**Corresponding author. E-mail: ahmadali@mu.ac.in*

ABSTRACT

In this era of endless possibilities, we are continually challenged by several lifelong chronic diseases like diabetes, making it difficult to lead a healthy life. Diabetes has been described as a process for dysregulation of glucose metabolism, resulting in its toxic levels in circulation. This, in turn, results in the modification of biomolecules, especially proteins, with the help of several mechanisms including glycation. Glycation is a direct expression of the reductive nature of glucose, unlike other glucose-induced protein modifications. It is a nonenzymatic process where carbonyl groups of the sugar part exclusively react with the amino group of the proteins and the nucleic acids resulting in the formation of advanced glycation end products. These products are difficult to eliminate and keep accumulating inside the cell altering the normal function and structure of biomolecules. Currently, antidiabetic therapies include the use of recombinant or artificial insulin taken in intravenous form and synthetic hypoglycemic agents taken orally. However, many antidiabetic agents are known to cause severe side effects which also add to the problems of diabetes-related complications. The screening of plants for antidiabetic and antiglycation activity can provide a safer alternative to the problems related to glucotoxicity. This chapter

Biologically Active Small Molecules: Modern Applications and Therapeutic Perspectives.
Debarshi Kar Mahapatra, & Sanjay Kumar Bharti (Eds.)

describes, briefly, the need for screening of antiglycating agents from natural sources to surmount the problems associated with diabetes by using simple biochemical approaches.

2.1 INTRODUCTION

We all know the importance of DNA, RNA, and other biomolecules that exist inside a cell. However, mere presence of these components is not enough to drive a life process. It is the biomolecular interaction known as chemical reactions, also called the biochemical reactions, which allow energy generation from the raw materials as requisite to perform the work. Biochemical reactions that occur inside the cell are key to almost all the life processes. Every biochemical reaction is linked to a cellular function, and any error in it can lead to fatal consequences for the organism. These consequences may be due to toxic accumulation of reaction products in the cell, damage due to oxidation, deficiency of an enzyme or protein, overstimulation of certain reaction pathways, and so on.

One such biochemical reaction, known as glycation, is a spontaneous nonenzymatic amino-carbonyl reaction between sugar components and lipids, nucleic acids, and proteins. It is among the posttranslational modification processes that occur between the reducing sugars with the protein's free amino groups.[50] In the initial period of glycation, the intermediate synthesis leads to the development of Schiff bases and Amadori compounds and in the later stage, advanced glycation end products (AGEs) are formed permanently due to the complex reaction cascade.[12] These products are deleterious and accumulate inside the cell in a time-dependent manner. If not eliminated or metabolized by the body, they can alter the normal function and structure of biomolecules.

Glycation-induced damage to biomolecules is believed to be a contributing factor in several metabolic and physiological disorders, such as Alzheimer's, Parkinson's, atherosclerosis, cataracts. The most common being diabetes and damage due to reactive oxygen species (ROS) production.[17,27,44,88] It is also linked to the natural physiological process of aging.[74] According to the AGE concept, the pathogenesis of diabetic complications is due to the glycation-mediated structural alteration and cross-linking of lipids, nucleic acids, and proteins.[16]

In order to prevent the complication of these disorders, it is extremely important to understand the mechanism of the biochemical events that lead the cell toward its fatal destination. We have no control over the random mutations that may take place at the cellular level leading to disasters like cancer or chromosomal disorders. Treatment, as well as the management of such disorders, is also difficult. Moreover, due to the complex and diverse cellular mechanisms that lead to mutations, developing a hypothesis for effective prevention becomes extremely difficult. On the contrary, most of the metabolic disorders like diabetes or enzyme deficiency diseases are comparatively easier to manage. The understanding of these disorders is also easier due to the fact that its progression can be linked to a single metabolic event, step or pathway. The treatment of these diseases, however, is currently a topic of medical research.

To understand the process of glycation, it is necessary that we know the fundamentals of basic metabolism and how it is linked to diabetes and its related complications. The food we eat primarily comprises of carbohydrates, fats, and proteins. The energy is mainly obtained from the digestion of carbohydrate (i.e., complex sugar) components of food. The common carbohydrates available as food are disaccharide sugars like sucrose (table sugar), starch (vegetables), and lactose (milk). They are broken down into simple sugars, such as glucose, galactose, and fructose with the help of digestive enzymes.

Glucose is the most abundant monosaccharide available as food and it is also the most important molecule for living organisms. Accordingly, the balanced metabolism of this molecule is of primary importance. A minor fluctuation in the glucose metabolism impairs the functions of many organs and tissues leading to a diseased condition. The two conditions of glucose accessibility and utilization, hypoglycemia and hyperglycemia, have been well described in the literature for many centuries. Hyperglycemia has been accepted as a primary factor accountable for the commencement and promotion of diabetic complications.[29,32]

2.2 DIABETES

Diabetes has become a major public health concern globally because of a tremendous increase in the number of people affected by this disease and health system cost.[71] Recently, NCD Risk Factor Collaboration[58] has reported that there is a fourfold rise in the number of diabetics—from 108 million in 1980 to 422 million in 2014 with half of them living in India, China, USA, Brazil, and Indonesia. The scenario in India is more alarming as the number of diabetics has climbed from 11.9 million in 1980 to 64.5 million in 2014, almost more than fivefold increase. This has put an enormous burden on the socioeconomic as well as the healthcare systems at both the national and international levels.

Under normal circumstances, our body maintains a balance by regulating the metabolism of food in order to maintain the optimum concentration of nutrients required for cellular activity. Diabetes has been described as a process that either originates from or manifests the dysregulation of glucose metabolism. Type 2 diabetes mellitus characterized either due to a decrease in insulin production or insulin resistance has become a common disorder in recent times. Type 1 diabetes mellitus, on the other hand, is a less common autoimmune disorder commonly occurring in children.

As the concentration of glucose amplifies in blood, a lot of transformations are observed. Since the fundamental pathway of glucose metabolism fails in diabetes, several alternative attempts are made by the body to overcome the defects. First, the water is drawn outside the cell into the bloodstream, which allows excretion of excess glucose. This is manifested in the form of symptoms like excessive thirst, dehydration, polyuria, and glucosuria. Next, the body is forced to look for alternative sources of energy like fatty acids due to lack of glucose inside the cells, which in turn occurs due to lack of insulin. This leads to the generation of ketone bodies which is a fair source of energy by the body in times of emergency, but its accumulation over time leads to ketosis which is manifested in the form of ketonuria and presents as a complication of diabetes. The consequence of diabetes complications is increased in morbidity and mortality due to its characteristic increase in susceptibility to infection and poor wound healing.[72] Finally, the enhanced level of glucose leads to the modification of biomolecules especially proteins. There can be several ways by which glucose modifies the proteins: glycation, carbonylation, nitration, nitrosylation, acetylation, ADP-ribosylation, and succination.[97] After modification, the proteins tend to lose their structure and function. This is manifested in the form of several degenerative disorders and aging.[42,62]

2.3 CURRENT TREATMENT OF DIABETES AND ITS COMPLICATIONS

Currently, antidiabetic therapies include the use of recombinant or artificial insulin taken in intravenous (IV) form, and synthetic hypoglycemic agents taken orally. The key factors in designing treatment plans for diabetes include the following:[2]

1. Inhibiting glucose absorption from the intestine.

2. Increasing insulin secretion from the pancreas.
3. Inhibiting glucose production from hepatocytes.
4. Enhancing glucose uptake by adipose and muscle tissues.

Apart from using drugs, inhibition of AGEs is believed to be a valuable curative strategy in the management of diabetes.

The most widely used drugs for diabetes are α-glucosidase inhibitors, such as acarbose, metformin, glimepiride, and so on, which regulates blood glucose level by inhibiting digestion of complex sugars, such as maltose, dextrins or sucrose into glucose. They have no effect on pancreas stimulation or insulin resistance. Similarly, other synthetic hypoglycemic agent works on any one of the above-mentioned aspects in managing diabetes. The combination of oral drugs and/or insulin has been found to effectively contribute to diabetes management. For instance, combination therapy with rosiglitazone and repaglinide has been verified to be safe and effective in diabetes treatment. Similarly, the combination of voglibose with glibenclamide or gliclazide; or miglitol with metformin offer better glycemic control.[73] However, long-term use of these agent results in the development of complications related to diabetes, nutrition absorption, as well as certain other metabolic disorders.[18,24,37] In addition, the use of several therapeutic agents has shown to aid in oxidative stress, which further increases diabetes-related complications.[3,38,60] Along with ROS, reactive nitrogen species (RNS) are also responsible for oxidative stress and have a deleterious effect on the cell by DNA damage, membrane lipids peroxidation and increased inflammatory responses. All these interactions lead to diabetic micro- and macrovascular complications.[76]

In addition to the metabolic and oxidative stress, high levels of fructose in the diet are also related to complications in diabetes. Fructose diet is linked to augments efflux of glucose and glycogen depletion in the liver that results in decreased activity of glycogen synthase and glucokinase or downregulation of gene expression of these enzymes. This entire phenomenon is, in turn, linked to insulin resistance.[80] A study demonstrated that high levels of fructose feeding in rats decrease insulin-induced insulin receptor substrates (IRS-1/2) phosphorylation in both the liver and muscles.[80] This leads to AGEs and polyol formation, which leads to the abundant complications as observed in diabetes. HbA1c due to hyperglycemia is recognized to originate additional oxidative stress via free radical reactions that are iron catalyzed. Furthermore, the augmentation of plasma AGEs are linked with enhancement in the serum creatinine level.[14]

2.4 GLYCATION

Glycation is the most significant process of protein modification mechanisms in diabetes. It is a direct manifestation of the reductive nature of glucose, unlike other glucose-induced protein modifications. This process is irreversible and is subdivided into three stages: early, intermediate, and late (Figure 2.1).

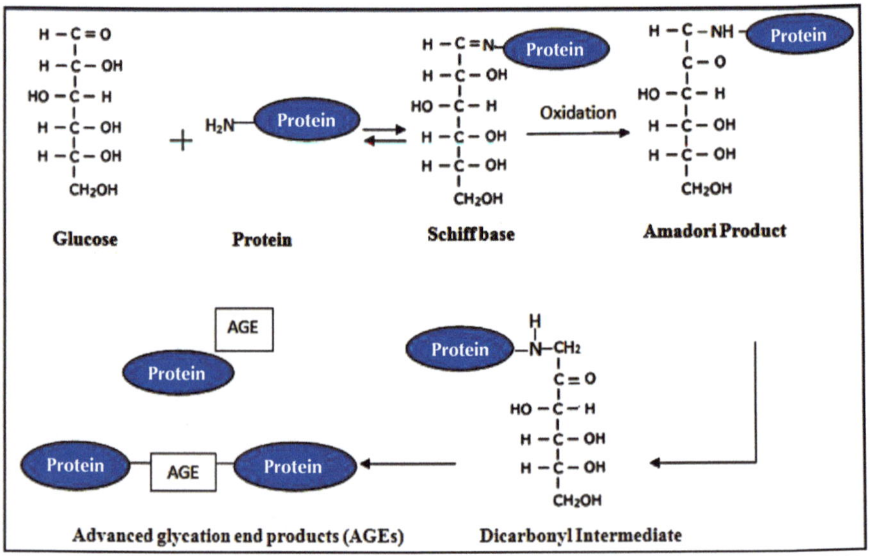

FIGURE 2.1 Glycation of proteins.

A spontaneous nonenzymatic reaction between sugars and proteins leads to the formation of Schiff bases which undergo chemical rearrangements to form irreversible Amadori products. These products further rearrange themselves forming highly reactive dicarbonyl compounds which lead to stable end products known as AGEs.

In the early stage, the reaction starts with the interaction between the aldehyde group of glucose and ε-amino group of lysine/arginine or N-terminal amino group of the proteins resulting in the Schiff base adduct formation.[57] These unstable adducts undergo chemical rearrangements to form Amadori products that are also known as the early glycation products.[6] In the intermediate stage, some of these Amadori products and Schiff bases lead to the production of reactive dicarbonyls, such as glyoxal, methyl glyoxal, and deoxyglucosone.[74] In the third stage, the Amadori products reorganize themselves via complexed chemical rescheduling, such as reductions, hydration, and oxidation, and form cross-linked proteins.[81] This process is irreversible and takes place in days, weeks, or months. The final brownish products formed are called the AGEs.

AGEs are a heterogeneous group of molecules.[28] As indicated in Figure 2.1, there are three main routes for the formation of AGEs.[74]

1. Autoxidation of sugars
2. Amadori rearrangements
3. Schiff adducts

Ingested food is another source of AGEs in humans. The intestinal absorption of exogenous AGEs is affected by digestive enzymes and microbial flora in the gut.[57] Irrespective of their sources, AGEs share some common functional features like cross-linking and fluorescence and have been identified using these features.[6] Accordingly, they are grouped into three different categories of AGEs: fluorescent, nonfluorescent, and cross-linking AGEs. Examples of some well-known AGEs are carboxyethyl lysine (CEL), carboxymethyl lysine (CML), pyrralline, pentosidine, and so on.[74] The rate of AGEs formation is raised in the presence of reactive oxygen species, oxygen, and redox active transition metals.[28] Cerami and coworkers[22] have also shown that environmental factors like diet and smoking influence the production of AGEs.

Proteins are typically glycated through the lysine residues.[11] In the human body, lysine content is highest for the histones situated in the cell nucleus, and therefore produces the glycated protein N (6)- CML.[11] A receptor for AGE known as RAGE is found on smooth muscle, cells of the immune system, and endothelial cells.[89] AGEs bind to these receptors, initiate reactive oxygen species formation, and downstream signaling basically leads to initiation of pro-inflammatory response. This contributes to diabetes and age-related chronic inflammatory diseases, such as asthma, atherosclerosis, myocardial infarction, arthritis, retinopathy, nephropathy, neuropathy, and periodontitis.[89] The pathogenesis of this process assumes regarding the foundation of the nuclear factor kappa B (NF-κB) following the binding of AGE. NF-κB controls several genes which are involved in inflammation.

Although glucose-related complications in the formation of AGEs are well documented and scientifically proven, biochemically, it is the least reactive sugars.[19] The presence of other reactive sugars in diet like fructose is inevitably linked to Mallard reaction and diabetes complications. Schalkwijk and coworkers[70] have reported fructose-induced intracellular AGE production in diabetes.[70]

The glycation process takes place incessantly in the body, yet at the usual blood glucose levels, and the damage caused by the Amadori and AGE products keeps on accrued gradually over time. When the blood glucose level enhances (suh as hyperglycemic conditions), the harmful consequences of these products are further considerably detected.

Consequently, an widespread research endeavor has been applied to develop the anti-AGE therapeutics. These include

1. Preventing the formation of AGEs by using inhibitors, such as ascorbic acid, aspirin, metformin, and so on.
2. Deglycation of Amadori products/Schiff bases involving transglycation approach using drugs like hydralazine.
3. Reversal of AGE-induced modifications like cross-links with the help of phenacylthiazolium bromide (PTB).
4. Preventing the body from the ill effects of AGE formation by using natural components like resveratrol and curcumin.

2.5 PHARMACOLOGICAL INTERVENTION OF DELETERIOUS REACTIONS

Based on the pathophysiological roles of deleterious reactions described in the current chapter, it can be seen that considerable benefits can be achieved by inhibiting these reactions and thereby causing damages to biomolecules. Of the many, the most common feature among AGE inhibitors known to date is the presence of nucleophilic groups such as amine or hydrazine capable of reacting with intermediate carbonyl compounds formed during glycation and preventing their progression to AGEs.[77]

Various approaches have been employed to combat the damaging effects of AGEs. First is by inhibition of glycation pathway using synthetic as well natural compounds, and the second approach is by repairing the damage. Although many compounds may be useful as an anti-AGE drug, the multi-cascade nature of the glycation reaction imposes severe limitations to the use of only one class of inhibitors. Several classes of inhibitors such as metal ion chelators and radical-trapping antioxidants exhibit their activities by preventing oxidative acceleration of glycation.[55] Further, it is possible that an inhibitor may exert its effect by more than one mechanism. Overall, the mechanisms of action of AGE inhibition are complex and depend on the inhibitor under study. Attempts have been made to develop novel multi-functional AGE inhibitors designed to function as a carbonyl trap, metal ion chelator, and/or radical-trapping antioxidant.[78]

2.6 PREVENTION OF GLYCATION

Reasons and causes for the onset of the diabetes are well documented. However, the drug development for the therapy has not been accomplished. One of the approaches is to design a drug which can inhibit the AGEs formation. While doing this, it should be kept in mind that an ideal AGE inhibitor should be capable of preventing the AGE formation by blocking all the pathways which lead to its formation.

AGE inhibitors have been classified by several researchers on the basis of their mechanism of action. The preventive strategies are outlined in Figure 2.2.

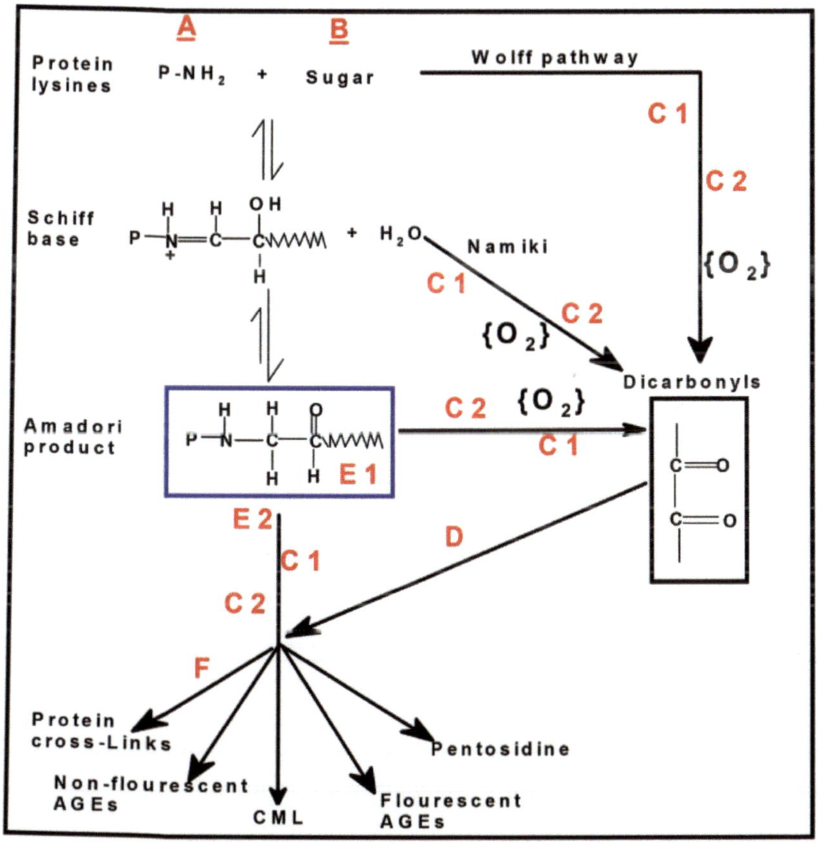

FIGURE 2.2 Inhibitors of AGE formation.

The diagram represents the pathway for the formation of AGEs and its possible inhibition sites. Type A inhibitors (pyridoxal-5-phosphate, aspirin, etc.) act by modifying free amino groups in order to prevent sugar attachment. Type B inhibitors (gentamicin) react with the ketose sugars and aldose sugars that redirect them from the Maillard reaction. Type C inhibitors (DETAPAC, phytate, etc.) act as chelators to inhibit glycoxidation by sequestering all free metal ions in vivo. Type C2 inhibitors (vitamin C, vitamin E, etc.) act as antioxidants. Type D inhibitors (glyoxal, glucoaldehyde, aminoguanidine, L-arginine, etc.) trap reactive dicarbonyl intermediates of AGE formation. Type E inhibitors (aminoguanidine, antiglycated monoclonal antibodies) act on the formed Amadori products. Type F inhibitors (phenacylthiazolium bromide) breaks the already-formed AGE cross-links rather than preventing AGE formation.

The inhibitory compounds generally inhibit the reactants or enzymes of a pathway to stop the production of end products.[1,35] Based on the literature, inhibition of AGEs can be achieved by the following mechanisms.

2.6.1 INHIBITION OF SUGAR ATTACHMENT WITH PROTEINS

It can be achieved by the modification of sugar or protein molecules. Some compounds may have the capacity to contend for the amino groups on the protein. Others can express binding to the glycation intermediates or proteins to impede the AGE formation progress. Otherwise, they may have the characteristics to eradicate the open-chain structure of glycating sugars. Theoretically, it is the most effective strategy to inhibit glycation and AGE formation. However, they are not promising therapeutic agents, since it is not practical to block all protein amino groups. Moreover, very few inhibitors are identified to impede with the primary connection of reducing sugars to the amino groups of proteins. Aspirin hinders the glycation process by acetylating the free-protein amino groups; therefore, it can obstruct the connection of the reducing sugars with the amino groups.[25,66] Diclofenac covalently bonds with proteins, thus blocking the attachment of sugars with proteins.[85] Similarly, other compounds, such as pyridoxal-5-phosphate, inositol, arginine, arginine-lysine complex, and α-glucosidase inhibitors, such as metformin, pioglitazone, and pentoxifylline inhibit the early stage of glycation by inhibiting sugar attachment with proteins.[96]

2.6.2 INHIBITION BY USING ANTIOXIDANTS (RADICAL SCAVENGERS)

It is already known that AGE formation and glycation are attended and further hastened by oxidative stress, therefore, antioxidant molecules may be gifted agents for averting the AGE formation and glycation. Some compounds can suppress AGE formation by attenuating glycol-oxidation and preventing oxidative stress because of their probable radical scavenging attributes which can trap glycation intermediate products including RNS, dicarbonyls, and ROS. Examples include calcium antagonists,[75] amlodipine,[8] kinetin,[87] quinine,[34] and some anti-inflammatory drugs like acetylsalicylic acid, ibuprofen, indomethacin, and diclofenac.[20]

2.6.3 INHIBITION BY METAL CHELATORS

TH-specific iron chelators, such as DETAPAC, desferoxamine, and phytate are also useful in preventing AGE formation.

2.6.4 INHIBITION OF DICARBONYL INTERMEDIATES

Few inhibitors of synthetic origin have the capability to lurk reactive free radicals and reactive carbonyls that are prepared during the glycation process or can directly obstruct the Amadori products formation. It is the most important target for the prevention of AGE formation since the ketone group of the 1-amino-1-deoxy fructose residue in the Amadori product is the precursor for the subsequent AGE formation. Thus, the chemical inactivation of the ketone group would prevent AGE formation. One of the popular drugs studied is aminoguanidine. It is a low molecular weight, highly nucleophilic compound that readily reacts with ketones and aldehydes. It has been shown to avert the in vivo AGEs formation in a wide variety of systems and tissues. Aminoguanidine shows its effect by targeting the α-dicarbonyl intermediates formed during the glycation reaction. Another example includes pyridoxamine, a known carbonyl trapping compound modified by the addition of 6-dimethylamino substituent (Culbertson et al, 2003). The modification resulted in enhanced radical-trapping ability in addition to carbonyl trapping. Other examples of such inhibitors include thiamine pyrophosphate, carnosine, and buformin.[96] The inhibitors mentioned above have a major limitation, as they cannot reverse already formed AGEs particularly cross-links. Thus, a compound that could break already formed cross-links can restore damaged tissue to its original functionality.

2.6.5 INHIBITION OF AMADORI PRODUCT FORMATION

Some antiglycating compounds can hinder the formation of Amadori products. For example, tenilsetam and ethanol react with the sugar-derived moieties of glycated proteins and Amadori products, respectively,[9,53] thus, blocking further polymerization reactions. In other instances, antibodies against Amadori products are found to be useful in its inhibition. Blocking of AGE-receptors, that is, RAGE, may also be a successful step in the prevention of AGE formation.[96] However, in-depth studies are required to practically materialize the claims.

2.6.6 INHIBITION OF AGE AND PROTEIN CROSS-LINKS

Certain compounds are known to break cross-linking in the formed AGEs. They are hence not a preventive method, but rather a curative approach toward the problem of AGEs. The cross-linking between AGEs and proteins is the basis following the hardening of vessels and cardiovascular injury.[98] Hence, rupture of AGE cross-links are brilliant method to avert diabetic hurdles caused by cross-linking. Examples include Alagebrium (ALT-711), TRC4186, N-phenacylthiazolium bromide, and so on.[14,23,86]

2.7 NEED FOR NATURAL SOURCES OF INHIBITORS

So far, no inhibitor has been identified which can completely block the formation of AGEs. A combination of different inhibitors acting at different levels may be a solution. In the market, some synthetic AGE inhibitors, like aminoguanidine, are available, but they have not become a potential drug because of their side effects.[64] On the other hand, some natural products have shown promising results but they are still to be marketed at a commercial level.[10] One such compound is resveratrol, a natural phytoestrogen found in grapes.[48] Another natural compound, curcumin, an active principle isolated from turmeric, has also shown some promising results.[68] The possible approaches for the pharmacotherapeutics of diabetes include the screening of natural products, taking care of the side effects, and understanding their mechanism of action of these products.

Although synthetic inhibitors prevent AGE formation strongly but these antiglycating agents wield stern adverse effects. For example, the clinical trials of aminoguanidine were ceased due to safety concerns due to precipitation of adverse effects, such as anemia, gastrointestinal disturbances, flu-like symptoms, and rare vasculitis. Its intervention with the metabolism of vitamin B6 is a chief concern. Metformin, another synthetic antiglycating compound wields numerous adverse effects including diarrhea and.[96] Hence, there is a need to shift our focus toward a safer alternative to hypoglycemic and antiglycating agents, and plants can prove to be an effective source. Synergistic activities between varied biocomponents present in plants may provide beneficial therapeutic results for the treatment of metabolic disorders like diabetes. Moreover, plants are an ideal source that may exhibit antioxidant, antiglycating, as well as hypoglycemic activity.[83]

In the past few decades, "natural products" and "natural product mimics" have gained immense importance as sources of drugs. For centuries, humans have depended on herbal sources for effective remedy of simple as well as complicated disorders. This is evident from the modern literature as well as the historical documents, including religious scriptures which provide authentic evidence of plants as a source of medicine. Even today, around 80% of the rural population in India and more than 50% of the global population relies on plants-derived pharmaceutics for the purpose of therapy.[43] At least 40% of all medicines, drugs, and pharmaceuticals worth $100 billion per year owe their existence to the genetic resources of wild plants, mostly from the tropical forest of less-developed countries.[47] About 70% of the 3000 plants from tropical forests were identified by the National Cancer Institute (NCI) as sources of cancer-fighting chemicals between the year 1986–1991. The practice of traditional medicine is widespread in China, India, Japan, Pakistan, Sri Lanka, and Thailand. In China, 40% of the medicine comes from traditional sources. Natural products may lay the path to drug development and drug discovery in treating human diseases.[33]

2.8 ANTIDIABETIC AND ANTIGLYCATING ACTIVITIES IN PLANTS

Although many studies report antidiabetic activity of plants, very little is known about the exact mode of action of these plants. Administration of Triphala for a period of 45 days shows promising results in a significant reduction of blood glucose levels.[26] It is an ayurvedic drug consisting of equal portions of three medicinal herbs, viz. Indian Gooseberry (*Emblica officinalis*), Chebulic Myrobalan (*Terminalia chebula*) and Beleric Myrobalan (*Terminalia bellirica*). Triphala extract also inhibits protein glycation. This activity is mediated by the presence of tannins.[26] In another study, turmeric extracts have proved to be a potent antioxidant as well as antiglycating agent.[40] Tupe et al.[83] have screened the extracts of several Indian antidiabetic plant species for their antiglycating activities.

Synthetic flavonoid analogs have shown to possess antioxidant, antidiabetic, and antiglycation activities in rats.[54] *Azadirachta indica* (Neem), *Syzygium cumini* (Jambul), and *Terminalia bellirica* (Behada) are commonly used antidiabetic plants in India. They are believed to aid in the averting the diabetes and related complications. These claims have also been indicated in the animal as well as human models.[49,52,] A total of 76 medicinal plants are reviewed by Rahimi[65] for their antidiabetic potential. These plants include onion, garlic, neem, tulsi, and so on to name a few. *Ocimum* species is another plant considered as a boon for medicinal chemists. Leaf extracts of several species of these genera have a long and successful history of been used in ancient folk medicine having antioxidant, antistress, anticancer, radiation protection, antifungal, insecticidal, and several other bioactivities. *Ocimum tenuiflorum* leaf extracts exhibited hypoglycemic effects by initiating secretion of insulin from the pancreas, isolated pancreatic islets, and clonal pancreatic β-cells. However, the principal compound and its molecular mechanism of action are inadequately understood. Singh and coworkers have shown that eugenol isolated from *O. gratissimum* has a potential dual effectors role in diabetes control. It acts as an effective α-glucosidase inhibitor as well as a glycation inhibitor, thus imitating the effect of mixture therapy. Although both natural and synthetic antidiabetic agents are commercially available, the primary seems to be the a noticeable option due to its low-toxicity profile and much lesser side effects.[73]

As discussed earlier, supplementation of antioxidants with diet is an effective strategy for inhibiting AGE formation. Antioxidants can be classical molecules, such as vitamins, β-carotene, and lipoic acid. Carotene is found to be protective against hyperglycemia and suppresses lipid peroxidation and glucose tolerance ability.[77] Lipoic acid and dihydrolipoic acid have AGEs inhibitory activity through free radical scavenging and metal chelating properties. Certain thiazolium compounds, like phanacylthiazolium, similar to thiamine, which is found to inhibit AGEs formation and have antioxidative effects. Tamoxifen, which exerts antioxidant effects, has been reported to suppress H_2O_2 production.[41] The tamoxifen metabolite 4-hydroxytamoxifen is a more powerful inhibitor of lipid peroxidation than tamoxifen in lipoprotein and in the nuclear membrane.[94] Sulfasalazine may also act as a free-radical scavenger and is an excellent scavenger of several ROS, especially HOCl.[93] A recent study showed that bhasma also possesses antioxidant and free radical scavenging property, when given orally.[82]

Pepper et al.[61] have reported inhibitory activity of carnosine, folic acid, and aminoguanidine against glycation in prokaryotes. He further proposed that these agents may also prove to be beneficial to eukaryotes since the chemical processes of glycation are similar in these two domains of life. Another investigation revealed that the purified extract of *Centella asiatica* has both antiaging as well as antiglycating property.[45] *Artocarpus heterophyllus*, possessing phytochemical constituents, such as flavonoids, xanthones, and prenylated stilbenes, has also shown a strong antiglycation activity. Other studies have shown antiglycating effects of polyphenols,[79] Canadian berries (p4), *Fagopyrum tataricum*, Alligator pepper, ginger, and nutmeg, to name a few.

The continued investigation of naturally occurring antiglycating agents can prove to be a promising opportunity for complementary intervention that may be beneficial, and in fact more acceptable, to high-risk populations in search of nonpharmaceutical alternatives with reduced or no side effects. Table 2.1 represents several antidiabetic plants depending on their mode of action.

TABLE 2.1 AntiDiabetic Plants.

S. No.	Mechanism	Plants	Common Name
1	Enhancing glucose uptake by adipose and muscle tissues	*Cinnamomum zeylanicum*	Cinnamon
2		*Momordica charantia*	Bitter guard
3		*Andrographis paniculata*	Echinacea
4		*Cecropia obtusifolia*	Snakewood
5		*Amomum xanthioides*	Black cardamom
6	Inhibition of glucose absorption from the intestines	*Cornus officinalis*	Japanase cornell
7		*Alismatis Rhizoma*	Plantain
8		*Mangifera indica*	Mango
9		*Lagerstroemia speciosa*	Queens crepe
10		*Terminalia chebula*	Black myrobalan
11	Increasing insulin secretion from the pancreas	*Abies pindrow*	West Himalayan Fir
12		*Aegle marmelos*	Bael/ Golden apple
13		*Agrimony eupatoria*	Agrimony/ church steeples/ sticklewort
14		*Aloe barbadensis*	Aloe vera
15		*Annona squamosal*	Sugar apple
16	Inhibiting glucose production from hepatocytes	*Ipomoea batatas*	Sweet potato
17		*Fagopyrum tataricum*	Buckwheat
18		*Cucurbita ficifolia*	Winter squash
19		*Artemisia dracunculus*	Tarragon
20		*Agave tequilana*	Blue agave

2.9 INTRINSIC DEFENSE MECHANISMS FOR PREVENTION OF GLYCATION

Cells and tissues have an elaborate system of controlling the production of AGEs and their degradation. This is very much similar to the control mechanism of oxidative stress by a combination of antioxidant enzymes and antioxidants. There are some well-characterized molecules which help in the prevention of AGEs indirectly or directly. Glutathione, carnosine, and taurine have been shown to reduce the production of AGEs.[56,59,95] Serum albumin is another very good scavenging molecule because of the presence of hydroxyl radical scavenging activity of Cys-34.[21]

Amadori rearrangement of Schiff bases leads to the formation of ketoamines. Fructosamines are one of the best examples of a ketoamines produced in the body due to the reaction between glucose and the amino groups. A carbon and nitrogen need to be broken down to separate the ketoamine attached to amino acids or proteins. This is achieved by a diverse group of enzymes commonly known as amadoriases or deglycases: oxidases, isomerases, and kinases[69] represented in Figure 2.3. These enzymes vary from each other with respect to their site of action, substrate specificity, and location.[63]

Oxidases are found in microorganisms and act on the C-N bond of the fructosamines releasing a dicarbonyl. These flavoenzymes catalyze irreversible reactions leading to the creation of H_2O_2 from O_2. Accordingly, the eukaryotic oxidases have been located in peroxisomes. Monnier and Wu[51] have reported the occurrence of two different types of amadoriases, I and II in *Aspergillus* sp. Both the enzymes show different specificity toward the size and charge of the substrate.

Isomerases have been reported from bacteria.[91,92] These enzymes act on phosphorylated fructosamines, fructosamines 6-phosphate, generated by the action of either an ATP-dependent kinase or phosphotransferase system. Their role in deglycation is not significant because they catalyze the reversible reaction for the Amadori rearrangement through Schiff bases.[90]

The third class of deglycases is an ATP-dependent kinase, which catalyzes the phosphorylation of ketoamines. The resultant destabilization induced by phosphorylation leads to the separation of the amino group from the carbon moiety. Although the role of phosphorylation is exactly not clear, but the consumption of ATP makes the reaction irreversible.[69] The evidence that these kinases belong to the protein kinase family has helped in establishing their role as the major enzyme for protein deglycation.

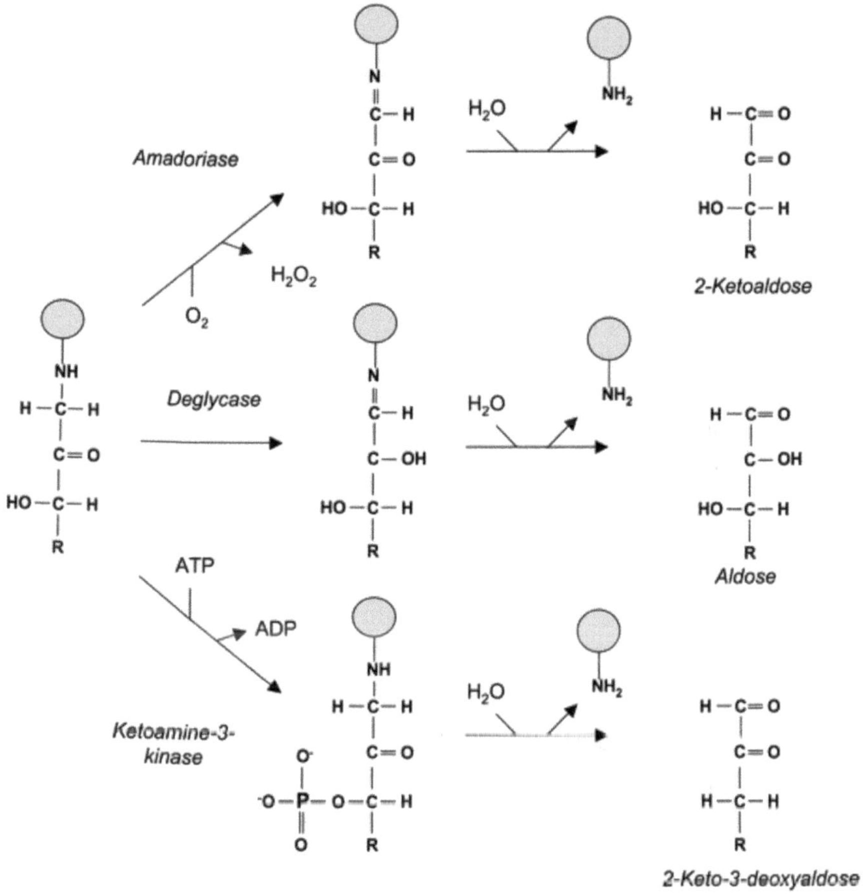

FIGURE 2.3 Types of deglycases and their mechanism of action.

Oxidases break the C-N bond in the ketoamine. Isomerases catalyze a reverse Amadori reaction and lead to the production of aldose. The third enzyme, a kinase, phosphorylates the ketoamine and results in the formation of a Dicarbonyl.

Source: Adapted from Ref. [69].

2.10 BIOCHEMICAL AND ANALYTICAL TECHNIQUES TO STUDY ANTIGLYCATION PROPERTIES OF PLANTS

Analytical techniques for the detection of biochemical parameters are of highest importance in the field of biochemistry. Many factors are considered in making a choice of detection methods. This choice depends on the advantages as well as disadvantages of a technique. Given the fact that any possible analysis will have at least one disadvantage when it comes to precision and cost-effectiveness, it is the list of our requirements that is weighed along in order to strike a balance and make a choice. For example, simple techniques which may show low precision or reproducibility may still be preferred for demonstration and teaching in practical sessions given its cost-effectiveness and ease of handling. However, every minute detail should be peered upon in designing a technique that may predict health risks or is anyhow associated with the community.

As we overview the concept of diabetes, it is believed that simply detecting the glucose levels in the blood is the ultimate statement of health in diabetic patients. If this was true, then there would be no disorders termed as "diabetes-related disorders" which is a result of glycation of proteins and RBCs not detected by routine pathological tests. Moreover, recent studies have put forward that AGEs can form not only from glucose, but also from other sugars like fructose, and carbonyl compounds derived from the sugars autoxidation as well as through other metabolic pathways. In vitro studies show clearly that fructose is a more potent glycating agent than glucose and is as much as 10-fold more efficient at forming AGEs. Among different AGEs, there is substantiation that glyceraldehyde-derived AGEs are associated with cytotoxicity.[5] Table 2.2 provides a list of biochemical and analytical techniques used for the detection of glycation and AGEs.

Simple biochemical techniques described by McPherson et al.[46] can be used for detection of glycating properties of proteins. Briefly, BSA samples can be incubated in dark with fructose prepared in phosphate-buffered saline, at 37°C for a few hours to several days. A parallel set up can be made for the detection of antiglycation activity of plants or other natural sources by incubating BSA along with herbal extracts and fructose. After incubation, the comparative analysis of unbound fructose in both the samples will give an overview of the extent of antiglycating activity of plants. This technique, however, may not be reliable for the screening of antiglycating activities of plant extracts, due to the unspecific reactivity of sugars with proteins and/or other compounds. It is for this reason that all the unbound fructose is removed from the system using dialysis technique and the remaining extracts can be tested for the presence of several reaction intermediates using colorimetric, spectroscopic as well as analytical techniques.

Since direct detection of molecules in a system may not be possible owing to their extremely small size as well as concentration, biochemical assays exploit the reactivity and color intensity of the formed products to detect the concentration of desired compounds. Several such indirect detection protocols are available for the study of antiglycation potential of plant extracts. For instance, detection of fructosamines- the Amadori product in glycated proteins can be attained with the help of a simple colorimetric assay using NBT. In this assay, the NBT solution is prepared in carbonate buffer and incubated with glycated samples at 37°C for 30 min. After which the absorbance can be measured and fructosamine concentration can be calculated using standard 1-deoxy-1-moepholinofructose curve. The NBT reduction method is also commonly used for the detection of glycated hemoglobins in diabetic patients. It is a widely accepted method used both for medical as well as nonmedical purpose since it does not require any sample preparation and is not affected by any chemical interference. Moreover, it is easily available, simple and easy to store for over 6 months. This makes it an ideal reagent for performing simple instrumental techniques like colorimeter and spectrophotometer.[83] Similarly, other methods listed in Table 2.2 like DNPH assay and DTNB assay can be used for detection of the products which contain carbonyl and thiol groups, respectively, and hence is useful to monitor the gathering of early glycation end products.

TABLE 2.2 Summary of Biochemical and Analytical Techniques for the Detection of Glycation.

S. No.	Name of the method	Uses
	Biochemical techniques	
1	Nitroblue tetrazolium (NBT) assay	Detection of Amadori products like fructosamine
2	Dinitrophenyl hydrazine (DNPH) assay	Detection of protein carbonyl group in Amadori products
3	5,5′-Dithiobis 2-nitrobenzoic acid (DTNB) assay	Detection of the thiol group in Amadori products
4	Congo Red/thioflavin T assay	Detection of amyloid cross-β-structure
5	Thiobarbituric acid colorimetric assay	Detection of Amadori products like fructosamine
	Analytical techniques	
6	ELISA	Detection of fluorescent as well as nonfluorescent AGEs and AGE antibodies.
7	HPLC	Detection of Amadori products and AGEs
8	LC-MS	Detection of Amadori products and AGEs
9	UV-visible spectroscopy	Qualitative detection of fluorescent AGEs
10	Fluorescence spectroscopy	Detection of fluorescent AGEs
11	Boronate affinity chromatography	Separation of glycated and unglycated proteins in a sample

The process of glycation induces protein conformational changes by boosting the level of amyloid cross-β-structure, which plays a pivotal role in the formation of AGEs and protein aggregation. Protein aggregation can be evaluated with the help of two β-amyloid specific dyes, viz., Congo Red and Thioflavin T. Thioflavin T connects in the channels running similar to the fibril long axis in either a monomeric form or dimeric form, whereas Congo red might attach in grooves produced along the β-sheets as a planar molecule in either a supramolecular form or monomeric form. Furthermore, fluorescence of Thioflavin T can be bias as polyphenolics are chiefly inter-relate with Thioflavin T, and hence, another method using Congo Red is advocated by Hudson et al.[31] However, both the methods are extensively used for the detection of amyloid cross-β-structures.[83]

Several analytical techniques have also been standardized for the detection of AGEs. The designing of these techniques is based on the type of AGE to be detected. AGEs are divided into three main groups[4,15,67]:

1. Cross-linking AGEs which fluoresce such as pentosidine.
2. Noncross-linking AGEs like CML.
3. Nonfluorescent cross-linking AGEs like imidazolium dilysine cross-links.

Fluorescence spectroscopic techniques can be exploited for detection and measurement of glycation reaction. Fluorescent AGEs have a characteristic fluorescent spectrum at 440 nm subsequent to excitation at 370 nm.[7] This property of AGEs has been classically used to determine their accumulation at the tissue level.[84] Detection of fluorescence is simple and noninvasive and can be used for instantaneous analysis. Excitation at 370 nm/ emission at 440 nm in a sample can be directly measured with a 96-well plate spectrophotofluorimeter.

Most widely used techniques for nonfluorescent AGEs, for example, HPLC, gas chromatography, and LC-MS have been found beneficial in its detection. The HPLC method is the gold standard for the HbA1c determination as it measures HbA1c rather than total glycosylated hemoglobin. In contrast to the previously used double-column ion-exchange HPLC, advanced single-column reverse-phase HPLC provides clear chromatograms with stable retention times. LC-MS is an ideal method for the detection of biomolecules in the majority of the cases. However, the limited database information for glycation adducts makes it unsuitable in various cases. Moreover, the cost of handling and expertise required for using chromatographic techniques limits its use to a great extent.[13]

Simpler techniques like ELISA allow rapid detection and evaluation of AGE-protein adducts. The AGE adducts quantification in protein samples is completed by evaluating its absorbance with a known AGE-BSA standard curve. In the primary step of AGE competitive ELISA, an AGE conjugate is coated onto an ELISA plate. Then the unidentified AGE-protein samples or identified AGE-BSA standards are adjoined to the AGE conjugate preabsorbed ELISA plates. The plate remains for a brief period in incubation, after which anti-AGE antibody is added, followed by a conjugated secondary antibody.[13] This technique, however, presents several drawbacks like varied antigen specificity, nonreproducibility of results, and so on. It is for this reason that the focus is now slowly shifting toward alternatives for detection of AGEs.

One such example is boronate affinity chromatography (BAC) which is based on the application of "biological affinity" for separating analytes within a sample. Much attention has been given to both boric acid and boronic acid because of their prospective applicability in recognizing sugars. Glycated proteins differ from the unglycated proteins by the adding carbohydrate moiety through ketoamine bond. Glycated proteins like glycated hemoglobin and glycated albumin contain 1,2- and 1,3-cis-vicinal diol moieties that are not present in nonglycated proteins. A robust attraction remains between boronic acid and peptide containing cis–diol groups. As a result of this, stable esters are produced between boronic acid and the glycated proteins containing a cis–diol part.[36] Thus, the diol groups in the glycated proteins offer the foundation of their separation from unglycated proteins by BAC.

2.11 CONCLUSION

In the last several decades, diabetes has become one of the major causes of death in both developed and developing countries. Hyperglycemia is accountable for many of the diabetic hurdles. One of the outcomes of hyperglycemia is the rise in the formation of AGEs. These products have been associated in several pathophysiological states which include cataract, dialysis-related amyloidosis, Alzheimer's, Parkinson's atherosclerosis, and physiological aging. The deterrence of AGEs formation is becoming a new tool in the management of diabetes. Several approaches have been used to avert the development and amassing of these products. Some artificial molecules have been tested and commercialized for their potential in inhibiting the generation of glycation products. However, many of them have been reported to have an adverse effect on the physiological systems. Because of these reasons attempts have been made to isolate the natural compounds from the plants and test their efficacy in preventing the AGEs formation. Some of these natural products have shown promising results. They can be used as a good and safe alternative for the treating diabetes and its related problems.

ACKNOWLEDGMENT

The authors thank Suji George for the proof reading of the article. This work was supported by grants awarded to AA by the University Grants Commission, New Delhi and University of Mumbai, Mumbai, India.

KEYWORDS

- **ages**
- **antiglycating agents**
- **hyperglycemia**
- **glycation**
- **natural products**

REFERENCES

1. Abbas, G.; Al-Harrasi, A. S.; Hussain, H.; Hussain, J.; Rashid, R.; Choudhary, M. I. Antiglycation Therapy: Discovery of Promising Antiglycation Agents for the Management of Diabetic Complications. *Pharma. Biol.* **2016,** *54* (2), 198–206.
2. Abhar, H. S.; Schaalan, M. F. Phytotherapy in Diabetes: Review on Potential Mechanistic Perspectives. *World J. Diabetes* **2014,** *5* (2), 176–197.
3. Abiko, T.; Abiko, A.; Clermont, A. C.; Shoelson, B.; Horio, N.; Takahashi, J.; Adamis, A. P.; King, G. L.; Bursell, S. E. Characterization of Retinal Leukostasis and Hemodynamics in Insulin Resistance and Diabetes: Role of Oxidants and Protein Kinase C Activation. Diabetes **2003,** *352*, 829–837.
4. Ahmad, S.; Siddiqui, Z. Protein Glycation: A Firm Link to Cause Metabolic Disease and Their Complications. *J. Glycomics Lipidomics* **2015,** *4*, 127.
5. Ahmed, N.; Furth, A. J. Failure of Common Glycation Assays to Detect Glycation by Fructose. Clin. Chem. **1992,** *38* (7): 1301–1303.

6. Ahmed, N. Advanced Glycation Endproducts- Role in Pathology of Diabetic Complications. *Diabetes Res. Clin. Pract.* **2005,** *67*, 3–21.
7. Akhter, F.; Khan, M. S.; Shahab, U.; Moinuddin, A. S. Bio-Physical Characterization of Ribose Induced Glycation: A Mechanistic Study on DNA Perturbations. *Int. J. Biol. Macromol.* **2013,** *58*, 206–210.
8. Akira, K.; Amano, M.; Okajima, F.; Hashimoto, T.; Oikawa, S. Inhibitory Effects of Amlodipine and Fluvastatin on the Deposition of Advanced Glycation End Products in Aortic Wall of Cholesterol and Fructose-Fed Rabbits. *Biol. Pharm. Bull.* **2006,** *29*, 75–81.
9. Al-Abed, Y.; Mitsuhashi, T.; Li, H.; Lawson, J. A.; FitzGerald, G. A.; Founds, H.; Donnelly, T.; Cerami, A.; Ulrich, P.; Bucala, R. Inhibition of Advanced Glycation Endproduct Formation by Acetaldehyde: Role in the Cardioprotective Effect of Ethanol. *Proc. Natl. Acad. Sci. USA* **1999,** *96*, 2385–2390.
10. Ali, A.; Sharma, R.; Sivakami, S. Role of Natural Compounds in the Prevention of DNA and Proteins Damage by Glycation. *Bionano Front.* **2014,** *7*(12), 25–30.
11. Ansari, N. A.; Moinuddin, A. R. Glycated Lysine Residues: A Marker for Non-Enzymatic Protein Glycation in Age-Related Diseases. *Dis. Mark.* **2011,** *30* (6), 317–324.
12. Arai, K.; Maguchi, S.; Fujii, S.; Ishibashi, H.; Oikawa, K.; Taniguchi, N. Glycation and Inactivation of Human Cu-Zn-Superoxide Dismutase. Identification of the Glycated Site In Vitro. *J. Biochem.* **1987,** *262*, 16969–16972.
13. Ashraf, J. M.; Ahmad, S.; Choi, I.; Ahmad, N.; Farhan, M.; Tatyana, G.; Shahab, U. Recent Advances in Detection of AGEs: Immunochemical.; Bioanalytical and Biochemical Approaches. *Int. Union Biochem. Mol. Biol.* **2015,** *67* (12), 897–913.
14. Asif, M.; Egan, J.; Vasan, S.; Jyothirmayi, G. N.; Masurekar, M. R.; Lopez, S.; Williams, C.; Torres, R. L.; Wagle, D.; Ulrich, P.; Cerami, A.; Brines, M.; Regan, T. J. An advanced Glycation Endproduct Cross-Link Breaker Can Reverse Age-Related Increases in Myocardial Stiffness. *Proc. Natl. Acad. Sci. USA* **2000,** *97*, 2809 2813.
15. Basta, G.; Lazzerini, G.; Massaro, M.; Simoncini, T.; Tanganelli, P. Advanced Glycation End Products Activate Endothelium through Signal-Transduction Receptor RAGE: A Mechanism for Amplification of Inflammatory Responses. *Circulation* **2002,** *105*, 816–822.
16. Bierhaus, A.; Hofmann, M. A.; Ziegler, R., Nawroth, P. P. AGEs and Their Interaction with AGE-Receptors in Vascular Disease and Diabetes Mellitus. I. The AGE Concept. *Cardiovasc. Res.* **1998,** *37*, 586–600.
17. Bras, I. D.; Colitz, C. M.; Kusewitt, D. F.; Chandler, H.; Lu, P.; Gemensky Metzler, A. J.; Wilkie, D. A. Evaluation of Advanced Glycation End Products in Diabetic and Inherited Canine Cataracts. *Graefes Arch. Clin. Exp. Ophthalmol.* **2007,** *245*, 249–257.
18. Buchanan, T. A.; Xiang, A. H.; Peters, R. K.; Kjos, S. L.; Marroquin, A.; Goico, J.; Ochoa, C.; Tan, S.; Berkowitz, K.; Hodis, H. N.; Azen, S. P. Preservation of Pancreatic Beta-Cell Function and Prevention of Type 2 Diabetes by Pharmacological Treatment of Insulin Resistance in High-Risk Hispanic Women. *Diabetes* **2002,** *51*, 2796–2798.
19. Bunn, H. F.; Gabbay, K. H.; Gallop, P. M. The Glycosylation of Hemoglobin: Relevance to Diabetes Mellitus. *Science* **1978,** *200* (4337), 21–27.
20. Caballero, F.; Gerez, E.; Batlle, A.; Vazquez, E. Preventive Aspirin Treatment of Streptozotocin Induced Diabetes: Blockage of Oxidative Status and Revertion of Heme Enzymes Inhibition. *Chem. Biol. Interact.* **2000,** *126*, 215–225.
21. Carballal, S.; Alvarez, B.; Turell, L.; Botti, H.; Freeman, B. A.; Radi, R. Sulfenic Acid in Human Serum Albumin. *Amino Acids* **2007,** *32*, 543–551.

22. Cerami, C.; Founds, H.; Nicholl, I.; Mitsuhashi, T.; Giordano, D.; Vanpatten, S.; Lee, A.; Al-Abed, Y.; Vlassara, H.; Bucala, R.; Cerami, A. Tobacco Smoke Is a Source of Toxic Reactive Glycation Products. *Proc. Natl. Acad. Sci. USA* **1997,** *94*, 13915–13920.
23. Chandra, K. P.; Shiwalkar, A.; Kotecha, J.; Thakkar, P.; Srivastava, A.; Chauthaiwale, V.; Sharma, S. K.; Cross, M. R.; Dutt, C. Phase I Clinical Studies of the Advanced Glycation End-Product (AGE)-Breaker TRC4186: Safety, Tolerability and Pharmacokinetics in Healthy Subjects. *Clin. Drug. Investig.* **2009,** *29*, 559–575.
24. Chiasson, J. L.; Josse, R. G.; Gomis, R.; Hanefeld, M.; Karasik, A.; Laakso, M. Acarbose for Prevention of Type 2 Diabetes Mellitus: The STOP-NIDDM Randomized Trial. *Lancet* **2002,** *359*, 2072–2077.
25. Crompton, M.; Rixon, K. C.; Harding, J. J. Aspirin Prevents Carbamylation of Soluble Lens Proteins and Prevents Cyanate-Induced Phase Separation Opacities In Vitro: A Possible Mechanism by Which Aspirin Could Prevent Cataract. *Exp. Eye Res.* **1985,** *40*, 297–311.
26. Ganeshpurkar, A.; Jain, S.; Agarwal, S. Experimental Studies on Glycolytic Enzyme Inhibitory and Antiglycation Potential of Triphala. *Ayu* **2015,** *36* (1), 96–100.
27. Geroldi, D.; Falcone, C.; Emanuele, E. Soluble Receptor for Advanced Glycation End Products, from Disease Marker to Potential Therapeutic Target. *Curr. Med. Chem.* **2006,** *13*, 1971–1978.
28. Gkogkolou, P.; Bohm, M. Advanced Glycation End Products Key Players in Skin Aging? *Dermato-Endocrin.* **2012,** *4* (3), 259–270.
29. Hanssen, K. F. Blood Glucose Control and Microvascular and Macrovascular Complications. *Diabetes* **1997,** *46* (2), 101–103.
30. Harris, C. S.; Cuerrier, A.; Lamont, E.; Haddad, P. S.; Arnason, J. T.; Bennett, S. A.; Johns, T. Investigating Wild Berries as a Dietary Approach to Reducing the Formation of Advanced Glycation Endproducts: Chemical Correlates of In Vitro Antiglycation Activity. Plant Foods Hum. Nutr. **2014,** *69*, 71–77.
31. Hudson, S. A.; Ecroyd, H.; Kee, T. W.; Carver, J. A. The Thioflavin T Fluorescence Assay for Amyloid Fibril Detection Can be Biased by the Presence of Exogenous Compounds. *FEBS J.* **2009,** *276*, 5960–5972.
32. Jakuš, V.; Rietbrock, N. Advanced Glycation End-Products and the Progress of Diabetic Vascular Complications. *Physiol. Res.* **2004,** *53*, 131–142.
33. Julian, K. C. *Forests: More Than Just Trees*; Infobase Publishing, 2007.
34. Jung, Y. S.; Joe, B. Y.; Cho, S. J.; Konishi, Y. 2,3-Dimethoxy-5-methyl-1,4-benzoquinones and 2-methyl-1,4-naphthoquinones: Glycation Inhibitors with Lipid Peroxidation Activity. *Bioorg. Med. Chem. Lett.* **2005,** *15*, 1125–1129.
35. Khalifah, R. G.; Baynes, J. W.; Hudson, B. G. Amadorins: Novel Post-Amadori Inhibitors of Advanced Glycation Reactions. *Biochem. Biophys. Res. Commun.* **1999,** *257* (2), 251–258.
36. Klenk, D. C.; Hermanson, G. T.; Krohn, R. I.; Fujimoto, E. K.; Mallia, A. K. Determination of Glycosylated Hemoglobin by Affinity Chromatography: Comparison with Colorimetric and Ion-Exchange Methods, and Effects of Common Interferences. *Clin. Chem.* **1982,** *28*, 2088–2094.
37. Knowler, W. C.; Barrett, C. E.; Fowler, S. E.; Hamman, R. F.; Lachin, J. M.; Walker, Diabetes Prevention Program Research Group. Reduction in the Incidence of Type 2 Diabetes with Lifestyle Intervention or Metformin. *N. Engl. J. Med.* **2002,** *346*, 393–403.
38. Kowluru, R. A.; Kenned, A. Therapeutic Potential of Anti-Oxidants and Diabetic Retinopathy. *Expert. Opin. Invest. Drugs* **2001,** *10*, 1665–1676.

39. Lee, C. C.; Lee, B. H.; Lai, Y. J. Antioxidation and Antiglycation of *Fagopyrum tataricum* Ethanol Extract. *J. Food. Sci. Technol.* **2015,** *52* (2), 1110–1116.
40. Lekshmi, P. C.; Arimboor, R.; Nisha, V. M.; Nirmala, A. M.; Raghu, K. G. In Vitro Antidiabetic and Inhibitory Potential of Turmeric (*Curcuma longa* L.) Rhizome against Cellular and LDL Oxidation and Angiotensin Converting Enzyme. *J. Food. Sci. Technol.* **2014,** *51* (12), 3910–3917.
41. Lim, J. S.; Frenkel, K.; Troll, W. Tamoxifen Suppresses Tumor Promoter Induced Hydrogen Peroxide Formation by Human Neutrophils. *Cancer Res.* **1992,** *52* (18), 4969–4972.
42. Luo, X.; Wu, J.; Jing, S.; Yan, L. J. Hyperglycemic Stress and Carbon Stress in Diabetic Glucotoxicity. *Aging Dis.* **2016,** *7* (1), 90–110.
43. Madana, N. G.; Tejaswinin, J.; Mantry, S.; Kumar, S. A. International Standards of Medicinal Plants. *Int. J. Innov. Pharm. Sci. Res.* **2014,** *2* (10), 2498–2532.
44. Makita, Z.; Vlassara, H.; Rayfield, E.; Cartwright, K.; Friedman, E.; Rodby, R.; Cerami, A.; Bucala, R. Hemoglobin-AGE, a Circulating Marker of Advanced Glycosylation. *Science* **1992,** *258*, 651–653.
45. Maramaldi, G.; Togni, S.; Franceschi, F.; Lati, E. Anti-Inflammaging and Antiglycation Activity of a Novel Botanical Ingredient from African Biodiversity (Centevita™). *Clin. Cosm. Invest. Derm.* **2014,** *7*, 1–9.
46. McPherson, J. D.; Shilton, B. H.; Walton, D. J. Role of Fructose in Glycation and Cross-Linking of Proteins. *Biochemistry* **1988,** *27*, 1901–1907.
47. Miller, G.; Tyler, Jr. *Living in the Environment*; Wadsworth Publishing Company: Belmont, CA, 1996.
48. Mizutani, K.; Ikeda, K.; Yamori, Y. Resveratrol Inhibits AGEs-Induced Proliferation and Collagen Synthesis Activity in Vascular Smooth Muscle Cells from Stroke-Prone Spontaneously Hypertensive Rats. *Biochem. Biophys. Res. Commun.* **2000,** *274* (1), 61–67.
49. Modak, M.; Dixit, P.; Londhe, J.; Ghaskadbi, S.; Thomas, P.; Devasagaya, A. Indian Herbs and Herbal Drugs Used for the Treatment of Diabetes. *J. Clin. Biochem. Nutr.* **2007,** *40*, 163–173.
50. Monnier, V. M.; Cerami, A. Nonenzymatic Browning In Vivo: Possible Process for Aging of Long-Lived Proteins. *Science* **1981,** *211*, 491–493.
51. Monnier, V. M.; Wu, X. Enzymatic Deglycation with Amadoriase Enzymes from *Aspergillus* sp. as a Potential Strategy against the Complications of Diabetes and Aging. *Biochem. Soc. Trans.* **2003,** *31*, 1349–1353.
52. Mukherjee, P. K.; Maiti, K.; Mukherjee, K.; Houghton, P. J. Leads from Indian Medicinal Plants with Hypoglycemic Potentials. *J. Ethnopharmacol.* **2006,** *106*, 1–28.
53. Munch, G.; Taneli, Y.; Schraven, E.; Schindler, U.; Schinzel, R.; Palm, D.; Riederer, P. J. The Cognition-Enhancing Drug Tenilsetam Is an Inhibitor of Protein Crosslinking by Advanced Glycosylation. *Neural Park. Dis. Dement. Sect.* **1994,** *8*, 193–208.
54. Naik, Y.; Venkatachalam, H.; Daroji, V. K.; Mathew, G.; Jayashree, B. S.; Unnikrishnan, M. K. Antidiabetic Activity of 3-Hydroxy Flavone Analogues in High Fructose Fed Insulin Resistant Rats. *EXCLI* **2014,** *13*, 1055–1074.
55. Nagai, R.; Murray, D. B.; Metz, T. O.; Baynes, J. W. Chelation: A Fundamental Mechanism of AGE Inhibitors, AGE Breakers and Other Inhibitors of AGE Complications. *Diabetes* **2012,** *61*, 549–559.
56. Nandhini, A. T.; Thirunavukkarasu, V.; Anuradha, C V. Taurine Prevents Collagen Abnormalities in High Fructose-Fed Rats. *Indian J. Med. Res.* **2005,** *122*, 171–177.

57. Nass, N.; Bartling, B.; Navarrete Santos, A.; Scheubel, R. J.; Borgermann, J.; Silber, R. E.; Simm, A. Advanced Glycation End Products, Diabetes and Ageing. Z. *Gerontol. Geriatr.* **2007,** *40* (5), 349–356.
58. NCD Risk Factor Collaboration (NCD-RisC). Worldwide Trends in Diabetes Since 1980: A Pooled Analysis of 751 Population-Based Studies with 4.4 Million Participants. *Lancet*. April 06, 2016.
59. Osawa, T.; Kato, Y. Protective Role of Antioxidative Food Factors in Oxidative Stress Caused by Hyperglycemia. *Ann NY Acad. Sci.* **2005,** *1043*, 440–451.
60. Packer, L.; Kraemer, K.; Rimbach, G. Molecular Aspects of Lipoic Acid in the Prevention of Diabetes Complications. *Nutrition* **2001,** *17*, 888–895.
61. Pepper, E. D.; Farrell, M. J.; Nord, G.; Finkel, S. E. Antiglycation Effects of Carnosine and Other Compounds on the Long-Term Survival of *Escherichia coli. Appl. Env. Microbiol.* **2010,** *76* (24), 7925–7930.
62. Piero, M. N.; Nzaro, G. M.; Njagi, J. M. Diabetes Mellitus—A Devastating Metabolic Disorder. *Asian J. Biomed. Pharm. Sci.* **2014,** *40*, 1–7.
63. Popova, E. A.; Mironova, R. S.; Odjakova, M. K. Non-Enzymatic Glycosylation and Deglycating Enzymes. *Biotech Biotech Equip.* **2010,** *24* (3), 1928–1935.
64. Rahbar, S.; Yerneni, K. K.; Scott, S.; Gonzales, N.; Lalezari, I. Novel Inhibitors of Advanced Glycation End Products. *Biochem. Biophys. Res. Commun.* **1999,** *262*, 651–656.
65. Rahimi, M. A Review: Anti Diabetic Medicinal Plants Used for Diabetes Mellitus. *Bull. Environ. Pharmacol. Life Sci.* **2015,** *4* (2), 163–180.
66. Rao, G. N.; Lardis, M. P.; Cotlier, E. Acetylation of Lens Crystallins: A Possible Mechanism by Which Aspirin Could Prevent Cataract Formation. *Biochem. Biophys. Res. Commun.* **1985,** *128*, 1125–1132.
67. Rietbrock, J. V. Advanced Glycation End-Products and the Progress of Diabetic Vascular Complications. *Physiol. Res.* **2004,** *53*, 131–142.
68. Sajithlal, G. B.; Chithra, P.; Chandrakasan, G. Effect of Curcumin on the Advanced Glycation and Cross-Linking of Collagen in Diabetic Rats. *Biochem. Pharmacol.* **1998,** *56* (12), 1607–1614.
69. Schaftingen, E. V.; Collard, F.; Wiame, E.; Cunha, M. V. Enzymatic Repair of Amadori Products. *Amino Acids* **2012,** *42* (4), 1143–1150.
70. Schalkwijk, C. G.; Stehouwer, C. D.; Van Hinsbergh, V. W. Fructose Mediated Non Enzymatic Glycation : Sweet Coupling or Bad. *PubMed Commons* **2016,** *20* (5), 1–2.
71. Seuring, T.; Archangelidi, O.; Suhrcke, M. The Economic Costs of Type 2 Diabetes: A Global Systematic Review. *Pharmacoecon.* **2015,** *33*, 811–831.
72. Shim, E.; Babu, J. P. Glycated Albumin Produced in Diabetic Hyperglycemia Promotes Monocyte Secretion of Inflammatory Cytokines and Bacterial Adherence to Epithelial Cells. *J. Periodont. Res.* **2015,** *50*, 197–204.
73. Singh, P.; Jayaramaiah, R. H.; Agawane, S. B.; Vannuruswamy, G.; Korwar, A. M.; Anand, A.; Dhaygude, V. S.; Shaikh, M. L.; Joshi, R. S.; Boppana, R.; Kulkarni, M. J.; Thulasiram, H. V.; Giri, A. P. Potential Dual Role of Eugenol in Inhibiting Advanced Glycation End Products in Diabetes: Proteomic and Mechanistic Insights. *Sci. Rep.* **2016,** *6*, 18798.
74. Singh, V. P.; Bali, A.; Singh, N.; Jaggi, A. S. Advanced Glycation End Products and Diabetic Complications. *Korean J. Physiol. Pharmacol.* **2014,** *18*, 1–14.
75. Sobal, G.; Menzel, E. J.; Sinzinger, H. Calcium Antagonists as Inhibitors of In Vitro Low Density Lipoprotein Oxidation and Glycation. *Biochem. Pharmacol.* **2001,** *61*, 373–379.

76. Son, S. M. Reactive Oxygen and Nitrogen Species in Pathogenesis of Vascular Complications of Diabetes. *Diabetes Metab. J.* **2012,** *36*, 190–198.
77. Suji, G.; Sivakami, S. Glucose, Glycation and Aging. *Biogeront* **2004,** 5, 365–373.
78. Suji, G.; Sivakami, S. Approaches to the Treatment of Diabetes Mellitus: An Overview. *Cell. Mol. Biol.* **2003,** *49* (4), 635–639.
79. Sun, C.; McIntyre, K.; Saleem, A.; Haddad, P. S.; Arnason, J. T. The Relationship between Antiglycation Activity and Procyanidin and Phenolic Content in Commercial Grape Seed Products. *Can. J. Physiol. Pharmacol.* **2012,** *90*, 167–174.
80. Tappy, L.; Le, K. A. Metabolic Effects of Fructose and the Worldwide Increase in Obesity. *Physiol. Rev.* **2010,** *90*, 23–46.
81. Thorpe, S. R.; Baynes, J. W. Maillard Reaction Products in Tissue Proteins: New Products and New Perspectives. *Amino Acids* **2003,** *25*, 275–281.
82. Tripathi, B.; Ajita, P.; Pandey, V. S.; Singh, R. S.; Kumar, B. K.; Dixit, M.; Yamini, S. K. Toxicology and Free Radicals Scavenging Property of Tamra Bhasma. *Ind. J. Clin. Biochem.* **2003,** *18* (2), 181–189.
83. Tupe, R. S.; Kemse, N. G.; Khaire, A. A.; Shaikh, S. A. Attenuation of Glycation-Induced Multiple Protein Modifications by Indian Antidiabetic Plant Extract. *Pharma. Biol.* **2016,** *55* (1), 68–75.
84. Uribarri, J.; Peppa, M.; Cai, W.; Goldberg, T.; Lu, M. Dietary Glycotoxins Correlate with Circulating Advanced Glycation End Product Levels in Renal Failure Patients. *Am. J. Kidney Dis.* **2003,** *42*, 532–538.
85. Van Boekel, M. A. M.; Van den Bergh, P. J.; Hoenders, H. J. Glycation of Human Serum Albumin: Inhibition by Diclofenac. *Biochim Biophys Acta.* **1992,** *1120*, 201–204.
86. Vasan, S.; Zhang, X.; Kapurniotu, A.; Bernhagen, J.; Teichberg, S.; Basgen, J.; Wagle, D.; Shih, D.; Terlecky, I.; Bucala, R.; Cerami, A.; Egan, J.; Ulrich, P. An Agent Cleaving Glucose-Derived Protein Crosslinks In Vitro and In Vivo. *Nature* **1996,** *382*, 275–278.
87. Verbeke, P.; Siboska, G. E.; Clark, B. F.; Rattan, S. I. Kinetin inhibits protein oxidation and glycoxidation in vitro. Biochem. Biophys. Res. Commun. **2000,** *276*, 1265–1270.
88. Webster, J.; Wilke, M.; Stahl, P.; Kientsch-Engel, R.; Munch, G. Maillard Reaction Products in Food as Pro-Inflammatory and Pro-Arteriosclerotic Factors of Degenerative Diseases. *Z. Gerontol. Geriatr.* **2005,** *38*, 347–353.
89. Wells-Knecht, K. J.; Zyzak, D. V.; Litchfield, J. E.; Thorpe, S. R.; Baynes, J. W. Mechanism of Autoxidative Glycosylation: Identification of Glyoxal and Arabinose as Intermediates in the Autoxidative Modification of Proteins by Glucose. *Biochemistry* **1995,** *34* (11), 3702–3709.
90. Wiame, E.; Delpierre, G.; Collard, F.; Van Schaftingen, E. Identification of a Pathway for the Utilization of the Amadori Product Fructoselysine in *Escherichia coli*. *J. Biol. Chem.* **2002,** *277*, 42523–42539.
91. Wiame, E.; Duquenne, A.; Delpierre, G.; Van Schaftingen, E. Identification of Enzymes Acting on Alpha-Glycated Amino Acids in *Bacillus subtilis*. *FEBS Lett.* **2004,** *577*, 469–472.
92. Wiame, E.; Lamosa, P.; Santos, H.; Van Schaftingen, E. Identification of Glucoselysine-6-Phosphate Deglycase, an Enzyme Involved in the Metabolism of the Fructation Product Glucoselysine. *Biochem J.* **2005,** *392*, 263–269.
93. Wiseman, H.; Halliwell, B. Damage to DNA by Reactive Oxygen and Nitrogen Species: Role in Inflammatory Disease and Progression to Cancer. *Biochem J.* **1996,** *313* (1), 17–29.

94. Wiseman, H.; Halliwell, B. Tamoxifen and Related Compounds Protect against Lipid Peroxidation in Isolated Nuclei: Relevance to the Potential Anticarcinogenic Benefits of Breast Cancer Prevention and Therapy with Tamoxifen? *Free Rad. Biol. Med.* **1994,** *17*, 485–488.
95. Yan, H.; Guo, Y.; Zhang, J.; Ding, Z.; Ha, W.; Harding, J. J. Effect of Carnosine, Aminoguanidine and Aspirin Drops on the Prevention of Cataracts in Diabetic Rats. *Mol. Vis.* **2008,** *14*, 2282–2291.
96. Younus, H.; Anwar, S. Prevention of Non-Enzymatic Glycosylation (Glycation): Implication in the Treatment of Diabetic Complication. *Int. J. Health Sci.* **2016,** *10* (2), 261–277.
97. Zheng, H.; Wu, J.; Jin, Z.; Yan, L. J. Protein Modifications as Manifestations of Hyperglycemic Glucotoxicity in Diabetes and Its Complications. *Biochem. Insights* **2016,** *9*, 1–9.
98. Zieman, S. J.; Melenovsky, V.; Kass, D. A. Mechanisms, Pathophysiology, and Therapy of Arterial Stiffness. *Arterioscler. Thromb. Vasc. Biol.* **2005,** *25*, 932–943.

CHAPTER 3

Perspectives of Chalcone-Based Nf-Kβ Inhibitors as Anti-Inflammatory Agents

DEBARSHI KAR MAHAPATRA[1*], VIVEK ASATI[2], and SANJAY KUMAR BHARTI[3]

[1]*Department of Pharmaceutical Chemistry, Dadasaheb Balpande College of Pharmacy, Nagpur 440037, Maharashtra, India*

[2]*Department of Pharmaceutical Chemistry, NRI Institute of Pharmacy, Bhopal 462021, Madhya Pradesh, India*

[3]*Institute of Pharmaceutical Sciences, Guru Ghasidas Vishwavidyalaya (A Central University), Bilaspur 495009, Chhattisgarh, India*

[*]*Corresponding author. E-mail: mahapatradebarshi@gmail.com*

ABSTRACT

Inflammation is an important process in the human body which is broadly mediated by two main enzymes which are cyclooxygenase-2 (COX-2) and 5-lipoxygenase (5-LOX). The former plays a crucial role in the biosynthesis of prostanoids like thromboxanes (TXs) and prostacyclins (PCs) while the latter plays a vital role in the biosynthesis of leukotrienes (LTs). Attenuating inflammatory responses by the inhibition of the proinflammatory cytokines and its symptoms may have an insightful role toward the treatment of chronic inflammatory diseases, such as inflammatory bowel disease (IBD), asthma, rheumatoid arthritis (RA), and chronic obstructive pulmonary disease (COPD). The suppression of inflammation based on the molecular targets remained an emerging perspective. NF-κB is a proinflammatory target, its activation, NF-κB-luciferase activity, NF-κB

Biologically Active Small Molecules: Modern Applications and Therapeutic Perspectives.
Debarshi Kar Mahapatra, & Sanjay Kumar Bharti (Eds.)

target genes, κBα, κB kinase-α/β, NF-κB p65, and so on. aggravates serious inflammatory conditions. Chalcones of both natural (hesperidin, isoliquiritigenin, butein, and xanthohumol) and synthetic (dihydrotriazine-chalcones, carboxamide-linked chalcone, E-α-p-OMe-C_6H_4-TMC, and L6H9) origin have been found to profoundly inhibit the above targets associated with the NF-κB pathway.

3.1 INTRODUCTION

Inflammation has remained a key concern among human civilization for centuries. Inflammation can be both external as well as internal in response to immunological and nonimmunological stimuli.[1] It is a process in which white blood cells and some important chemical mediators are produced which protect us from pathogenic organisms, such as bacteria and virus. It is a vital part of the body's immune response which is caused by a number of physical reactions triggered by the immune system in response to a physical injury or an infection.[2] It is often considered as a self-protection mechanism of the body to remove harmful stimuli, begin the healing process, and start an imperative essential process for the existence of life. Although, the process is necessary for survival and has its own importance in biology, however, the aggravation of this biological process leads to extreme conditions of discomfort, pain, and suffering.[3] Therefore, researchers argued that inflammation can persist longer than necessary, causing more harm than benefit. The etiology and treatment approaches dates back to several civilizations, which reveals the prevailing continuous obstacles to mankind.[5] It is often characterized by redness, swelling, warmth, and sometimes pain and some immobility. Inflammatory diseases involve a series of disorders, such as ankylosing spondylitis, osteoarthritis, systemic lupus erythematosus, rheumatic fever, polyarthritis nodosa, rheumatoid arthritis, Alzheimer's disease, cardiac problems, diabetic vascular complication, asthma, allergic rhinitis, psoriasis, colitis ulcerosa, inflammatory bowel disease (IBD), and so forth, which leads to granuloma formation, edema, leukocyte infiltration, respiratory conditions.[6] In the case of autoimmune diseases like arthritis, the patients' sufferings reaches several folds higher than it is expected, since it attacks the body's own tissue. Therefore, to reduce the pain and further responses, anti-inflammatory agents are seldom required.[7]

3.2 INFLAMMATORY MEDIATORS

Inflammation is an important process in the human body which is broadly mediated by two main enzymes which are cyclooxygenase-2 (COX-2) and 5-lipoxygenase (5-LOX). The former plays a crucial role in the biosynthesis of prostanoids like thromboxanes (TXs) and prostacyclins (PCs) while the latter plays vital role in the biosynthesis of leukotrienes (LTs).[8] The biochemical processes release proteins called cytokines as emergency signals that bring in the body's immune cells, hormones, and nutrients to fix the problem and brings out responses in the body milieu. Various inflammatory mediators implicated in the pathological process include prostaglandins (PG), thromboxanes, and leukotrienes.[9] The production of various PG is directed by the coordinated activity of eicosanoid forming enzymes named cyclooxygenase (COX). COX-1 functions as a housekeeping isoform of cyclooxygenase and are constitutively expressed to serve functions such as control of renal blood flow, imparting protection to the stomach against ulcers, production of prostaglandin E_2 (PGE_2) to maintain coherence and structure of the gastric mucosal surface, and production of prostanoid thromboxane in platelets.[10] These include lipopolysaccharide (LPS), interleukin-1 (IL-1), tumor necrosis factor (TNF), epidermal growth factor (EGF), platelet activating factor (PAF), serum, endothelin, and arachidonic acid.[11] The cyclooxygenase-2 (COX-2) is a primary inflammatory target, when inhibited, produces relief from pain, fever, and swelling caused by inflammatory conditions. COX-2 is an inducible early response gene and is activated in response to various extracellular or intracellular physiological stimuli.[12] By specifically only blocking COX-2 enzymes, COX-2 inhibitors relieve inflammation and pain with less adverse gastrointestinal effects than nonsteroidal anti-inflammatory drugs (NSAIDs) that inhibit both COX-1 and COX-2 enzymes.[13]

Lipoxygenases (LOXs; EC1.13.11.12) are nonheme-containing dioxygenases widely distributed in fungi, plants, and animals. The nature of its activity and its relation to unsaturated lipids in soybeans were first recognized in 1932 by Andre and Hou.[14] The biochemistry of LOX and its related products now span a period of nearly 70 years. In human, there are five lipoxygenases, the 5*S*-(arachidonate: oxygen 5-oxidoreductase), 12*R*-(arachidonate 12-lipoxygenase, 12*R*-type), 12*S*-(arachidonate: oxygen 12-oxidoreductase), and two distinct 15*S*-(arachidonate: oxygen 15-oxidoreductase). The most common analogs are 5-LOX, 8-LOX, 12-LOX, and 15-LOX, with the number indicating at which carbon the oxygen is inserted.[15] LOX is a vital enzyme that catalyzes several biochemical reactions in the inflammatory pathway, producing leukotrienes, a more potent inflammatory mediator, which plays an essential role in acute inflammation. This noniron-containing enzyme is found inside the biologically active cells, such as basophils, neutrophils, leukocytes, mast cells, macrophages, mastocytoma cells, and eosinophils which have an imperative role in the biosynthesis of LTs.[16] In several human ailments, such as psoriasis, asthma, rheumatoid arthritis, colitis ulcerosa, and allergic rhinitis, a very high level of LTs have been observed. Lipoxygenase reactions may initiate the synthesis of a signaling molecule or be involved in inducing structural or metabolic changes in the cell.[17] LOXs induces structural and metabolic changes in the cell in a number of pathophysiological conditions. LOXs catalyze the formation of hydroperoxy eicosatetraenoic acids (HPETEs) from arachidonic acid by specifically catalyzing the peroxidation steps of polyunsaturated fatty acids with a 1,4-cis, cis-pentadiene motif by inserting the oxygen moiety into the 5-carbon of arachidonic acid (stereospecific oxygenation reactions) arrested in the myeloid cells.[18] These HPETEs are subsequently reduced and transformed to form eicosanoids, which are signaling molecules that play an important regulatory role in the immune responses and other physiological processes. The produced LOX-mediated leukotriene B_4 (LTB_4) and peptidoleukotrienes leads to chemotaxis and potent bronchoconstriction.[19] The LOXs converts arachidonic acid into thromboxane family and is the most potent inflammatory and allergic mediator and enhances inflammatory responses and often precipitate unwanted chemical reactions which ultimately results in the aggravation of disease conditions, such as arthritis, bronchial asthma, IBD, and so on.[20]

3.3 TREATMENT APPROACHES

With the progress of time, new inhibitors have been developed, but the majority of them demonstrated similar complications which restricted their long-duration therapy due to the manifestation through the identically associated biochemical pathway(s). For the suppression of these inflammatory conditions, primarily 5-LOX, cyclooxygenase-1/2 (which catalyzes the formation of arachidonic acid to prostaglandins H2), and thromboxane A2 are the key targets.[21] NSAIDs block the COX enzymes and reduce the production of PG. The best way of reducing the inflammation is the complete termination of LT production, which can be achieved directly or indirectly by the inhibition of the lipoxygenase pathway.[22] Therefore, inflammation, pain, and fever are reduced by all COX inhibitors. The availability of starting material, ease of synthesis, easy separation techniques for purification, and multifarious pharmacological activities (anti-inflammatory, analgesic, anti-pyretic, etc.) made them popular among the scientific community.[23]

The modern therapeutics have seen several complications in long-term usage. Particularly in treating inflammatory conditions for a long-duration range, many problems, such as stomach irritation (aspirin), bleeding in the stomach (ibuprofen), cardiac complications (celecoxib), etc. are frequently encountered among the patients.[24] Like all NSAIDs, a higher risk of stroke and heart attack has been associated with COX-1/2 mixed inhibitors. COX-2 inhibitors may increase the risk of serious, even fatal, stomach and intestinal adverse reactions, such as bleeding, ulcers, and perforation of the stomach or intestines.[25] These events occur at any time during treatment and without warning symptoms. Elderly patients are at greater risk for these types of reactions.[26] Dated historically, several natural plant products have been seen in the available ancient manuscript which has numerous claims and prejudices. Though, all the experimentally and clinically approved molecules do have one or the other complications which have hindered long-term applications.[27] However, attempts are continuously being made to overcome these associated issues of anti-inflammatory drugs by diverse rational approaches to achieve an ideal drug.

3.4 DRUG DESIGN, DISCOVERY, AND DEVELOPMENT

The drug discovery is an endless process and its evidenced progress has endeavored several leads which are of immense importance for mankind. Several unexplored classes of the chemical have been explored through sophisticated techniques and a lot more is yet more to discover and need to be explored.[28] Over the years, with the discovery of aspirin, the wonder blockbuster drug, several eligible candidates appeared in the market and nearly all heterocycles have been preclinically studied for any sort of anti-inflammatory activity. The heterocyclic compounds have emerged as potent medicinal agents in the 21st century.[29] In the market, nearly 75% of the modern drugs are of heterocyclic origin. The combination of heteroatoms in the low molecular weight compounds offers strong binding with the biological targets which may be a receptor, enzyme, or a channel by forming hydrogen bonding, Van der Waals forces, or may even covalent bonding in extreme cases.[30] A great pace has been observed in the area of rational drug discovery of novel nonsteroidal anti-inflammatory candidates (NSAIC) where the pharmacological potentials of the multiple fused rings (four to five rings) need to explore and further optimize them by rational approaches. A number of anti-inflammatory drugs of both steroidal and nonsteroidal nature (comprising of nitrogen and oxygen heterocycles) with attributes of inhibiting different mediators, such as cyclooxygenase-1/2, lipoxygenase,and so on have received attention for the pharmacotherapeutics, but due to some or the other factors like compromised pharmacokinetic, unwanted side effects.[31] People have often shifted to herbal therapeutics for a better patient compliance, but the biological activity comprised significantly no considerable advantages in side effects. The emerging world has seen numerous drug discoveries over the past few decades.[32] In the modern concept of drug design, target-based development of new drug molecules is the most promising approach.

3.5 CHALCONES

Chalcone is a natural product present abundantly in nature. It consists of a benzylideneacetophenone component, often referred to as prop-2-ene-1-one where the two aromatic nuclei are joined by a three-carbon α, β unsaturated carbonyl bridge.[33] The term "chalcone" was first termed by Tambor and Kostanecki in the year 1899 for designating these natural chromophoric products. They are the precursor of flavonoids and isoflavonoids, open-chain intermediates in aurones synthesis, isomeric forms of flavonones, and later interconvert easily into flavonone. They are also Michael acceptors in Michael addition reactions.[34] This scaffold is known to exert multiple pharmacological activities, such as antifungal, antiprotozal, antiobesity, hypnotic, antidiabetic, antigout, antihistaminic, hepatoprotective, antioxidant, anti-invasive, antiplatelet, antiretroviral, antiarrhythmic, osteogenic, antitubercular, antimetastatic, sedative, hypolipidemic, antifibrinogenic, antitrypanosomal, antileishmanial, anti-inflammatory, antihypertensive, analgesic, antimalarial, immunosuppressive, cardioprotective, anxiolytic, antiulcer, antimicrobial, antinociceptive, artificial sweetener, etc.[35–40]

The scaffold has received global attention in this century owing to the simplicity of the structure, noncomplicated synthetic steps, ease of computational studies, the presence of a number of replaceable hydrogen, and a variety of promising biological activities.[41] The inhibitory perspectives of the scaffold include critical inhibition of the enzymes, channels, and receptors involved in hyperactive states in any biomolecular route and play a crucial role in the exaggeration of the diseases. The profile involved inhibition of Cystic Fibrosis Transmembrane Conductance Regulator (CFTR), tyrosine kinase, epoxide hydroxylase, quinone reductase, xanthine oxidase, along with LTD_4 antagonism, increasing p53 (tumor suppressor protein) level, preventing ovarian cancer cell proliferation, modulating multidrug resistance offered by P-gp (P-glycoprotein), etc.[42–44]

In the due course, numerous nontherapeutic applications are well known. The scaffold played an imperative role in the structural elucidation of a variety of natural products, such as flavonoid, coumarin, flavanone, chromane, chromanochromane, tannin, etc.[45] The derivatives have been reported to be applied as insecticide, scintillator, organic brightening agent, fluorescent whitening agent, polymerization catalyst, fluorescent polymers, analytical receptor for iron determination.[46–47]

3.6 NF-κβ

NF-κB is a component in a proinflammatory signaling pathway which stimulates the release of cytokines, chemokines, and adhesion molecules by promoting the expression of proinflammatory genes.[48] The factor is often found to be highly activated in chronic inflammatory diseases, such as chronic obstructive pulmonary disease (COPD), IBD, rheumatoid arthritis (RA), and asthma.[49] A complex role of this component has been found in the pathogenesis of both inflammation and cancer. The potential role of NF-κB has also been perceived in the immune system.[50] The hyperactive state of NF-κB has been implicated as the expression of IL-1 and TNFα signaling pathway and attracted the pharmaceutical scientists toward the development of anti-inflammatory drugs.[51]

3.7 CHALCONES AS NF-KB INHIBITORS

The dose-dependent inhibitory potential of butein **(1)** against TNF-α-induced NF-κB activation was studied by Pandey et al. (Figure 3.1). The natural product also inhibited the proliferative components, such as IκBα kinase (IKK) and NF-κB regulated gene products that are actively involved in the proliferation by directly interacting with the Cys179 residue.[52]

The NF-κB inhibitory attribute of isoliquiritigenin **(2)**, a chalcone isolated from the leaves of *Glycyrrhiza inflate* have been studied by the researchers. The natural chalcone significantly reduced the adipose tissue inflammation by inflammasome-independent and well as inflammasome-dependent pathways which lead to acute suppression of IκBα phosphorylation and also decreased the macrophage stimulation, ultimately leading to enhanced anti-inflammatory response.[53]

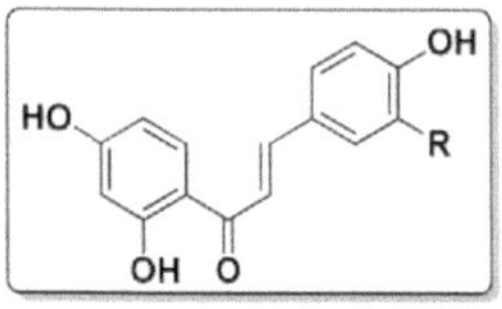

(1) R = OH; (2) R = H

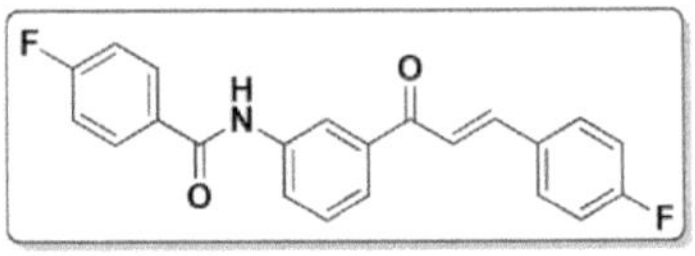

(3)

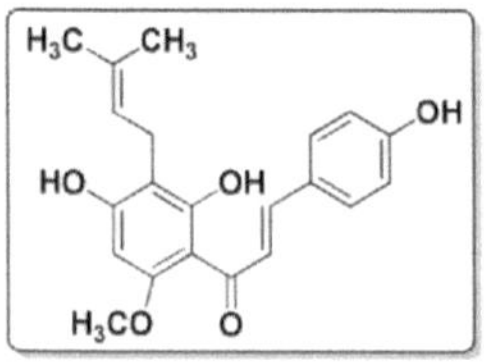

(4)

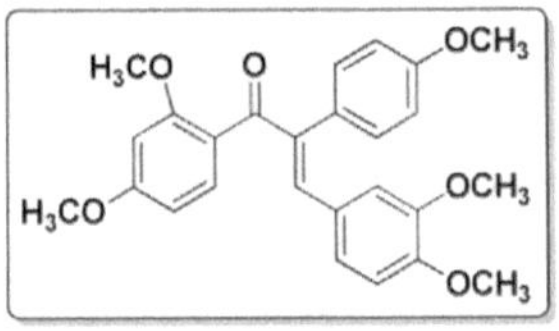

(5)

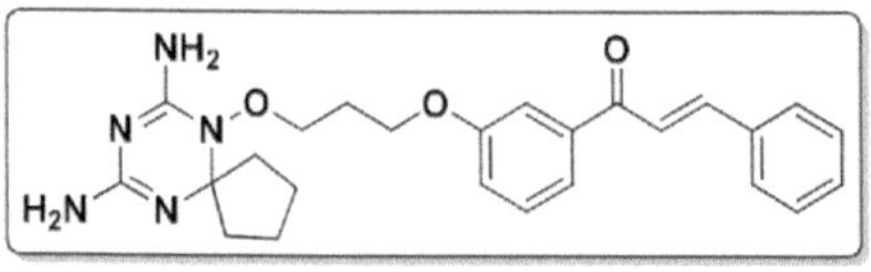

(6)

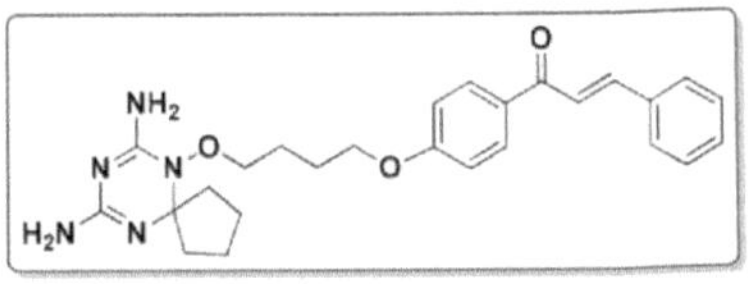

(7)

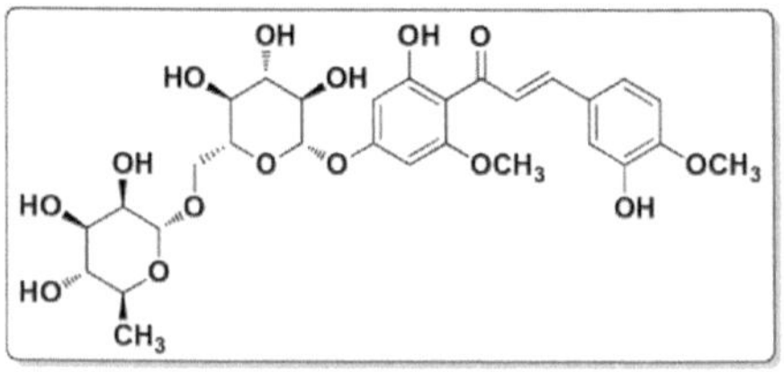

(8)

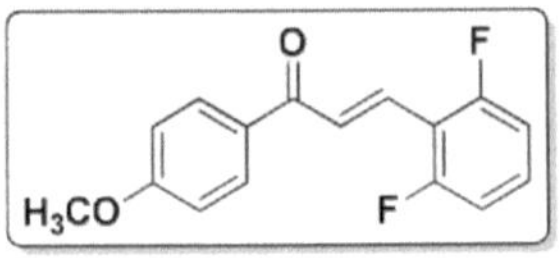

(9)

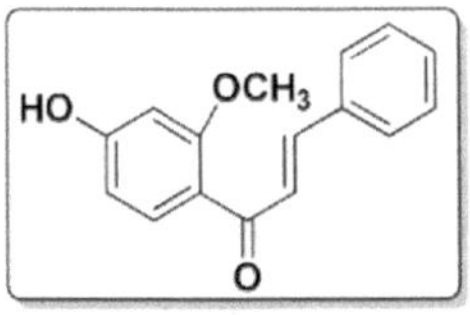

(10)

FIGURE 3.1 List of some chalcone-based NF-κβ inhibitors.

The inhibitory characteristics of carboxamide-linked chalcone **(3)** against LPS-stimulated NF-κB activation was studied. The compound suppressed the NF-κB activation with IC_{50} of 1.12 μM along with considerable inhibition of mediators, such as PG and interleukins.[54]

A prenylated natural chalcone obtained from *Humulus lupus* L., xanthohumol **(4)** has been isolated by Gao et al. and was found to express a noteworthy immunosuppressive response by profoundly inhibiting the anti-inflammatory target NF-κB as well as interleukins, T-cells, lymphocytes, cytokines -mediated immune responses.[55]

E-α-p-OMe-C_6H_4-TMC **(5)**, which is a novel compound has been screened for the dose-dependent NF-κB suppression in LPS-treated RAW264.7 macrophages and inhibition of proinflammatory factors in HK-2 and Jurkat cells with no substantial toxicity.[56]

The synthetically hybridized dihydrotriazine-chalcones **(6–7)** scaffold have been screened for anti-inflammatory prospective by Gan and coworkers. The hybrid compounds inhibited the NF-κB pathway in a dose-dependent manner by targeting the cysteine residue in IKKa/β along with LPS-induced iNOS suppression. The hybrids were found to be much more active than the parent compounds, such as chalcone and dihydrotriazine.[57]

Interestingly, Felipe et al. presented the anti-inflammatory potential of water-soluble hesperidin methyl chalcone **(8)** in Swiss albino rats in the dose of 30 mg/kg b.w. The chalcone compound suppressed the edema by the mixed mechanism (TRPV1 and NF-κB inhibition) along with prevention of carrageenan-induced cytokine (IL-1, IL-6, and IL-10) release.[58]

A noteworthy NF-κB blockade has been produced by a novel fluorinated chalcone derivative L6H9 **(9)** which led to a tremendous protection against the high glucose-induced anti-inflammatory activity. The chalcone drastically reduced the cardiac cytokine expression, Nrf2 levels, and decrease diabetes-induced inflammation.[59]

Ren et al. reported the potential of cardamonin **(10)** in mediating NF-κB p65 blockade, exhibiting the inhibition of NF-κB-luciferase activity, downregulating NF-κB target genes, κBα inhibition, and suppression of κB kinase-α/β activity in dextran sulfate sodium-induced mouse which marked the improved anti-inflammatory conditions.[60]

3.8 CONCLUSION

The discovery and rational development of new anti-inflammatory drugs based on molecular targets represent a rising approach. Attenuating inflammatory responses and its symptoms by the suppression of NF-κB activation, NF-κB-luciferase activity, NF-κB target genes, κBα, κB kinase-α/β, NF-κB p65, remained an emerging perspective. Chalcones of both natural (hesperidin, isoliquiritigenin, butein, and xanthohumol) and synthetic (dihydrotriazine-chalcones, carboxamide-linked chalcone, E-α-p-OMe-C_6H_4-TMC, and L6H9) origin have been found to profoundly inhibit the above targets associated with the NF-κB pathway. The inhibition of the proinflammatory cytokines by the repression of NF-κB may have an insightful role toward the treatment of chronic inflammatory diseases, such as IBD, asthma, RA, and COPD.

KEYWORDS

- **chalcone**
- **inflammation**
- **anti-inflammatory**
- **NF-κB**
- **inhibitor**
- **molecular target**

REFERENCES

1. Joseph, T. M.; Mahapatra, D. K. A Nascent Step towards the Discovery of New Generation Non-Steroidal Anti-inflammatory Agents (NSAIDs): Induced-Fit Molecular Docking and Pharmacokinetic Prediction Studies of Some Chromene Derivatives as Potential Cyclooxygenase-2 (COX-2) and Lipoxygenase (LOX) Inhibitors. *Res. Rev.: J. Bioinfo.* **2018,** *5* (2), 30–36.
2. Joseph, T. M.; Mahapatra, D. K. In Silico Induced-Fit Molecular Docking and Pharmacokinetic Investigations of Pentacyclic Compounds as Promising Inhibitors of Cyclooxygenase-2 (COX-2) and Lipoxygenase (LOX). *Res. Rev.: J. Drug Design Discov.* **2018,** *5* (2), 13–17.
3. Joseph, T. M.; Mahapatra, D. K. In Silico Molecular Docking of Cuminaldehyde and Its Bioconverted Ligands as Lipoxygenase and Cyclooxygenase-2 Inhibitors. *Inventi. Impact: Mol. Model.* **2017,** *2017* (3), 118–121.
4. Joseph, T. M.; Mahapatra, D. K. 5-Lipoxygenase and Phospholipase A2 Inhibitory Potentials of Alcoholic Extract of *Cyperus rotundus*: In Vitro and in Silico Study. *Res. Rev.: J. Pharmacol.* **2018,** *8* (2), 1–5.
5. Joseph, T. M.; Sharanya, C. S.; Mahapatra, D. K.; Suresh, K. I.; Sabu, A.; Haridas, M. In Vitro Assessment of Selected Benzoic Acid Derivatives as Anti-inflammatory Compounds. *J. Sci. Indus. Res.* **2018,** *77* (6), 330–336.
6. Mahapatra, D. K.; Chhajed, S. S.; Shivhare, R. S. Halogen Substituted Murrayanine Based Chalcone as Emerging Non-Steroidal Anti-Inflammatory Drug. *ARC J. Pharm. Sci.* December **2018,** *4* (4), 29–33.

7. Mahapatra, D. K.; Shivhare, R. S. Substituting Tert-butyl Group on Murrayanine-Chalcone Scaffold Produced Tremendously High Anti-oxidant Activity than the Individual Components. *Int. J. Analy. Med. Chem.* **2018,** *1* (1), 1–5.
8. Mahapatra, D. K.; Shivhare, R. S.; Kumar, P. Murrayanine-Chalcone Transformed Into Novel Pyrimidine Compounds Demonstrated Promising Anti-Inflammatory Activity. *Asian J. Pharm. Res.* **2018,** *8* (1), 6–10.
9. Mahapatra, D. K.; Shivhare, R. S.; Haldar, A. G. M. Novel Schiff's Base Containing Murrayanine–1,3,4-Thiadiazole Hybrids as Potential Anti-Inflammatory Agents. *Asian J. Chem. Pharm. Sci.* **2017,** *2* (2), 27–32.
10. Mahapatra, D. K.; Shivhare, R. S. 3′,4′-Methylenedioxy Moiety Containing Murrayanine Based Chalcone as Emerging Anti-inflammatory Agent. *J. Modern Chem. Chem. Technol.* **2018,** *9* (1), 12–16.
11. Mahapatra, D. K.; Dadure, K. M.; Haldar, A. G. M. Exploring the Anti-Inflammatory Potentials of N-((5-(((1,3-dioxoisoindolin-2-yl)methyl)amino)-1,3,4-thiadiazol-2-yl) methyl)benzamide. *Drug Discov.* **2018,** *12*, 42–45.
12. Mahapatra, D. K.; Shivhare, R. S. Anti-inflammatory Potential of a Novel Imidazole Containing Murrayanine Based Chalcone. *Modern App. Pharm. Pharmacol.* **2018,** *2* (2), MAPP.000533. 2018 (1–4 pp).
13. Mahapatra, D. K.; Dadure, K. M.; Shivhare, R. S. Edema Reducing Potentials of Some Emerging Schiff's bases of Murrayanine. *MOJ Bioorg. Org. Chem.* **2018,** *2* (4), 172–175.
14. Mahapatra, D. K.; Shivhare, R. S.; Ugale, V. G. Anti-Inflammatory Potentials of Some Novel Murrayanine Containing 1,3,4-Oxadiazole Derivatives. *Asian J. Pharm. Technol.* **2018,** *8* (1), 47–51.
15. Mahapatra, D. K.; Dadure, K. M.; Shivhare, R. S. Murrayanine-1,3,4-Thiadiazole-Uracil Hybrid as Emerging Anti-inflammatory Candidate. *Madridge J. Novel Drug Res.* **2018,** *2* (1), 75–78.
16. Mahapatra, D. K.; Ugale, V. G.; Shivhare, R. S. Murrayanine-1,3,4-Oxadiazole-Uracil Hybrid: The Emerging Anti-inflammatory Candidate. *Int. J. Pharma Sci. Res.* **2018,** *9* (8), 99–102.
17. Mahapatra, D. K.; Thote, L. T.; Dhabarde, D. M. The Emerging Anti-Inflammatory Potentials of Some Novel Schiff's Base Derivatives of 3,4 dioxymethylene Acetophenone. *J. Appl. Pharm. Sci. Res.* **2018,** *1* (3), 1–4.
18. Mahapatra, D. K.; Dadure, K. M.; Shivhare, R. S. Exploring the Site-Specific Influence of Hydroxyl group in Ring-B of Murrayanine-Chalcone on Edema Reducing Potential. *MOJ Drug Design Dev. Ther.* **2018,** *2* (4), 191–194.
19. Mahapatra, D. K.; Shivhare, R. S.; Bhat, A. R. Piperidine Containing Murrayanine-Chalcones as Emerging Bactericidal and Fungicidal Agents. *J. Pharm. Pharm.* **2018,** *5* (2), 88–91.
20. Mahapatra, D. K.; Shivhare, R. S.; Bhat, A. R. Piperazine Containing Murrayanine-Chalcones as Emerging Anti-microbial Agents. *J. Adv. Res. Pharm. Sci. Pharmacol. Interv.* **2018,** *2* (2), 12–15.
21. Craig, C. R.; Stitzel, R. E. *Modern Pharmacology with Clinical Applications*; Lippincott Williams & Wilkins: Philadelphia, 2004.
22. Brunton, L.; Parker, K.; Blumenthal, D.; Buxton, I. *Goodman & Gilman's Manual of Pharmacology and Therapeutics*; The McGraw-Hill: New York, 2008.
23. Colledge, N. R.; Walker, B. R.; Ralston, S. H. *Davidson's Principles and Practice of Medicine*; Elsevier Ltd.: Amsterdam, 2010.

24. Katzung, B. G.; Masters, S. B.; Trevor, A. J. *Katzung's Basic and Clinical Pharmacology*; McGraw-Hill Education: New York, 2012.
25. Lemke, T. L.; Williams, D. A. *Foye's Principles of Medicinal Chemistry*; Lippincott Williams & Wilkins: Philadelphia, 2012.
26. Beale, J. M.; Block, J. H. *Wilson and Gisvold's Textbook of Organic Medicinal and Pharmaceutical Chemistry*; Lippincott Williams & Wilkins: Philadelphia, 2011.
27. Golan, D. E. *Principles of Pharmacology: The Pathophysiologic Basis of Drug Therapy*; Lippincott Williams & Wilkins: Philadelphia, 2008.
28. Mahapatra, D. K.; Bharti, S. K., Eds. *Handbook of Research on Medicinal Chemistry: Innovations and Methodologies*; Apple Academic Press: New Jersey, 2017.
29. Mahapatra, D. K.; Bharti, S. K., Eds. *Medicinal Chemistry with Pharmaceutical Product Development*; Apple Academic Press: New Jersey, 2019.
30. Chhajed, S. S.; Bastikar, V.; Bastikar, A. V.; Mahapatra, D. K. *Computer Aided Drug Design*. Everest Publishing House: Pune, 2019.
31. Chhajed, S. S.; Upasani, C. D.; Wadher, S. J.; Mahapatra, D. K. *Medicinal Chemistry*; Career Publications Private Limited: Nashik, 2017.
32. Mahapatra, D. K.; Bharti, S. K. *Drug Design*; Tara Publications Private Limited: New Delhi, 2016.
33. Singh, P., Anand, A., Kumar, V. Recent Developments in Biological Activities of Chalcones: A Mini Review. *Eur. J. Med. Chem.* **2014,** *85*, 758–777.
34. Sahu, K. N.; Balbhadra, S.; Choudhary, J.; Kohli, V. D. Exploring Pharmacological Significance of Chalcone Scaffold: A Review. *Curr. Med. Chem.* **2012,** *19* (2), 209–225.
35. Mahapatra, D. K.; Bharti, S. K.; Asati, V. Anti-Cancer Chalcones: Structural and Molecular Target Perspectives. *Eur. J. Med. Chem.* **2015,** *98*, 69–114.
36. Mahapatra, D. K.; Bharti, S. K.; Asati, V. Chalcone Scaffolds as Anti-Infective Agents: Structural and Molecular Target Perspectives. *Eur. J. Med. Chem.* **2015,** *101*, 496–524.
37. Mahapatra, D. K.; Bharti, S. K.; Asati, V. Chalcone Derivatives: Anti-inflammatory Potential and Molecular Targets Perspectives. *Curr. Topics Med. Chem.* **2017,** *17* (28), 3146–3169.
38. Mahapatra, D. K.; Asati, V.; Bharti, S. K. Chalcones and Their Therapeutic Targets for the Management of Diabetes: Structural and Pharmacological Perspectives. *Eur. J. Med. Chem.* **2015,** *92*, 839–865.
39. Mahapatra, D. K.; Bharti, S. K. Therapeutic Potential of Chalcones as Cardiovascular Agents. *Life Sci.* **2016,** *148*, 154–172.
40. Mahapatra, D. K.; Bharti, S. K.; Asati V. Recent Perspectives of Chalcone Based Molecules as Protein Tyrosine Phosphatase 1B (PTP-1B) Inhibitors. In *Medicinal Chemistry with Pharmaceutical Product Development*; Mahapatra, D. K., Bharti, S. K., Eds.; Apple Academic Press, 2018: New Jersey.
41. Zhou, B.; Xing, C. Diverse Molecular Targets for Chalcones with Varied Bioactivities. *Med. Chem.* **2015,** *5* (8), 388.
42. Mathew, B.; Suresh, J.; Anbazghagan, S.; Paulraj, J.; Krishnan, G. K. Heteroaryl Chalcones: Mini Review About Their Therapeutic Voyage. *Biomed. Prev. Nutr.* **2014,** *4* (3), 451–458.
43. Zhuang, C.; Zhang, W.; Sheng, C.; Zhang, W.; Xing, C.; Miao, Z. Chalcone: A Privileged Structure in Medicinal Chemistry. *Chem. Rev.* **2017,** *117* (12), 7762–7810.

44. Nasir Abbas Bukhari, S.; Jasamai, M.; Jantan, I. Synthesis and Biological Evaluation of Chalcone Derivatives (Mini Review). *Mini Rev. Med. Chem.* **2012,** *12* (13), 1394–1403.
45. Matos, M. J.; Vazquez-Rodriguez, S.; Uriarte, E.; Santana, L. Potential Pharmacological Uses of Chalcones: A Patent Review (from June 2011–2014). *Expert Opin. Therap. Patents* **2015,** *25* (3), 351–366.
46. Patil, C. B.; Mahajan, S. K.; Katti, S. A. Chalcone: A Versatile Molecule. *J. Pharm. Sci. Res.* **2009,** *1* (3), 11.
47. Nasir Abbas Bukhari, S.; Jasamai, M.; Jantan, I.; Ahmad, W. Review of Methods and Various Catalysts Used for Chalcone Synthesis. *Mini-Rev. Org. Chem.* **2013,** *10* (1), 73–83.
48. Lenardo, M. J.; Baltimore, D. NF-κB: A Pleiotropic Mediator of Inducible and Tissue-Specific Gene Control. *Cell* **1989,** *58* (2), 227–229.
49. Schreck, R.; Albermann, K. A. J.; Baeuerle, P. A. Nuclear Factor kB: An Oxidative Stress-Responsive Transcription Factor of Eukaryotic Cells (a Review). *Free Radical Res. Commun.* **1992,** *17* (4), 221–237.
50. Serasanambati, M.; Chilakapati, S. R. Function of Nuclear Factor Kappa B (NF-kB) in Human Diseases-a Review. *South Indian J. Biol. Sci.* **2016,** *2* (4), 368–387.
51. Yadav, V. R.; Prasad, S.; Sung, B.; Aggarwal, B. B. The Role of Chalcones in Suppression of NF-κB-Mediated Inflammation and Cancer. *Int. Immunopharmacol.* **2011,** *11* (3), 295–309.
52. Pandey, M. K.; Sandur, S. K.; Sung, B.; Sethi, G.; Kunnumakkara, A. B.; Aggarwal, B. B. Butein, a Tetrahydroxychalcone, Inhibits Nuclear Factor (NF)-κB and NF-kB-Regulated Gene Expression through Direct Inhibition of IkB Kinase β on Cysteine-179 Residue. *J. Biol. Chem.* **2007,** *282* (24), 17340–17350.
53. Watanabe, Y.; Nagai, Y.; Honda, H.; Okamoto, N.; Yamamoto, S.; Hamashima, T.; Ishii, Y.; Tanaka, M.; Suganami, T.; Sasahara, M.; Miyake, K. Isoliquiritigenin Attenuates Adipose Tissue Inflammation In Vitro and Adipose Tissue Fibrosis through Inhibition Of Innate Immune Responses in Mice. *Sci. Rep.* **2016,** *6*, 23097.
54. Chu, J.; Guo, C. L. Design and Discovery of Some Novel Chalcones as Antioxidant and Anti-Inflammatory Agents via Attenuating NF-κB. *Archiv der Pharmazie* **2016,** *349* (1), 63–70.
55. Gao, X.; Deeb, D.; Liu, Y.; Gautam, S.; Dulchavsky, S. A.; Gautam, S. C. Immunomodulatory Activity of Xanthohumol: Inhibition of T Cell Proliferation, Cell-Mediated Cytotoxicity and Th1 Cytokine Production through Suppression of NF-κB. *Immunopharmacol. Immunotoxicol.* **2009,** *31* (3), 477–484.
56. Kaufmann, K. B.; Gothwal, M.; Schallner, N.; Ulbrich, F.; Ruecker, H.; Amslinger, S.; Goebel, U. The Anti-Inflammatory Effects of E-α-(p-methoxyphenyl)-2′, 3, 4, 4′-Tetramethoxychalcone Are Mediated via HO-1 Induction. *Int. Immunopharmacol.* **2016,** *35*, 99–110.
57. Gan, F. F.; Zhang, R.; Ng, H. L.; Karuppasamy, M.; Seah, W.; Yeap, W. H.; Ong, S.M.; Hadadi, E.; Wong, S.C.; Chui, W.K.; Chew, E. H. Novel Dual-Targeting Anti-Proliferative Dihydrotriazine-Chalcone Derivatives Display Suppression of Cancer Cell Invasion and Inflammation by Inhibiting the NF-κB Signaling Pathway. *Food Chem. Toxicol.* **2018,** *116*, 238–248.
58. Pinho-Ribeiro, F. A.; Hohmann, M. S.; Borghi, S. M.; Zarpelon, A. C.; Guazelli, C. F.; Manchope, M. F.; Casagrande, R.; Verri, Jr., W. A. Protective Effects of the Flavonoid

Hesperidin Methyl Chalcone in Inflammation and Pain in Mice: Role of TRPV1, Oxidative Stress, Cytokines and NF-κB. *Chem. Biol. Interact.* **2015,** *228*, 88–99.
59. Zhong, P.; Wu, L.; Qian, Y.; Fang, Q.; Liang, D.; Wang, J.; Zeng, C.; Wang, Y.; Liang, G. Blockage of ROS and NF-κB-Mediated Inflammation by a New Chalcone L6H9 Protects Cardiomyocytes from Hyperglycemia-Induced Injuries. *Biochimica et Biophysica Acta* **2015,** *1852* (7), 1230–1241.
60. Ren, G.; Sun, A.; Deng, C.; Zhang, J.; Wu, X.; Wei, X.; Mani, S.; Dou, W.; Wang, Z. The Anti-Inflammatory Effect and Potential Mechanism of Cardamonin in DSS-Induced Colitis. *Am. J. Physiol.-Gastrointestinal Liver Physiol.* **2015,** *309* (7), G517–G527.

CHAPTER 4

Coronaviruses: Molecular Modeling and Docking Studies of Potential Drug Candidates Showing Anti-Covid-19 Activity

SEFA CELIK[1*], ALIYE DEMET DEMIRAG[2], SEVIM AKYUZ[3], and AYSEN E. OZEL[1]

[1]Istanbul University, Faculty of Science, Department of Physics, Vezneciler, Istanbul, 34134 Turkey

[2]Institute of Graduate Studies in Sciences, Istanbul University, Istanbul 34452, Turkey

[3]Istanbul Kultur University, Faculty of Science and Letters, Department of Physics, Bakirkoy-Istanbul 34156, Turkey

[*]Corresponding author. E-mail: scelik@istanbul.edu.tr

ABSTRACT

The coronaviruses, which have threatened human health since 1937, have today brought the global epidemic to a different dimension with new mutations. With the development of technology, scientists from all over the world have started to look for ways to destroy or reduce this virus, both by vaccines and existing drugs. Treatment for this viral infection include either to develop an antiviral molecule that stops protein synthesis or to produce antibodies against pathogens. Especially the vaccines that are developed now have been carried out by producing antibodies against pathogens. The general rationale in drug research is to prevent the virus from attaching to

Biologically Active Small Molecules: Modern Applications and Therapeutic Perspectives.
Debarshi Kar Mahapatra, & Sanjay Kumar Bharti (Eds.)

the cell by creating a change in the cell surface where the drug molecule attaches with the COVID-19 virus. For this purpose, it is necessary to know the genome sequences of the virus well. The emergence of new mutations of the new type of coronavirus has intensified the research of scientists toward determining the nucleotide sequences of new mutations well and developing more effective vaccines and drug molecules. Molecular simulations play a key role in the development of effective vaccines and drugs. In this chapter, the structure of the coronavirus, the drugs used in the treatment process, and the molecular modeling and molecular docking studies on the discovery of the effective molecule against the coronavirus are discussed.

4.1 INTRODUCTION: THE DISEASE THAT SHOOK THE WORLD IN 2019

This epidemic gripped the world in December 2019 and became everyone's nightmare resulting in the death of millions of people worldwide and has turned into a catastrophe threatening global public health. All coronaviruses that cause infections in humans are found in alpha and beta genera. Although this disease, which is often confused with colds and flu, shows the same symptoms, it can be very severe in the later stages. The human coronavirus (HCoV) causes upper respiratory tract infections (flu, cold, cold, etc.).[1] The novel coronavirus is closely related to SARS-CoV that emerged in China in 2002. Therefore, it is identified as SARS-CoV-2, which comes from the beta family of coronavirus.[2] Globally, efforts are underway to develop vaccines and drugs to both find an effective treatment for the new coronavirus disease and prevent future spread.

4.2 TYPES OF CORONAVIRUS THREATS AND THREATS HUMANITY FROM PAST TO PRESENT

Coronaviruses are a large family of viruses that have been around for a long time and are known to cause illness. The first coronaviruses were detected in humans and animals (camel, bat, cat, etc.) in 1937.

The seven HCoVs that have been identified since 1937 are as follows[3]:

1. 229E (alpha coronavirus).
2. NL63 (alpha coronavirus).
3. OC43 (beta coronavirus).

4. HKU1 (beta coronavirus).
5. MERS-CoV (beta coronavirus).
6. SARS-CoV (beta coronavirus).
7. 2019 Novel Coronavirus (2019-nCoV).

The reason why these viruses have reached such a global size is that they are zoonotic. Zoonotic means that the virus is transportable between animals and humans.[4]

The coronavirus type SARS-CoV, which the world met in 2002, caused human-to-human transmission and resulted in 30% deaths. In this virus, transmission from person to person occurred via the ACE2 protein receptor.[5] The coronavirus type, which came into our lives as MERS-CoV in June 2012, was an epidemic with a low rate of human-to-human transmission, but that had caused great number deaths. In this type of virus, the virus has been transported from person to person via the DPP4 protein receptor.[5]

Many sources have shown that the new type of coronavirus is transported human-to-human by binding to the human ACE2 receptor through the receptor-binding domain (RBD) of the S protein of the virus.[6] The vaccines developed for COVID-19 are antibody-enhancing vaccines that recognize the RBD polypeptide to prevent the virus from entering the cell. Sources say that this disease spreads from bats and is transmitted from person to person.[7]

Recent studies aimed to design a peptide-structured inhibitor that inhibits the interaction between the RBD fragment in the structure of the S protein of this virus and the ACE2 protein receptor.[8] These, of course, are mainly preclinical trials. There are some ways for identifying and characterizing drug molecules. These are designing drugs by determining the structures and chemical characterizations of intracellular molecules, and then determining the activities of designed drug molecules with in vitro and in vivo experiments, investigating their mechanism of action, calculating toxicity, and investigating their biosafety and bioavailability.[9]

SARS and MERS coronavirus receptors are found in cells of various organs in the body, particularly the lung cells. When the new type of coronavirus binds to ACE2 protein receptors in lung cells, it causes cytokine storms due to systemic immune overactivation and this situation is especially striking in severely ill patients.[10,11]

The ability of coronaviruses to enter the cells depends on the activation of the envelope glycoproteins of the viruses by the host cell proteases. Thus, it is thought that targeting viral entry, and more specifically, the endosomal proteolysis step of entry, is an attractive strategy to discover a new antiviral agent.[12] Many of these trials aim to reduce the duration of the new type of coronavirus disease by examining the effect of drugs used to treat other diseases (rheumatism, malaria, etc.) on SARS-CoV-2, allowing patients to have milder disease. Some experts have revealed that some antiviral drugs developed to destroy the Ebola virus are effective against the new type of coronavirus. They found that drugs known as protease inhibitors help in improving the immune system against coronavirus.[12]

A vaccine takes between 2 and 5 years to produce. It is critical to thoroughly characterize the virus's activity and properties in the intermediate host.[13] To carry out these procedures, animal tests are needed. To properly activate our immune system and avoid any harm to it, it is important to examine the properties of vaccines produced after this stage.

To treat viral infections caused by SARS, MERS, and a new form of coronavirus, as well as produce drugs and vaccines, there are two options. The first is to create an antiviral enzyme that inhibits protein synthesis, and the second is to create antiviral antibodies. Our bodies develop antibodies to battle the virus on their own after around 2 weeks. Antibodies are inserted into the bloodstream during vaccines so that the body can resist the virus. As a result, the infection is expelled from the body.[14] Tiny antiviral compounds, on the other hand, have the disadvantage of binding with other molecules in the body. Disruption of our body's balance leads to the emergence of other diseases.[14]

4.3 NOVEL CORONAVIRUS

Coronavirus RNA is defined as a virus, which is positively sensitive, single-stranded, and enveloped. Because the glycoprotein is "pointed," its membrane has a crown-like appearance.[2] There are three categories of epidemics that affect the entire world and are classified as global epidemics by the World Health Organization (WHO). SARS, MERS, and SARS-CoV-2 are all beta genome subgroups.[15] Two polyproteins called replicases, pp1a and pp1ab, are encoded by two-third of the novel coronavirus genome. These polyproteins are cleaved into 16 nonstructural proteins by two major viral proteases 3C-like proteases (3CLpro) and PLpro. These viruses copy their genomes in the host's cytoplasm, thanks to the nonessential protein RdRp, or RNA-linked RNA polymerase gene. The new coronavirus genome encodes four major structural proteins and a series of helper proteins, all of which thwart the immune response of the host.[16,17] There are four structural proteins in the new coronavirus. These are given below.

Spike(S) glycoproteins: Glycoproteins are responsible for attachment to the host receptor, mediating membrane fusion, and entry of coronavirus (coronavirus attaches to the ACE2 protein (ligand) on the surface of the host cell with S spike proteins). It needs the TMPRSS2 protease to achieve this attachment. This process is followed by the replication and translation (polypeptide-protein synthesis) of the RNA sequence.

The virus then multiplies and spreads inside the host cell, amplifying the infection's power. The immune system takes action against the virus, and the macrophage-like antigen-producing cells break down the proteins in the virus's structure and introduce them into T cells as peptides. Thus, the T-cell begins to differentiate and activate. T-cells activate immunity with cytokine production. Where there is an excess of cytokine production, the situation is known as a cytokine storm. The priority in drug and vaccine development is to prevent the S protein from binding to the ACE2 receptor. The entry of the virus into the target cell, its proliferation can be prevented, or neutralizing antibodies can be developed.[18]

Small envelope (E) proteins: It is responsible for the coronavirus's enveloped structure. CoV enveloped proteins have a variety of functions, ranging from viral growth during infection and cell differentiation to changes in the structure and function of the cell caused by chemical, physical, and microbiological agents in respiratory syndrome. These proteins are small hydrophobic viroporins.[19] They form ion channels as oligomerized.[18]

Membrane/matrix (M) proteins: It is the most abundant triple envelope protein with a short amino-terminal domain and a suitable carboxyl-terminal.[20] These terminals are outside and inside the virus. Membrane proteins function as receptors and are a type of protein that allows immune system cells to communicate with foreign structures or with each other.[18] This protein is a mobile, amphipathic molecule.

Nucleocapsid (N) proteins: N protein is a helix and flexible phosphoprotein. These proteins take part in the replication and transcription of coronaviruses.[18]

4.3.1 STRUCTURE AND GENOME OF THE NOVEL CORONAVIRUS

The genome of the novel coronavirus most often encodes ORF1ab nonstructural polyproteins. Replicase pp1a and pp1ab are also cleaved into the 16 nonstructural proteins by 3C-like protease (3CLpro) and PLpro.[21] It encodes spike glycoprotein, envelope (E) glycoprotein, membrane (M) glycoprotein, and nucleocapsid (N) phosphoprotein, which are the main structural proteins in a certain part of the viral genome. Nonstructural proteins are also encoded by the ORF3a, ORF6, ORF7a, ORF7b, ORF8, and ORF10 genes.[21,22]

4.3.2 NOVEL CORONAVIRUS PROTEINS

The structural and nonstructural proteins of the new coronaviruses are classified into two categories. Allowing viral cell entry is the first step in the development of a new coronavirus disease. The binding of a large amount of glycosylated S protein to the ACE2 enzyme allows the virus to enter viral cells.[23] The serine protease (TMPRSS2) in the host cell membrane aids virus entry. These proteins are potential targets for the survival of the virus and drug therapies. The S protein of the new type of coronavirus plays an important role in viral binding in the HCoV.[24] The virus cellular receptor ACE2 protein, like the novel coronavirus proteins, is being studied as a therapeutic agent (antivirus drugs) to treat and prevent Coronavirus disease. To prevent the virus from entering target cells, the interaction of the S receptor binding domain with ACE2 must be blocked.[25,26] The S1 subunit of S protein is responsible for the binding of SARS-CoV-2 by ACE2 and plays a role in viral entry into the target cell. This subunit protein has an N-terminal domain (S1-NTD) and a C-terminal domain (S1-CTD). It mediates sugar-binding of S1-NTD and S1-CTD allows the virus to bind to ACE2 via the RBD.[27,28]

4.3.2.1 THE S (SPIKE) PROTEIN

S (spike) protein is a transmembrane protein with three regions: the external region (extracellular N-terminal), transmembrane sequence, and internal region (short-intracellular C-terminal).[29] The external region gives the virus antigenic property and forms a receptor for cell surface receptor binding. They use a membrane-bound aminopeptidase N as the coronavirus receptor. The internal region of the S protein is responsible for membrane fusion during the binding phase of the virus to the host cell. Antibodies against this protein are neutralized.[30,31]

4.3.2.2 THE S1 SUBUNIT POLYPEPTIDE

The S1 subunit polypeptide contains two domains: the C-terminal RBD in the external domain and the internal domain that interacts with host receptors.

It facilitates virus infection by interacting with the host receptor. The inner RDB binds to the cell receptor ACE2 in the N region. Because of this feature, it is an important target in drug production.[31,32]

4.3.2.3 THE S2 SUBUNIT POLYPEPTIDE

S2 Subunit Polypeptide is responsible for viral fusion and viral entry. FP consists of HR1, HR2, TM domain, and cytoplasmic domain fusion (CT). The S2 subunit protein domain contains the HR domain responsible for the membrane fusion mechanism of the virus.[33]

4.3.3 REPLICATION CYCLE OF NOVEL CORONAVIRUS

Replications of virus genomes are carried out in several steps. Replication of the virus begins with the attachment of the virus onto the cell surface by finding a suitable host cell. There must be a specific binding site for the virus to interact with the host cell. The virus must then cross the host cell membrane and enter the cell. This phenomenon is called penetration. After the penetration into the host cell, the viral capsid is removed and degraded by viral enzymes or host enzymes releasing the viral genomic nucleic acid. This event is called uncoating (separation). By changing the biosynthetic mechanism of the host cell, it synthesizes polymerase enzymes that will carry out replication. This event is defined as early protein synthesis. Replication of viral nucleic acid is performed in the replication of the viral genome. In the process defined as the final protein synthesis, the synthesis of subunits of the structural protein takes place. The structural subunits and viral nucleic acid, which are defined as membrane components in enveloped viruses, are packaged into the nucleocapsid. In enveloped viruses, virions are released by budding. This event is defined as lysis.[34]

4.3.4 NEW TYPE CORONAVIRUS-RELATED INHIBITORS

The entry of the new type Coronavirus into the cell is completed with the preparation of the Spike protein by the TMPRSS2 protease. After the virus enters the host cell, the viral RNA nucleotide is copied there and surrounds the cell. In order for this event to occur, five helper proteins named ORF3a, ORF6, ORF7a, ORF7b, ORF8 take part in the new coronavirus genome.[35]

Recently, therapeutics have been studied to block and reduce expression of the SARS-CoV-2 virus to inhibit cellular entry and this virus in various tissues expressing ACE2.[36]

4.3.4.1 THE FUNCTIONAL PROTEINS OF SARS-COV-2

Antiviral drugs already developed or being developed for COVID-19 are targeted to prevent the virus from binding to the host cell or to prevent its growth and reproduction if the host cell has entered.

In order to understand the development of new antiviral drugs for COVID-19 and the effect of existing drugs on COVID-19, it is necessary to mention four important functional proteins in the novel coronavirus genome sequence. These are as follows:

- 3CLpro (chymotrypsin-like cysteine protease).
- PLpro (Papain protease).
- RdRp (RNA-dependent RNA-polymerase).
- Helicase.

It has been observed that the novel coronavirus, MERS-CoV, and SARS-CoV are highly conserved in their functional regions. The coronaviral proteases PLpro and 3CLpro activate the viral proteins pp1a and pp1ab in the release and mutation of nonstructural proteins.[37–39] PLpro's function blocks the production of cytokines by deglycosylation.[40] Thiopurine compounds, naphthalene-based inhibitors, zinc ion, and zinc conjugate inhibitors are FDA-approved SARS-CoV PLpro inhibitors.[24] Docking studies have shown that molecules such as ribavirin, chloramphenicol, and levodropropizine can have a high binding affinity with PLpro.[41]

By examining the crystal structure of the 3CLpro functional protein, α-ketoamide inhibitors were designed.[42] The task of RdRp is to catalyze RNA synthesis. In other words, the virus joins a chemical reaction during RNA synthesis, causing the reaction to proceed without modification, or to raise its speed by lowering its energy.[43]

Nucleoside analogs inhibit the activity of RdRp. Therapeutic drugs such as ribavirin, remdesivir, sosfosbuir, galidesevir, IDX-184, and tenofovir have been shown to have high binding affinity in a virtual screening study with RdRp.[43]

Helicase is an important CoV replication enzyme that can split double-stranded nucleic acid into single strands using ATP. Various inhibitors developed are responsible for inhibiting helicase activity and ATPase activity.[44,45]

As new drugs for the new coronavirus will take many years to develop; scientists have focused on the potential of antiviral drugs used in the treatment of certain diseases with pharmacokinetic and pharmacodynamic properties. Studies on toxicity properties through virtual studies prior to clinical trials are still going on.[41]

4.3.4.2 VARIOUS DRUGS UNDER INVESTIGATION FOR COVID-19

1. Direct-acting antivirals.
2. Antiviral drugs targeting host cells.
3. Immunoregulatory or inhibited therapy modalities.

4.3.4.2.1 Direct-Acting Antiviral Drugs

Remdesivir: It is an RdRp inhibitor developed against Ebola virus. It has broad-spectrum anticoronavirus activity and has no serious side effects.[46]

Ribavirin: It is a nucleoside analog. Its viral target is DNA synthesis. Provides inhibition of RNA viruses.[47]

Ribavirin, remdesivir, lopinavir, ritonavir, darunavir, cobicistat, emtricitabine, tenofovir, azvudine, saquinavir, nelfinavir: HIV-1 protease inhibitors and potential drug candidates.[48,49]

Nelfinavir: Inhibits SARS-CoV replication.[50]

Lopinavir, ritonavir, saquinavir: Strong interactions occurred in the active site of 3CLpro of novel coronavirus.[51]

Lopinavir, ritonavir, ribavirin: Clinical results for SARS patients are successful.[52]

Lopinavir, ritonavir, ribavirin, interferonalfa 2a: It has been successful in treating MERS.[53]

Danoprevir, ritonavir, sofosbuvir, ledipasvir, sofosbuvir, daclatasvir: The combination of these drugs is used to treat the hepatitis C virus. Studies for the novel coronavirus have shown improved clinical symptoms and decreased viral infection in patients.[54]

Favipiravir, umifenovir (Arbidol), triazavirin, baloxavir, and marboxil: Antiviral drugs designed against influenza disease.[54]

4.3.4.2.2 Host Cell Targeted Drugs

Hydroxychloroquine, chloroquine: It prevents novel coronavirus from binding to the ACE2 enzyme.[55]

Leflunomide and teriflunomide: They are known as dihydroorotate dehydrogenase inhibitors and inhibit novel coronavirus replication in cell culture.[56]

Dipyridamole: Anticoagulant agent.[57]

Chlorpromazine: Anti-MERS-CoV and anti-SARS-CoV agent capable of regulating neurotransmitters. If this drug is used with clathrin, it inhibits endocytosis (the introduction of molecules large enough to pass through the membrane, i.e., polymers).[58] Estrogen therapy is a strategic form of treatment in COVID-19[59].

4.3.4.2.3 Immunomodulatory or Inhibitory Therapy Modalities

Corticosteroids, intravenous immunoglobulin, and interleukin-6 antagonists: The immune modulator or immunosuppressive agent is a potential therapeutic agent for COVID-19.[60]

Cyclosporine A: By binding to cellular cyclophilin, its antiviral activity is broad spectrum. Blocks replication, including SARS-CoV and avian flu.[61]

Interferons: It restores innate immune responses to the host cell.[62] SARS-CoV and MERS-CoV are potent inhibitors of replication.[63]

Combination of interferon-α ribavirin with lopinavir/ritonavir: It has shown good results for COVID-19 patients.[64]

Interferon + ribavirin and/or lopinavir-ritonavir: It is used to treat patients with SARS, MERS, and COVID-19.[64,65]

The combination of interferon with other effective antivirals is still under investigation.

Corticosteroids, intravenous immunoglobulin, and interleukin-6 antagonists: Immunomodulator or immunosuppressive agent is a potential therapeutic agent for COVID-19.[66]

Meplazumab (anti-CD147): Inhibits viral entry.[67]

In certain antibody tests, the nucleocapsid (N) or S (S1, RBD) protein is used. Antibodies against the virus (IgM, IgG, and IgA) began to rise in the second week after the onset of symptoms in the first trials.[68–70] In the following weeks, it was observed that all patients had IgM and IgG seroconversion and IgM decreased in the fifth week and even disappeared in the seventh week.[69–71]

Vaccines and drugs used in the treatment of COVID-19 and their mechanism of action are tabulated in Table 4.1.

TABLE 4.1 Vaccines and Drugs Used in the Treatment of COVID-19 and Their Mechanism of Action.

Drug name	Mechanism of action	Functional protein and/or its effectiveness	Pharmaceutical grade	FDA and WHO approved
Remdesivir	It inhibits RNA polymerase[72]	Envelope (E) protein[73] Nsp12[73] It is a RdRps inhibitor with antiviral activity[72]	Antiviral drug belonging to the class of nucleoside analogs[74]	Approved for coronavirus
Dexamethasone	Suppresses inflammation[75]	It is used in oxygen deficiency[75]	Anti-inflammatory drug[75]	Approved for coronavirus (WHO)
Favipiravir	Inhibits RNA polymerase[76]	Selective inhibition of viral RNA-dependent RNA polymerase[76] Purine nucleoside analogue[76]	It is an antiviral drug developed against the RNA virus[76]	Approved for coronavirus
Molnupiravir (EIDD-2801)	It inhibits RNA replication[77]	It has broad-spectrum antiviral activity against all epidemics, from influenza virus to new coronavirus[78]	Reduces the replication of novel coronavirus and while doing so does not show cytotoxic effect[78]	Approved for coronavirus
Convalescent plasma	It acts by strengthening the immune system[79]	Neutralizing antibody titer[79]	Immune plasma therapy, a classical adaptive immunotherapy[80]	Approved for coronavirus (WHO)
Monoclonal antibodies	It acts by strengthening the immune system[81]	Neutralizes SARS-CoV and MERS-CoV infection[81]	It is a pharmaceutical-grade drug[81]	Approved for coronavirus (WHO)
Interferons	It acts by strengthening the immune system. It inhibits human type 1 IFN production via the ORF3b gene[82]	COVID-19 and similar treatments provide antiviral activity with correct IFN application[82]	They can be inhibited by IFN-alpha and IFN-lambda[82] It is combined with antiviral drugs to detect the presence of viruses and reduce blood levels[82]	Approved for coronavirus

TABLE 4.1 *(Continued)*

Drug name	Mechanism of action	Functional protein and/or its effectiveness	Pharmaceutical grade	FDA and WHO approved
Cytokine inhibitors (such as Tocilizumab, anakinra)	It suppresses the cytokine storm[83]	Anakinra inhibits IL-1 cytokines Tocilizumab IL-6 cytokines[83]	Neutrophil macrophage[84] Plasma adsorption[84] CRRT (continuous renal replacement therapy)[84] High-dose SRRT with cell membrane[84]	Approved for coronavirus
Stem cell therapy	Cytokine storm is suppressed by producing anti-inflammatory molecules[85]	Minimizing tissue and organ damage in the body during cytokine storm[86] Organizing cells in the new coronavirus disease[86] Increasing growth factors[86]	Neo-angiogenic Anti-fibrotic Anti-inflammatory[87] Anti-apoptotic Anti-bacterial Chemoattractant Tissue Regeneration Immunosuppressive Immunomodulator[87]	Approved for coronavirus
Coagulants	It is used for increased coagulation problem due to COVID-19[88]	It also inhibits FXIa, FXa and FIXa[88] Thrombin inhibitors[88]	Defibrinating drugs[88]	Studies continue for the coronavirus.
Lopinavir	It is a 3CL protease inhibitor[89]	Nsp13[90] NRBD[90] Possible targets Nsp3b and Nsp3c[90]	Antiviral drug[89]	Studies continue for the coronavirus.

TABLE 4.1 *(Continued)*

Drug name	Mechanism of action	Functional protein and/or its effectiveness	Pharmaceutical grade	FDA and WHO approved
Chloroquine/ hydroxychloroquine	It inhibits the entry of the virus into the cell and endocytosis. Immunomodulatory effect: cytokine production, autophagy, and lysosomal activity inhibition in the cell[91]	Possible target Nsp3b[91]	Efficacy for virus-cell fusion[91] It is an antiviral drug[91]	Approved for coronavirus
Arbidol	It inhibits RNA polymerase[92]	Nsp7–Nsp8 Complex Nsp14 Nsp15 possible target[92]	It is an antiviral drug[92]	Studies continue for the coronavirus
Ritonavir	It is a 3CL protease inhibitor[93]	Possible targets Nsp3c[93]	It is an antiviral drug[93]	Studies continue for the coronavirus
Hesperidin	ACE2 complex[94]	It is an antiviral drug[94]	It is an antiviral drug[94]	Studies continue for the coronavirus

4.3.5 NEW TYPE CORONAVIRUS WITH A MOLECULAR APPROACH

In the study of the new type of coronavirus disease, scientists thought that it was an intermediate host and so they took viral samples from certain animals and looked at the nucleotide sequences of the viruses, which included adenine, guanine, cytosine, and thymine. These nucleotide sequences are genome sequences. By looking at the nucleotide sequences of the new type of coronavirus, it can be compared with the sequence of viruses found today, and their similarities and differences can be determined.[95] As a result of this determination, it is possible to find out whether the new type of coronavirus is transmitted directly from person to person or is transmitted from an intermediate host animal, whether the virus is enveloped or not, sequences encoding proteins and kinship degrees. The new type of coronavirus genome known as type 2 SARS-CoV consists of single-stranded RNA. This shows that they are more susceptible to mutations than DNA viruses. At the end of 2020, mutations of a new type of virus began to emerge. This means that the nucleotide sequence, variety, and number of the virus are changed.

Ongoing drug research for the new type of coronavirus is on preventing the attachment of the drug molecule by creating a change on the cell surface where the virus will attach and prevent the binding of the S protein with the ACE2 receptor.[96] Genome sequences must be precisely determined for this. With the mutation of the virus, the genome sequences will change, and the effect of vaccine developed today will decrease. Therefore, the nucleotide sequences of the mutated virus should be determined well, and the vaccine and drug molecule studies should be renewed. A vaccine is an antigenic preparation that makes it possible to induce a pathogenic agent specific immune response in the vaccinated person or animal that can protect them against natural infection or mitigate its consequences.[97]

4.3.6 CLINICAL FEATURES OF THE NOVEL CORONAVIRUS

Fever, cough, respiratory distress, and headache are all typical signs of this viral infection. In more extreme cases, pneumonia, major acute respiratory tract infection, kidney failure, and even death have been recorded.[98]

4.3.7 DRUG METABOLISM AND PHARMACOKINETICS OF THE NEW TYPE CORONAVIRUS

When examining the drug and pharmacokinetic structure of a new form of coronavirus, it is important to consider drug–drug interactions and whether the drug that will be formulated for this virus will damage the metabolic enzyme or its carriers. In terms of drug formulations for the treatment of patients with prophylactic or symptomatic conditions, oral inhaled aerosol types are difficult to use in patients with respiratory issues or acute illnesses, and vitamins, nutrients, amino acids, and enzymes can all be administered in liquid form.[99] To eradicate the pandemic from our lives, vaccinations that are both effective and safe are being developed. It is well known that pharmacological treatment of infectious diseases that have infected our lives in recent years has shown positive outcomes.[100] While AIDS which is caused by HIV resulted in death in the past centuries, with the development of antiviral drugs, it has turned into a chronic disease in which one in two people is treated today. As an additional example, despite its vaccine resistance, the hepatitis-C virus has evolved into a virus that is excreted from the body as a result of the production of antiviral drugs.[101,102] These examples may be expanded upon. The findings suggest that pharmacological therapeutic options for SARS-CoV-2 will arise in the coming years.

4.3.8 MOLECULAR MODELING AND MOLECULAR DOCKING APPLICATIONS ON COVID-19

After Heisenberg's article in the field of quantum mechanics theory modeling in 1925, James Hendrickson used the force field principle for the first time in 1961 to calculate the conformational energies of cycloheptanes using the IBM 709 computer. Cyrus Levinthal published a paper in 1966 outlining the use of molecular graphics to analyze the structures of proteins and nucleic acids, thus expanding the use of computer use.[103–106]

Molecular docking can be used to describe the relationship between the small ligand and the target receptor, as well as to see whether they will coexist at the binding site.

The data obtained with molecular docking techniques, which have been widely used since the 1980s, have increased rapidly ever since. Programs structured using different algorithms for molecular insertion analysis have significantly improved pharmaceutical research. The first algorithm used has been designed for protein–protein interactions.[107,108]

Molecular modeling and docking studies aid researchers in developing more reliable and accurate processes and identifying novel compounds. The pharmaceutical industry, which focuses on the development and design of new molecules that can be used in healthcare,[109,110] is the most significant supporter in this area.

The following are studies of molecular modeling and molecular docking applications in this field.

The thermal parameters, dipole moment, polarizability, molecular electrostatic potential, and Mulliken atomic charges of antiviral N-heterocycles molecules, as a COVID-19 drug, were determined by Hagar et al. by density functional theory (DFT) calculations. Docking studies were conducted on the SARS-CoV-2 parent protease to evaluate the binding affinity of these candidate drugs. According to docking studies, the amodiaquine molecule has the highest binding affinity of −7.77 kcal/mol and can be used as a strong SARS-CoV-2 inhibitor.[111]

As a result of DFT and molecular docking measurements, chloroquine derivatives were proposed as antivirals for the treatment of COVID-19 diseases in a report conducted by Noureddine et al. in 2021.[112]

Investigation of the structure of favipiravir and prediction of its activity against COVID-19 by in silico calculations were performed by Harismah et al.[113] The tautomers of favipiravir were identified by DFT calculations, and the results revealed that four of the tautomeric structures can be used as ligands. To assess the activity of ligands in molecular docking, protease and polymerase macromolecules associated with COVID-19 were chosen as target receptors. Docking revealed that these four ligands would interact with each of the targets.[113]

As a result of DFT and molecular docking studies conducted by Sagaama et al. on succinic acid, L-pyroglutamic acid, *N*-phenyl-thioacetamide, 2-amino-5-chloropyridine hydrogen succinate, epigallocatechin gallate, and selenomethionine molecules, which were potentially antiviral in the treatment of COVID-19, these have been suggested that can be used as candidates.[114]

In another study, El Hadki employed DFT and docking calculations to investigate whether natural molecules had an antiviral effect. Crocin has a binding energy of −8.1 kcal/mol with the "SARS-CoV-2 base protease" receptor, according to molecular docking calculations. This suggests that it may be a good CoV-2-SARS inhibitor.[115]

Moreover, Mumit et al. synthesized benzyl-3-*N*-(2,4,5-trimethoxyphenylmethylene) hydrazinecarbodithioate, which was investigated using DFT and molecular docking. The binding energy of this molecule, which is synthesized as a result of COVID-19 main protease calculations, is found to be higher than that of hydroxychloroquine. Furthermore, the calculated inhibition constant values show that the synthesized molecule has a higher potential to inhibit the target protein's biological process than hydroxychloroquine.[116]

Shabab et al. performed DFT calculations to determine the molecular structure of the Triazavirin molecule they synthesized. Using molecular docking calculations, it was suggested that this molecule could be used as an inhibitor for 2019-nCoV Mpro, that is, for potential applications against Coronavirus 2019-nCoV.[117]

El Kalai et al. synthesized molecule of (E)-4-(4-methylbenzyl) 6-styrylpyridazin-3(2H)-one and used DFT calculations to determine its molecular structure. Molecular docking was used to assess if this synthesized molecule is a possible COVID-19 therapeutic.[118]

Douche et al obtained the optimized structure of synthesized 5-((1H-imidazol-1-yl)methyl)quinolin-8-ol molecule by DFT calculations in the gas phase. Afterward, molecular docking was used to establish its interactions with COVID-19 receptors. As a result of the docking analysis, it was found that the molecule could be used as a possible agent against the COVID-19 receptor.[119]

Shahab et al. used DFT calculations to obtain the most stable conformation of the Atazanavir molecule and the optimized structure was subjected to molecular docking calculations against the Coronavirus 2019-nCoV receptor. The binding energy of Atazanavir with the target receptor has been determined to be −64.29 kcal/mol, and it has been suggested that it can be used to cure the coronavirus pandemic.[120]

Bouzian et al. have been calculated the molecular orbital calculations and molecular electrostatic potentials of 2-(2-oxo-1,3-oxazolidin-3-yl) ethyl 2-[2-(2-oxo-1,3-oxazo-lidin-3-yl) ethoxy] quinoline-4-carboxylate molecule using by DFT theory. Molecular docking has been used to figure out how this molecule interacts with the COVID-19 primary protease receptor. Because of its ability to conform to the active site of the protein and its binding affinity, the molecule has been determined to be a strong agent as a result of the docking study.[121]

In a thesis, the optimized geometry of the 2–[(4,6-diaminopyrimidin-2-yl) sulfanyl]-*N*-(4-fluorophenyl) acetamide molecule, synthesized by Mary et al., was determined using DFT calculations. Afterward, molecular docking analysis was used to assess its relationship with SARS-COV-2. The binding energy of the molecule was determined to be 8.7 kcal/mol, meaning that it could interact with the SARS-CoV-2 protease.[122]

Celik et al. used the AM1 semiempirical quantum mechanical method to determine the most stable conformations of arbidol, colchicine, dexamethasone, favipiravir, galidesivir, hydroxychloroquine, and remdesivir molecules commonly used in the treatment of COVID-19. In the molecular docking analysis, the most stable conformations were used as the input data of the ligands. Molecular docking calculations were used to determine the interaction mechanism between these stable conformations and the apo- and holo-formes of the COVID-19 main protease enzyme (Mpro) and the spike glycoprotein of SARSCoV-2 receptors. These drugs may play an important role in the treatment of COVID-19, according to the interaction results with the apo form of COVID-19 Mpro and spike glycoprotein.[123]

Belhassan et al. studied the molecular interactions of 32 N-substituted oseltamivir derivatives with the SARS-CoV-2 main protease. As compared to chloroquine and hydroxychloroquine, certain derivatives were found to have higher binding affinity values with the SARS-CoV-2 main protease.[124]

Mpiana et al. performed a molecular docking analysis of 10 Aloe vera molecules with the main protease responsible for coronavirus replication (3CLpro). According to their binding affinities, the most stable complexes (feralolide, 9-dihydroxyl-2-O-(z) -sinnamoyl-7-methoxy-Aloesin, and Aloeresin) were determined to be −7.9, −7.7, and −7.7 kcal/mol, respectively.[125]

In the study of 2-(p-chloro-benzyl)-5-[3-(4-ethly-1-piperazynl)propionamido]- benzoxazole molecule, synthesized by Zeyrek et al., the basic molecular geometry was determined by DFT calculations. Molecular docking study between this molecule and the COVID-19 main protease receptor was also performed. The carbon–hydrogen bonding, Pi-sigma bonding, Pi-donor hydrogen bonding, and van der Waals interaction interactions with the enzyme were found as a result of the docking simulation.[126]

4.4 CONCLUSION

The COVID-19 pandemic that began in Asia has quickly spread across the globe. It is important to understand the structure of this emerging virus and find appropriate drugs for successful treatments. Modeling effective drugs and elucidating the mechanisms of interaction with COVID-19 receptors rely heavily on molecular modeling and molecular docking calculations. Targeted treatments can be calculated earlier and the therapy plan can be accelerated using these calculations.

KEYWORDS

- **coronavirus**
- **treatment**
- **clinical**
- **drug**
- **vaccine**
- **molecular modeling**
- **COVID-19**

REFERENCES

1. Zhang, Y. Z.; Holmes, E. C. (2020). A Genomic Perspective on the Origin and Emergence of SARS-CoV-2. *Cell* **2020,** *181* (2), 223–227.
2. Guo, Y. R.; Cao, Q. D.; Hong, Z. S.; Tan, Y. Y.; Chen, S. D.; Jin, H. J.; Yan, Y. The Origin, Transmission and Clinical Therapies on Coronavirus Disease 2019 (COVID-19) Outbreak–An Update on the Status. *Military Med. Res.* **2020,** *7* (1), 1–10.
3. Helmy, Y. A.; Fawzy, M.; Elaswad, A.; Sobieh, A.; Kenney, S. P.; Shehata, A. A. The COVID-19 Pandemic: A Comprehensive Review of Taxonomy, Genetics, Epidemiology, Diagnosis, Treatment, and Control. *J. Clin. Med.* **2020,** *9* (4), 1225.
4. Jalava, K. First Respiratory Transmitted Food Borne Outbreak? *Int. J. Hygiene Environ. Health* **2020,** *226*, 113490.
5. Rajendran, D. K.; Rajagopal, V.; Alagumanian, S.; Kumar, T. S.; Prabhakaran, S. S.; Kasilingam, D. Systematic Literature Review on Novel Corona Virus SARS-CoV-2: A Threat to Human Era. *VirusDisease* **2020,** *31* (2), 161–173.
6. Walls, A. C.; Park, Y. J.; Tortorici, M. A.; Wall, A.; McGuire, A. T.; Veesler, D. Structure, Function, and Antigenicity of the SARS-CoV-2 Spike Glycoprotein. *Cell* **2020,** *181* (2), 281–292.
7. Shi, Z.; Hu, Z. (2008). A Review of Studies on Animal Reservoirs of the SARS Coronavirus. *Virus Res. 2008*, *133* (1), 74–87.
8. Baig, M. S.; Alagumuthu, M.; Rajpoot, S.; Saqib, U. Identification of a Potential Peptide Inhibitor of SARS-CoV-2 Targeting Its Entry Into the Host Cells. *Drugs in R&D* **2020,** *20* (3), 161–169.
9. Xu, L.; Liu, J.; Lu, M.; Yang, D.; Zheng, X. Liver Injury during Highly Pathogenic Human Coronavirus Infections. *Liver Int.* **2020,** *40* (5), 998–1004.
10. Rosa-Calatrava, M.; Traversier, A.; Desuzinges-mandon, É.; Dejean, E. U.S. Patent No. 10,144,917. U.S. Patent and Trademark Office: Washington, DC, 2018.
11. Song, P.; Li, W.; Xie, J.; Hou, Y.; You, C. Cytokine Storm Induced by SARS-CoV-2. *Clinica Chimica Acta* **2020,** *509*, 280–287.

12. Zhou, Y.; Vedantham, P.; Lu, K.; Agudelo, J.; Carrion Jr, R.; Nunneley, J. W.; Simmons, G. Protease Inhibitors Targeting Coronavirus and Filovirus Entry. *Antiviral Res.* **2015,** *116*, 76–84.
13. Leroux-Roels, G.; Bonanni, P.; Tantawichien, T.; Zepp, F. Vaccine Development. *Persp. Vaccinol.* **2011,** *1* (1), 115–150.
14. Liu, C.; Zhou, Q.; Li, Y.; Garner, L. V.; Watkins, S. P.; Carter, L. J.; Albaiu, D. Research and Development on Therapeutic Agents and Vaccines for COVID-19 and Related Human Coronavirus Diseases, 2020.
15. Snijder, E. J.; van der Meer, Y.; Zevenhoven-Dobbe, J.; Onderwater, J. J. M.; van der Meulen, J.; Koerten, H. K.; Mommaas, A. M. Ultrastructure and Origin of Membrane Vesicles Associated with the Severe Acute Respiratory Syndrome Coronavirus Replication Complex. *J. Virol.* **2006,** *80* (12) 5927–5940.
16. Cui, J.; Li, F.; Shi, Z. L. Origin and Evolution of Pathogenic Coronaviruses. *Nat. Rev. Microbiol.* **2019,** *17* (3), 181–192.
17. Zhu, N.; Zhang, D.; Wang, W.; Li, X.; Yang, B.; Song, J.; Tan, W. A Novel Coronavirus from Patients with Pneumonia in China, 2019. *N Engl. J. Med*, 2020.
18. Fehr, A. R.; Perlman, S. Coronaviruses: An Overview of Their Replication and Pathogenesis. *Coronaviruses* **2015,** 1–23.
19. Sainz, Jr., B.; Mossel, E. C.; Gallaher, W. R.; Wimley, W. C.; Peters, C. J.; Wilson, R. B.; Garry, R. F. Inhibition of Severe Acute Respiratory Syndrome-Associated Coronavirus (SARS-CoV) Infectivity by Peptides Analogous to the Viral Spike Protein. *Virus Res.* **2006,** *120* (1–2), 146–155.
20. De Haan, C. A.; Kuo, L.; Masters, P. S.; Vennema, H.; Rottier, P. J. Coronavirus Particle Assembly: Primary Structure Requirements of the Membrane Protein. *J. Virol.* **1998,** *72* (8), 6838–6850.
21. Hoffmann, M.; Kleine-Weber, H.; Schroeder, S.; Krüger, N.; Herrler, T.; Erichsen, S.; Pöhlmann, S. SARS-CoV-2 Cell Entry Depends on ACE2 and TMPRSS2 and Is Blocked by a Clinically Proven Protease Inhibitor. *Cell* **2020,** *181* (2), 271–280.
22. Liu, C.; Zhou, Q.; Li, Y.; Garner, L. V.; Watkins, S. P.; Carter, L. J.; Albaiu, D. Research and Development on Therapeutic Agents and Vaccines for COVID-19 and Related Human Coronavirus Diseases, 2020.
23. Cai, Y.; Zhang, J.; Xiao, T.; Peng, H.; Sterling, S. M.; Walsh, R. M.; Chen, B. Distinct Conformational States of SARS-CoV-2 Spike Protein. *Science* **2002,** *369* (6511), 1586–1592.
24. Gussow, A. B.; Auslander, N.; Faure, G.; Wolf, Y. I.; Zhang, F.; Koonin, E. V. Genomic Determinants of Pathogenicity in SARS-CoV-2 and Other Human Coronaviruses. *Proc. Natl. Acad.f Sci.* **2020,** *117* (26), 15193–15199.
25. Donoghue, M.; Hsieh, F.; Baronas, E.; Godbout, K.; Gosselin, M.; Stagliano, N.; Acton, S. A Novel Angiotensin-Converting Enzyme–Related Carboxypeptidase (ACE2) Converts Angiotensin I to Angiotensin 1–9. *Circ. Res.* **2000,** *87* (5), e1–e9.
26. Çakal, B. COVID-19'da Antikor Bağımlı İmmünpataloji, Monoklonal Antikorlar ve Mutasyonlar. *Experimed* **2020,** *10* (2), 112–118.
27. Zhang, H.; Penninger, J. M.; Li, Y.; Zhong, N.; Slutsky, A. S. Angiotensin-Converting Enzyme 2 (ACE2) as a SARS-CoV-2 Receptor: Molecular Mechanisms and Potential Therapeutic Target. *Intens. Care Med.* **2020,** *46* (4), 586–590.
28. Lan, J.; Ge, J.; Yu, J.; Shan, S.; Zhou, H.; Fan, S.; Wang, X. Structure of the SARS-CoV-2 Spike Receptor-Binding Domain Bound to the ACE2 Receptor. *Nature* **2020,** *581* (7807), 215–220.

29. Bosch, B. J.; Van der Zee, R.; De Haan, C. A.; Rottier, P. J. The Coronavirus Spike Protein is a Class I Virus Fusion Protein: Structural and Functional Characterization of the Fusion Core Complex. *J. Virol.* **2003,** *77* (16), 8801–8811.
30. Huang, Y.; Yang, C.; Xu, X. F.; Xu, W.; Liu, S. W. Structural and Functional Properties of SARS-CoV-2 Spike Protein: Potential Antivirus Drug Development for COVID-19. *Acta Pharmacologica Sinica* **2020,** *41* (9), 1141–1149.
31. Wrapp, D.; Wang, N.; Corbett, K. S.; Goldsmith, J. A.; Hsieh, C. L.; Abiona, O.; McLellan, J. S. Cryo-EM Structure of the 2019-nCoV Spike in the Prefusion Conformation. *Science* **2020,** *367* (6483), 1260–1263.
32. Walls, A. C.; Park, Y. J.; Tortorici, M. A.; Wall, A.; McGuire, A. T.; Veesler, D. Structure, Function, and Antigenicity of the SARS-CoV-2 Spike Glycoprotein. *Cell* **2020,** *181* (2), 281–292.
33. Millet, J. K.; Whittaker, G. R. Physiological and Molecular Triggers for SARS-CoV Membrane Fusion and Entry Into Host Cells. *Virology* **2018,** *517*, 3–8.
34. Brian, D. A.; Baric, R. S. Coronavirus Genome Structure and Replication. Coronavirus Replication and Reverse Genetics **2005,** 1–30.
35. Zhao, L.; Jha, B. K.; Wu, A.; Elliott, R.; Ziebuhr, J.; Gorbalenya, A. E.; Weiss, S. R. Antagonism of the Interferon-Induced OAS-RNase L Pathway by Murine Coronavirus ns2 Protein Is Required for Virus Replication and Liver Pathology. *Cell Host Microbe* **2012,** *11* (6), 607–616.
36. Biswas, S. K.; Mudi, S. R. Genetic Variation in SARS-CoV-2 May Explain Variable Severity of COVID-19. Medical Hypotheses.
37. Gioia, M. Ciaccio, C.; Calligari, P.; De Simone, G.; Sbardella, D.; Tundo, G.; Fasciglione, G.F.; Di Masi, A.; Di Pierro, D.; Bocedi,A.; Ascenzi, P.; Coletta, M. Role of Proteolytic Enzymes in the COVID-19 Infection and Promising Therapeutic Approaches. *Biochem. Pharmacol.* **2020,** *182*, 114225. https://doi.org/10.1016/j.bcp.2020.114225
38. Báez-Santos, Y. M.; John, S. E. S.; Mesecar, A. D. The SARS-Coronavirus Papain-Like Protease: Structure, Function and Inhibition by Designed Antiviral Compounds. *Antiviral Res.* **2015,** *115*, 21–38.
39. Chen, Y.; Liu, Q.; Guo, D. Emerging Coronaviruses: Genome Structure, Replication, and Pathogenesis. *J. Med. Virol.* **2020,** *92* (4), 418–423.
40. Mielech, A. M.; Kilianski, A.; Baez-Santos, Y. M.; Mesecar, A. D.; Baker, S. C. (2014). MERS-CoV papain-like protease has deISGylating and Deubiquitinating Activities. *Virology* **2014,** *450*, 64–70.
41. Wu, C.; Liu, Y.; Yang, Y.; Zhang, P.; Zhong, W.; Wang, Y.; Li, H. (2020). Analysis of Therapeutic Targets for SARS-CoV-2 and Discovery of Potential Drugs by Computational Methods. *Acta Pharmaceutica Sinica B* **2020,** *10* (5), 766–788.
42. Zhang, L.; Lin, D.; Sun, X.; Rox, K.; Hilgenfeld, R. X-Ray Structure of Main Protease of the Novel Coronavirus SARS-CoV-2 Enables Design of α-Ketoamide Inhibitors. BioRxiv, 2020.
43. Elfiky, A. A. Ribavirin, Remdesivir, Sofosbuvir, Galidesivir, and Tenofovir against SARS-CoV-2 RNA Dependent RNA Polymerase (RdRp): A Molecular Docking Study. *Life Sci.* **2020,** *253*, 117592.
44. Kim, M. K.; Yu, M. S.; Park, H. R.; Kim, K. B.; Lee, C.; Cho, S. Y.; Chong, Y. 2, 6-Bis-arylmethyloxy-5-hydroxychromones with Antiviral Activity against Both Hepatitis C Virus (HCV) and SARS-Associated Coronavirus (SCV). *Eur. J. Med. Chem.* **2011,** *46* (11), 5698–5704.

45. Zaher, N. H.; Mostafa, M. I.; Altaher, A. Y. Design, Synthesis and Molecular Docking of Novel Triazole Derivatives as Potential CoV Helicase Inhibitors. *Acta Pharmaceutica* **2020,** *70* (2), 145–159.
46. Al-Tawfiq, J. A.; Al-Homoud, A. H.; Memish, Z. A. Remdesivir as a Possible Therapeutic Option for the COVID-19. *Travel Med. Infect. Dis*, 2020.
47. Chen, F.; Chan, K. H.; Jiang, Y.; Kao, R. Y. T.; Lu, H. T.; Fan, K. W.; Yuen, K. Y. In Vitro Susceptibility of 10 Clinical Isolates of SARS Coronavirus to Selected Antiviral Compounds. *J. Clin. Virol.* **2004,** *31* (1), 69–75.
48. Wensing, A. M.; van Maarseveen, N. M.; Nijhuis, M. Fifteen Years of HIV Protease Inhibitors: Raising the Barrier to Resistance. *Antiviral Res.* **2010,** *85* (1), 59–74.
49. Charpentier, C.; Camacho, R.; Ruelle, J.; Eberle, J.; Gürtler, L.; Pironti, A.; Obermeier, M. HIV-2EU—Supporting Standardized HIV-2 Drug-Resistance Interpretation in Europe: An Update. *Clin. Infect. Dis.* **2015,** *61* (8), 1346–1347.
50. Xu, Z.; Yao, H.; Shen, J.; Wu, N.; Xu, Y.; Lu, X.; Li, L. J. Nelfinavir Is Active against SARS-CoV-2 in Vero E6 Cells. *ChemRxiv*, 2020.
51. Ortega, J. T.; Serrano, M. L.; Pujol, F. H.; Rangel, H. R. Unrevealing Sequence and Structural Features of Novel Coronavirus Using in Silico Approaches: The Main Protease as Molecular Target. *EXCLI J.* **2020,** *19*, 400.
52. Chu, C. M.; Cheng, V. C. C.; Hung, I. F. N.; Wong, M. M. L.; Chan, K. H.; Chan, K. S.; Yuen, K. Y. Role of Lopinavir/Ritonavir in the Treatment of SARS: Initial Virological and Clinical Findings. *Thorax* **2004,** *59* (3), 252–256.
53. Min, C. K.; Cheon, S.; Ha, N. Y.; Sohn, K. M.; Kim, Y.; Aigerim, A.; Kim, Y. S. Comparative and Kinetic Analysis of Viral Shedding and Immunological Responses in MERS Patients Representing a Broad Spectrum of Disease Severity. *Sci. Rep.* **2016,** *6* (1), 1–12.
54. Nile, S. H.; Kai, G. Recent Clinical Trials on Natural Products and Traditional Chinese Medicine Combating the COVID-19. *Indian J. Microbiol.* **2020,** 1–6.
55. Yazdany, J.; Kim, A. H. Use of Hydroxychloroquine and Chloroquine during the COVID-19 Pandemic: What Every Clinician Should Know, 2020.
56. Liu, X.; Li, Z.; Liu, S.; Chen, Z.; Zhao, Z.; Huang, Y. Y.; Luo, H. B. Therapeutic Effects of Dipyridamole on COVID-19 Patients with Coagulation Dysfunction. *MedRxiv*, 2020.
57. Xiong, R.; Zhang, L.; Li, S.; Sun, Y.; Ding, M.; Wang, Y.; Xu, K. Novel and Potent Inhibitors Targeting DHODH, a Rate-Limiting Enzyme in De Novo Pyrimidine Biosynthesis, Are Broad-Spectrum Antiviral against RNA Viruses Including Newly Emerged Coronavirus SARS-CoV-2. *BioRxiv*, 2020.
58. Plaze, M.; Attali, D.; Petit, A. C.; Blatzer, M.; Simon-Loriere, E.; Vinckier, F.; Gaillard, R. Repurposing of Chlorpromazine in COVID-19 Treatment: The ReCoVery Study. *Encephale* **2020,** S35–S39.
59. Suba, Z. Prevention and Therapy of COVID-19 via Exogenous Estrogen Treatment for Both Male and Female Patients. *J. Pharm. Pharm. Sci.* **2020,** *23*, 75–85.
60. Tufan, A.; Güler, A. A.; Matucci-Cerinic, M. COVID-19, Immune System Response, Hyperinflammation and Repurposing Antirheumatic Drugs. *Turkish J. Med. Sci.* **2020,** *50* (SI–1), 620–632.
61. Pfefferle, S.; Schöpf, J.; Kögl, M.; Friedel, C. C.; Müller, M. A.; Carbajo-Lozoya, J.; von Brunn, A. The SARS-Coronavirus-Host Interactome: Identification of Cyclophilins as Target for Pan-Coronavirus Inhibitors. *PLoS Pathog* **2011,** *7* (10), e1002331.
62. Kuri, T.; Zhang, X.; Habjan, M.; Martínez-Sobrido, L.; García-Sastre, A.; Yuan, Z.; Weber, F. Interferon Priming Enables Cells to Partially Overturn the SARS Coronavirus-Induced Block in Innate Immune Activation. *J. Gen. Virol.* **2009,** *90* (Pt 11), 2686.

63. Mustafa, S.; Balkhy, H.; Gabere, M. N. Current Treatment Options and the Role of Peptides as Potential Therapeutic Components for Middle East Respiratory Syndrome (MERS): A Review. *J. Infect. Public Health* **2018,** *11* (1), 9–17.
64. Kim, U. J.; Won, E. J.; Kee, S. J.; Jung, S. I.; Jang, H. C. Case Report Combination Therapy with Lopinavir/Ritonavir, Ribavirin and Interferon-α for Middle East Respiratory Syndrome. *Antiviral Ther.* **2016,** *21*, 455–459.
65. Min, C. K.; Cheon, S.; Ha, N. Y.; Sohn, K. M.; Kim, Y.; Aigerim, A.; Kim, Y. S. Comparative and Kinetic Analysis of Viral Shedding and Immunological Responses in MERS Patients Representing a Broad Spectrum of Disease Severity. *Sci. Rep.* **2016,** *6* (1), 1–12.
66. Cinatl, J.; Morgenstern, B.; Bauer, G.; Chandra, P.; Rabenau, H.; Doerr, H. W. Treatment of SARS with Human Interferons. *Lancet* **2003,** *362* (9380), 293–294.
67. Bian, H.; Zheng, Z. H.; Wei, D.; Zhang, Z.; Kang, W. Z.; Hao, C. Q.; Zhu, P. Meplazumab Treats COVID-19 Pneumonia: An Open-Labelled, Concurrent Controlled Add-On Clinical Trial. *MedRxiv*, 2020.
68. To, K. K. W.; Tsang, O. T. Y.; Leung, W. S.; Tam, A. R.; Wu, T. C.; Lung, D. C.; Yuen, K. Y. Temporal Profiles of Viral Load in Posterior Oropharyngeal Saliva Samples and Serum Antibody Responses during Infection by SARS-CoV-2: An Observational Cohort Study. *Lancet Infect. Dis.* **2020,** *20* (5), 565–574.
69. Liu, R.; Han, H.; Liu, F.; Lv, Z.; Wu, K.; Liu, Y.; Zhu, C. Positive Rate of RT-PCR Detection of SARS-CoV-2 Infection in 4880 Cases from One Hospital in Wuhan, China, from Jan to Feb 2020. *Clin. Chim. Acta* **2020,** *505*, 172–175.
70. Yongchen, Z.; Shen, H.; Wang, X.; Shi, X.; Li, Y.; Yan, J.; Gu, B. Different Longitudinal Patterns of Nucleic Acid and Serology Testing Results Based on Disease Severity of COVID-19 Patients. *Emerg. Microb. Infect.* **2020,** *9* (1), 833–836.
71. Long, Q. X.; Liu, B. Z.; Deng, H. J.; Wu, G. C.; Deng, K.; Chen, Y. K.; Huang, A. L. Antibody Responses to SARS-CoV-2 in Patients with COVID-19. *Nat. Med.* **2020,** *26* (6), 845–848.
72. Sheahan, T. P.; Sims, A. C.; Leist, S. R.; Schäfer, A.; Won, J.; Brown, A. J.; Baric, R. S. Comparative Therapeutic Efficacy of Remdesivir and Combination Lopinavir, Ritonavir, and Interferon Beta against MERS-CoV. *Nat. Commun.* **2020,** *11* (1), 1–14.
73. Shannon, A.; Le, N. T. T.; Selisko, B.; Eydoux, C.; Alvarez, K.; Guillemot, J. C.; Canard, B. Remdesivir and SARS-CoV-2: Structural Requirements at Both nsp12 RdRp and nsp14 Exonuclease Active-Sites. *Antiviral Res.* **2020,** *178*, 104793.
74. Warren, T. K.; Jordan, R.; Lo, M. K.; Ray, A. S.; Mackman, R. L.; Soloveva, V.; Bavari, S. Therapeutic Efficacy of the Small Molecule GS-5734 against Ebola Virus in Rhesus Monkeys. *Nature* **2016,** *531* (7594), 381–385.
75. Chu, F. F.; Esworthy, R. S.; Shen, B.; Gao, Q.; Doroshow, J. H. Dexamethasone and Tofacitinib Suppress NADPH Oxidase Expression and Alleviate Very-Early-Onset Ileocolitis in Mice Deficient in GSH Peroxidase 1 and 2. *Life Sci.* **2019,** *239*, 116884.
76. Furuta, Y.; Komeno, T.; Nakamura, T. Favipiravir (T-705), a Broad Spectrum Inhibitor of Viral RNA Polymerase. *Proc. Japan Acad., Ser. B* **2017,** *93* (7), 449–463.
77. Abdelnabi, R.; Foo, C. S.; Kaptein, S. J.; Zhang, X.; Langendries, L.; Vangeel, L.; Neyts, J. Molnupiravir (EIDD-2801) Inhibits SARS-CoV2 Replication in Syrian Hamsters Model. *bioRxiv*, 2020.
78. Sheahan, T. P.; Sims, A. C.; Zhou, S.; Graham, R. L.; Pruijssers, A. J.; Agostini, M. L.; Baric, R. S. An Orally Bioavailable Broad-Spectrum Antiviral Inhibits SARS-CoV-2 in Human Airway Epithelial Cell Cultures and Multiple Coronaviruses in Mice. *Sci. Transl. Med.* **2020,** *12* (541).

79. Ciyiltepe, F.; Bilir, Y.; Bombacı, E.; Saraçoğlu, K. COVID-19'da Yoğun Bakım Hastalarında İmmün Plazma Tedavisi. Southern Clinics of Istanbul Eurasia, 2020.
80. Cheng, Y.; Wong, R.; Soo, Y. O. Y.; Wong, W. S.; Lee, C. K.; Ng, M. H. L.; Cheng, G. Use of Convalescent Plasma Therapy in SARS Patients in Hong Kong. *Eur. J. Clin. Microbiol. Infect. Dis.* **2005,** *24* (1), 44–46.
81. Schroeder, Jr, H. W.; Cavacini, L. Structure and Function of Immunoglobulins. *J. Allergy Clin. Immunol.* **2010,** *125* (2), S41–S52.
82. Schreiber, G. The Role of Type I Interferons in the Pathogenesis and Treatment of COVID-19. *Front. Immunol.* **2020,** *11*.
83. Figuero-Pérez, L.; Olivares-Hernández, A.; Escala-Cornejo, R. A.; Terán-Brage, E.; López-Gutiérrez, Á.; Cruz-Hernández, J. J. Anakinra as a Potential Alternative in the Treatment of Severe Acute Respiratory Infection Associated with SARS-CoV-2 Refractory to Tocilizumab. *Reumatología Clínica* (English Edition), 2020.
84. Smithuis, J. W.; Müller, M. C. A.; Smithuis, F. F.; Weenink, R. P. A Case of Cerebral and Coronary Air Embolism After Flushing of a Pleural Drain. *Netherlands J. Crit. Care* **2021,** 41.
85. Ciavarella, C.; Motta, I.; Valente, S.; Pasquinelli, G. (2020). Pharmacological (or Synthetic) and Nutritional Agonists of PPAR-γ as Candidates for Cytokine Storm Modulation in COVID-19 Disease. *Molecules* **2020,** *25* (9), 2076.
86. Weiss, A. R. R.; Dahlke, M. H. Immunomodulation by Mesenchymal Stem Cells (MSCs): Mechanisms of Action of Living, Apoptotic, and Dead MSCs. *Fronti. Immunol.* **2019,** *10*, 1191.
87. Rajarshi, K.; Chatterjee, A.; Ray, S. Combating COVID-19 with Mesenchymal Stem Cell Therapy. *Biotechnol. Rep.* **2020,** e00467.
88. Raskob, G. E.; Hull, R. D.; Pineo, G. F. *Heparin, Hirudin, and Related Agents. Williams Hematology*; McGraw-Hill: New York, 2001; pp 1793–1801.
89. Dayer, M. R.; Taleb-Gassabi, S.; Dayer, M. S. Lopinavir; a Potent Drug Against Coronavirus Infection: Insight from Molecular Docking Study. *Arch. Clin. Infect. Dis.* **2017,** *12* (4).
90. Patanè, S. Covid-19 and the Cardiologist: Drug Therapeutic Options. *J. Cardiol. Ther.*, *7* (1), 931–938.
91. Vankadari, N. Arbidol: A Potential Antiviral Drug for the Treatment of SARS-CoV-2 by Blocking Trimerization of the Spike Glycoprotein. *Int. J. Antimicrob. Agents* **2020,** *56* (2), 105998.
92. Zhu, Z.; Lu, Z.; Xu, T.; Chen, C.; Yang, G.; Zha, T.; Xue, Y. Arbidol Monotherapy Is Superior to Lopinavir/Ritonavir in Treating COVID-19. *J. Infect.* **2020,** *81* (1), e21–e23.
93. Mohamed, K.; Yazdanpanah, N.; Saghazadeh, A.; Rezaei, N. Computational Drug Discovery and Repurposing for the Treatment of Covid-19: A Systematic Review. *Bioorg. Chem.* **2021,** *106*, 104490.
94. Bellavite, P.; Donzelli, A. Hesperidin and SARS-CoV-2: New Light on the Healthy Function of Citrus Fruits. *Antioxidants* **2020,** *9* (8), 742.
95. Randhawa, G. S.; Soltysiak, M. P.; El Roz, H.; de Souza, C. P.; Hill, K. A.; Kari, L. Machine Learning Using Intrinsic Genomic Signatures for Rapid Classification of Novel Pathogens: COVID-19 Case Study. *Plos One* **2020,** *15* (4), e0232391.
96. Zhai, X.; Sun, J.; Yan, Z.; Zhang, J.; Zhao, J.; Zhao, Z.; Su, S. Comparison of Severe Acute Respiratory Syndrome Coronavirus 2 Spike Protein Binding to ACE2 Receptors from Human, Pets, Farm Animals, and Putative Intermediate Hosts. *J. Virol.* **2020,** *94* (15).

97. Mustafa, M. I.; Abdelmoneim, A. H.; Makhawi, A. M. Immunoinformatics Patterns and Characteristic of Epitope-Based Peptide Vaccine Candidates against COVID-19. *AIJR Preprints*, 2020.
98. Syed, A. Coronavirus: A Mini-Review. *Int. J. Curr. Res. Med. Sci.* **2020,** *6* (1), 8–10.
99. Bjornsson, T. D.; Callaghan, J. T.; Einolf, H. J.; Fischer, V.; Gan, L.; Grimm, S.; Wrighton, S. A. The Conduct of In Vitro and In Vivo Drug-Drug Interaction Studies: A Pharmaceutical Research and Manufacturers of America (PhRMA) Perspective. *Drug Metabol. Disposition* **2003,** *31* (7), 815–832.
100. Al-Bari, M. A. A. Targeting Endosomal Acidification by Chloroquine Analogs as a Promising Strategy for the Treatment of Emerging Viral Diseases. *Pharmacol. Res. Persp.* **2017,** *5* (1).
101. Mohs, R. C.; Greig, N. H. (2017). Drug Discovery and Development: Role of Basic Biological Research. *Alzheimer's & Dementia (New York, NY)* **2017,** *3* (4), 651–657.
102. Alexander, S. P.; Armstrong, J. F.; Davenport, A. P.; Davies, J. A.; Faccenda, E.; Harding, S. D.; Spedding, M. A Rational Roadmap for SARS-CoV-2/COVID-19 Pharmacotherapeutic Research and Development: IUPHAR Review 29. *Br. J. Pharmacol.* **2020,** *177* (21), 4942–4966.
103. Richon, A. B. An Early History of the Molecular Modeling Industry. *Drug Discov. Today* **2008,** *13* (15–16), 659–664.
104. Heisenberg, W. Quantum-Theoretical Re-interpretation of Kinematic and Mechanical Relations. *Z. Phys.* **1925,** *33*, 879–893.
105. Hendrickson, J. B. Molecular Geometry. I. Machine Computation of the Common Rings. *J. Am. Chem. Soc.* **1961,** *83* (22), 4537–4547.
106. Levinthal, C. Molecular Model-Building by Computer. *Sci. Am.* **1988,** *214* (6), 42–53.
107. Menchaca, T. M.; Juárez-Portilla, C.; Zepeda, R. C. Past, Present, and Future of Molecular Docking. In *Drug Discovery and Development-New Advances*; IntechOpen, 2020.
108. Phillips, M. A.; Stewart, M. A.; Woodling, D. L.; Xie, Z. R. Has Molecular Docking Ever Brought Us a Medicine. *Mol. Dock.*; IntechOpen: London, UK, 2018; pp 141–78.
109. Lakshmi Prabha, M.; Pradeepa, N. Role of Molecular Modelling in Drug Design. *Int. J. Innov. Res. Med. Sci.* **2016,** *1* (2), 39–41.
110. Smith, S.J.; Sutcliffe B. T. The Development of Computational Chemistry in the United Kingdom. *Rev. Comput. Chem.* **1997,** *10*, 271–316.
111. Hagar, M.; Ahmed, H. A.; Aljohani, G.; Alhaddad, O. A. Investigation of Some Antiviral N-heterocycles as COVID 19 Drug: Molecular Docking and DFT Calculations. *Int. J. Mol. Sci.* **2020,** *21* (11), 3922.
112. Noureddine, O.; Issaoui, N.; Al-Dossary, O. DFT and Molecular Docking Study of Chloroquine Derivatives as Antiviral to Coronavirus COVID-19. *J. King Saud Univ.-Sci.* **2021,** *33* (1), 101248.
113. Harismah, K.; Mirzaei, M. Favipiravir: Structural Analysis and Activity against COVID-19. *Adv. J. Chem., Sect. B: Nat. Products Med. Chem.* **2020,** *2* (2), 55–60.
114. Sagaama, A.; Brandan, S. A.; Issa, T. B.; Issaoui, N. Searching Potential Antiviral Candidates for the Treatment of the 2019 Novel Coronavirus Based on DFT Calculations and Molecular Docking. *Heliyon* **2020,** *6* (8), e04640.
115. El Hadki, H. DFT and Molecular Docking Study of Natural Molecules Proposed for COVID-19 Treatment. *Moroccan J. Chem.* **2021,** *9* (2).
116. Mumit, M. A.; Pal, T. K.; Alam, M. A.; Islam, M. A. A. A.; Paul, S.; Sheikh, M. C. DFT Studies on Vibrational and Electronic Spectra, HOMO–LUMO, MEP, HOMA, NBO

and Molecular Docking Analysis of benzyl-3-N-(2, 4, 5-trimethoxyphenylmethylene) Hydrazinecarbodithioate. *J. Mol. Struct.* **2020,** *1220*, 128715.
117. Shahab, S.; Sheikhi, M. Triazavirin-Potential Inhibitor for 2019-nCoV Coronavirus M Protease: A DFT Study. *Curr. Mol. Med*, 2020.
118. El Kalai, F.; Çınar, E. B.; Lai, C. H.; Daoui, S.; Chelfi, T.; Allali, M.; Benchat, N. Synthesis, Spectroscopy, Crystal Structure, TGA/DTA Study, DFT and Molecular Docking Investigations of (E)-4-(4-methylbenzyl)-6-styrylpyridazin-3 (2H)-one. *J. Mol. Struct.* **2021,** *1228*, 129435.
119. Douche, D.; Sert, Y.; Brandán, S. A.; Kawther, A. A.; Bilmez, B.; Dege, N.; Himmi, B. 5-((1H-imidazol-1-yl) methyl) quinolin-8-ol as Potential Antiviral SARS-CoV-2 Candidate: Synthesis, Crystal Structure, Hirshfeld Surface Analysis, DFT and Molecular Docking Studies. *J. Mol. Struct.* **2021,** *1232*, 130005.
120. Shahab, S.; Sheikhi, M.; Alnajjar, R.; Al Saud, S.; Khancheuski, M.; Strogova, A. DFT Investigation of Atazanavir as Potential Inhibitor for 2019-nCoV Coronavirus M Protease. *J. Mol. Struct.* **2021,** *1228*, 129461.
121. Bouzian, Y.; Baydere, C.; Dege, N.; Ahabchane, N. H.; Mague, J. T.; Abudunia, A.; Essassi, E. M. Crystal Structure, Hirshfeld Surface Analysis, DFT and Molecular Docking Investigation of 2-(2-oxo-1, 3-oxazolidin-3-yl) Ethyl 2-[2-(2-oxo-1, 3-oxazolidin-3-yl) Ethoxy] Quinoline-4-carboxylate. *Acta Crystallographica Sect. E: Crystallographic Commun.* **2021,** *77* (1), 28–33.
122. Mary, S. J.; Siddique, M. U. M.; Pradhan, S.; Jayaprakash, V.; James, C. Quantum Chemical Insight Into Molecular Structure, NBO Analysis of the Hydrogen-Bonded Interactions, Spectroscopic (FT–IR, FT–Raman), Drug Likeness and Molecular Docking of the Novel Anti COVID-19 Molecule 2-[(4, 6-diaminopyrimidin-2-yl) Sulfanyl]-N-(4-fluorophenyl) Acetamide-Dimer. *Spectrochim. Acta A: Mol. Biomol. Spectrosc.* **2021,** *244*, 118825.
123. Celik, S.; Demirag, A. D.; Ozel, A. E.; Akyuz, S. Interactions Mechanism of Commonly Used Drugs for the Treatment of COVID-19. *Bull. Chem. Soc. Ethiopia* **2020,** *34* (3), 613–623.
124. Belhassan, A.; Chtita, S.; Zaki, H.; Lakhlifi, T.; Bouachrine, M. Molecular Docking Analysis of N-Substituted Oseltamivir Derivatives with the SARS-Cov-2 Main Protease. *Bioinformation* **2020,** *16* (5), 404.
125. Mpiana, P. T.; Tshibangu, D. S.; Kilembe, J. T.; Gbolo, B. Z.; Mwanangombo, D. T.; Inkoto, C. L.; Tshilanda, D. D. Identification of Potential Inhibitors of SARS-CoV-2 Main Protease from Aloe Vera Compounds: A Molecular Docking Study. *Chem. Phys. Lett.* **2020,** *754*, 137751.
126. Zeyrek, C. T.; Arpacı, Ö. T.; Arısoy, M.; Onurdağ, F. K. Synthesis, Antimicrobial Activity, Density Functional Modelling and Molecular Docking with COVID-19 Main Protease Studies of Benzoxazole Derivative: 2-(p-chloro-benzyl)-5-[3-(4-ethly-1-piperazynl) Propionamido]-benzoxazole. *J. Mol. Struct.* **2021,** 130413.

PART II

Pharmacotherapeutically Active Classes

CHAPTER 5

Perspectives on Local Anesthetics in Medicine

T. N. V. GANESH KUMAR

Chebrolu Hanumaiah Institute of Pharmaceutical Sciences, Chandramoulipuram, Chowdavaram, Guntur 522019, Andhra Pradesh, India

E-mail: ganeshtnv@gmail.com

ABSTRACT

Local anesthesia is employed for operative practices and is distinguished by a lack of pain and muscle relaxation. This chapter deals with nerve conduction, mechanism of action, essential chemical features, structure–activity relationship, selection criteria for medical utility, systemic toxicity, techniques employed for administration, physicochemical characteristics, classification, pharmacokinetic, pharmacodynamics, side effects, adverse effects, pros and cons of numerous drug attributes such as ester-linked (benzocaine, procaine, piperocaine, propoxycaine, proparacaine, tetracaine, cyclomethycaine, chloroprocaine, oxybuprocaine, meprylcaine, butacaine, hexylcaine), amide-linked (lidocaine, prilocaine, etidocaine, ropivacine, dibucaine, bupivacaine, mepivacaine), and miscellaneous (diperidon, pramoxine, phenacaine, fomocaine, dimethisoquin, dyclonine, myrtecaine).

5.1 INTRODUCTION

The evolution in medical research has led us to find anesthetics from nonanesthetized operations in the olden days. The requirement of the most

Biologically Active Small Molecules: Modern Applications and Therapeutic Perspectives.
Debarshi Kar Mahapatra, & Sanjay Kumar Bharti (Eds.)

excellent molecule with good ADME properties and less toxicity is always a challenging requirement in the current situation. The use of general anesthetics (GA) is risky and complicated. For small and localized surgeries, the use of GA is not required. This made researchers invent local anesthetics (LAs) to overcome GA, where the pain receptors are blocked at a specific area without affecting the central nervous system (CNS). LA prevents or relieves pain by break off the nerve conduction. They connect to the exact receptors on the sodium (Na^+) channels present in the neuronal system and obstruct the ion traveling via these channels. The chemical and pharmacological properties of LA drugs establish their clinical properties. This chapter discusses the structural aspects, mechanism of action of LAs, and drugs with their properties. Anesthesia can be of three main categories: LA, regional anesthetics (RAs), and GA, all of which persuade the nervous system in concurrently and can be administered employing a range of methods and varied medications.

5.1.1 LOCAL ANESTHESIA

Local anesthesia is an anesthetic drug (which can be administered as a spray, ointment, shot, etc.) that anesthetizes only a diminutive, precise area of the body (e.g., hand, foot, or skin). With the application of LA, an individual is either sedated or awake, based on what is required. LA remains for a small duration of time and is habitually employed in negligible outpatient processes (when individuals arrive in for surgery and can walk off on the same day). The medicine used can freeze the area during the process and for a petite time subsequently to assist manage the postsurgery distress.

5.1.2 REGIONAL ANESTHESIA

Regional anesthesia is an anesthetic drug that is administered near the nerves cluster, which freezes a bigger region of the human body like below the waist region (like epidurals provide to women during labor). RA is typically utilized to make a person feel highly comfortable during the surgical practice as well as after the surgical practice.

5.1.3 GENERAL ANESTHESIA

General anesthesia attempts is to make and keep an entity absolutely insensible (or "asleep") through the operative functions, with no awareness or memory of the surgery. The general anesthesia can be injected via the IV route (that requires sticking a needle into a vein, characteristically in the arm) or by breathe in the vapors or gases by inhalation into a tube or a mask.

5.2 NERVE CONDUCTION

Nerve conduction is the promulgation of an electrical impulse produced by the speedy traveling of potassium (K^+) and sodium (Na^+) ions across the nerve cell membrane. The ionic gradient for K^+ (high intracellular and low extracellular) and Na^+ (high extracellular and low intracellular) is sustained by a Na^+–K^+ pump mechanism within the nerves. Under the latent state, the nerve membrane is more permeable to the K^+ ions as contrasted to the Na^+ ions, ensuing in the continual outflow of K^+ ions out of the nerve cell core. This outflow creates a negatively charged core comparative to the outer surface, which leads to the generation of electric potential of 60–70 mV across the nerve membrane. Sensory nerves have receptors, which serve as sensors through various mechanical, chemical, or thermal stimuli. Such stimuli are interpreted into the electric currents.

The electric potential in a nerve membrane subsequently to the receptor is customized due to stimuli, making it less negative. If the threshold potential is accomplished, an action potential is produced with an unexpected increase in the nerve membrane permeability to the Na^+ ions and an ensuing swift invasion of positively charged Na^+ ions. This results in depolarization, which creates an existing depolarization pattern that depolarizes the subsequent nerve segment in turn. This activates the nerve in transmitting a wave of sequential polarization across the nerve membrane.

The repolarization process takes place when the permeability of Na^+ declines and permeability of K^+ augments, resulting in an efflux of K^+ from within the cell and the restitution of the electrical balance. Consequently, both the ions are reinstated to their preliminary intracellular and extracellular concentrations by the Na^+–K^+ adenosine triphosphate pump mechanism. Because the rapid incursion of Na^+ ions occurs in reply to an alteration in the transmembrane potential, the Na^+ channels in the nerve are characterized as "voltage-gated." These channels are the protein structures with three subunits that penetrate the complete depth of the membrane bilayer and are in communication with both the extracellular surface of the nerve membrane and the axoplasm (interior) of the nerve. LA avert the creation and transmission of the nerve impulses by particularly binding with the α-subunit of the Na^+ channel and averting the spreading of Na^+ into the cell, thereby, halting the conduction across the nerve fiber of the progressive wave of depolarization. The resting potential of the nerve cells is about mV (Figure 5.1).

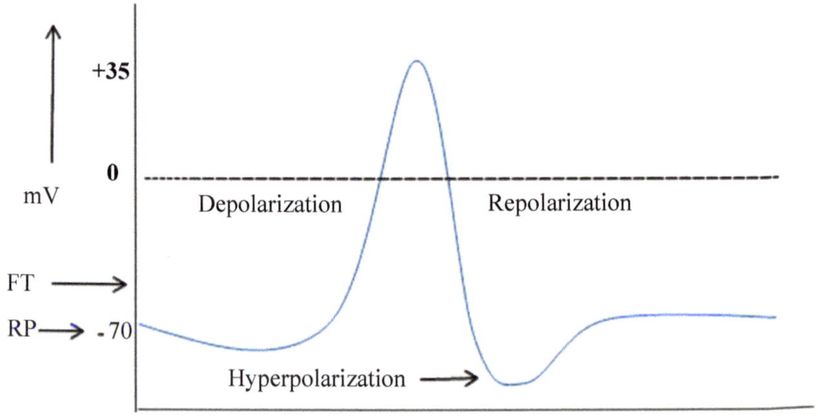

FIGURE 5.1 Electrical potential changes during nerve impulse. (mV) Millivolts, (FT) firing threshold, (RP) resting potential.

During a stimulus, a preliminary depolarization of the membrane occurs by about +20 mV. This is followed by a speedy increase of the membrane potential to a maximum of +35 mV, which is called depolarization. This is instantly trailed by repolarization. The repolarization exceeds the resting potential (hyperpolarization) previous to coming to resting potential. This speedy increase and descend of potential are called as spike potential and the point at which it commences is called as firing threshold of the axon. A resting nerve is less sensitive to a LA than a nerve that is continually stimulated. A superior frequency of stimulation and a more positive membrane potential cause an improved quantity of the transmission obstruction. The frequency- and voltage-dependent effects of LA occur because recurrent depolarization enhances the probability that a LA molecule will bump into a Na^+ channel that is in the activated or open form. Universally, the dissociation rate from the site of the receptor in the Na^+ channel pore is noteworthy for the frequency reliance of LA action.

5.3 MECHANISM OF ACTION OF LAS

LAs exceed through the nerve membrane in lipid-soluble base, nonionized form. When they stay within the nerve axoplasm milieu, they must equilibrate into an ionic form to put forth their action on the sodium channel. The drug in its unionized form surpass through the cell membrane by hydrophobic pathway, whereas the ionized drug goes into through the sodium channel and obstructs it. The unionized drug after inflowing the intracellular fluid converts to ionized form and goes into the sodium channel and obstructs it (Figure 5.2). In both the cases of equilibration, the drug obstructs the sodium channel and thus put forth its action of anesthesia by restraining the impulse.

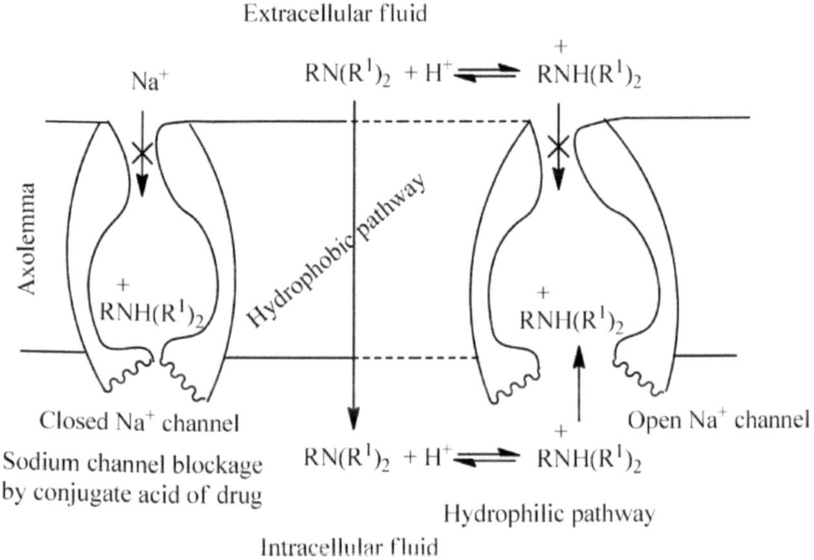

FIGURE 5.2 Mechanism of action of LAs in blocking the sodium channels.

5.4 ESSENTIAL CHEMICAL FEATURES OF LAS

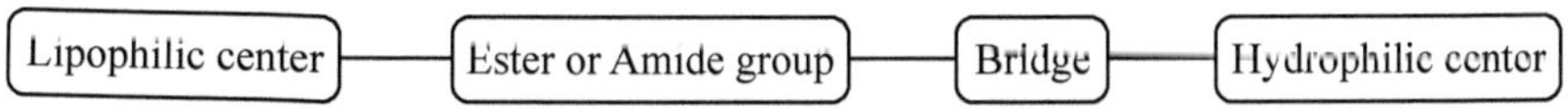

Basic Structural pattern of Local Anaesthetics

5.4.1 STRUCTURE-ACTIVITY RELATIONSHIP REQUIREMENTS OF LAS

The archetypal arrangement of a LA comprises hydrophobic and hydrophilic domains linked by a mediator amide linkage or ester linkage. The hydrophilic group is characteristically a tertiary amine, and the hydrophobic domain chiefly stays an aromatic component. The character of the linking group influences the pharmacological characteristics of LA agents. The physicochemical attributes of these agents primarily influence their potency and duration of action. For example, greater lipid solubility enhances the potency and duration of their action. This is owing to a higher attraction of the drug–to-lipid membranes and consequently enhanced closeness to its sites of action. The greater the drug's action on the sodium channel in the membrane, the longer the drug lingers in the proximity of the membrane. The higher lipid solubility also augments the toxicity by reducing the therapeutic index for the high hydrophobic drugs.

The p*K*a (the pH at which 50% is present as the base and 50% of the drug is ionized) of the LA is linked with pH and the concentrations of the cationic and base forms by the Henderson–Hasselbalch equation:

$$pH = pKa + \log([base]/[cation])$$

The p*K*a normally correlates with the pace of the beginning of action of most amide LA drugs; the nearer the p*K*a to the body pH, the earlier the commencement. The coexistence of the two forms of the drug (ionized and the unionized forms) is important because the LA requires the base (unionized) form to pass through the lipid nerve membrane and the cation (ionized) form creates a blockade of the sodium channel. An ester or an amide linkage is situated between the lipophilic end (benzene ring) and the hydrophilic end (amino group) in the molecule. The type of linkage resolves the position of metabolic degradation of the drug. Ester-linked LAs are principally metabolized in the plasma by enzyme pseudocholinesterase, whereas amide-linked drugs experience exclusive metabolism in the liver.

5.4.2 LOCAL ANESTHETIC DIFFUSION

An epineurium encircles the mixed peripheral nerve or nerve trunk consists of individual nerves. The LA in proximity to a peripheral nerve diffuses from the outer surface toward the core along a concentration gradient. Consequently, nerves situated in the outer shroud of the mixed nerve are blocked first. These fibers are typically dispersed to more proximal anatomic structures than those located near the center of the mixed nerve and often are motor fibers. When the volume and concentration of LA solution in the vicinity of the nerve are sufficient, the LA ultimately disseminates the interior to block the further centrally located fibers. This way, the block progresses from proximal structures to the distal organization. Lesser quantity and concentrations of a drug will only block the nerves in the layers and smaller and more sensitive central fibers.

5.4.3 ONSET OF BLOCKADE

The rate of diffusion of the drug transversely the nerve covering is resolute by concentration, its hydrophobicity, the physical attributes of the tissues contiguous to the nerve structures, and the degree of ionization (unionized LA diffuses faster). The time obligatory for a diffused LA drug to obstruct the receptor at nerve endings and stop the impulse is the commencement of the blockade (Figure 5.3).

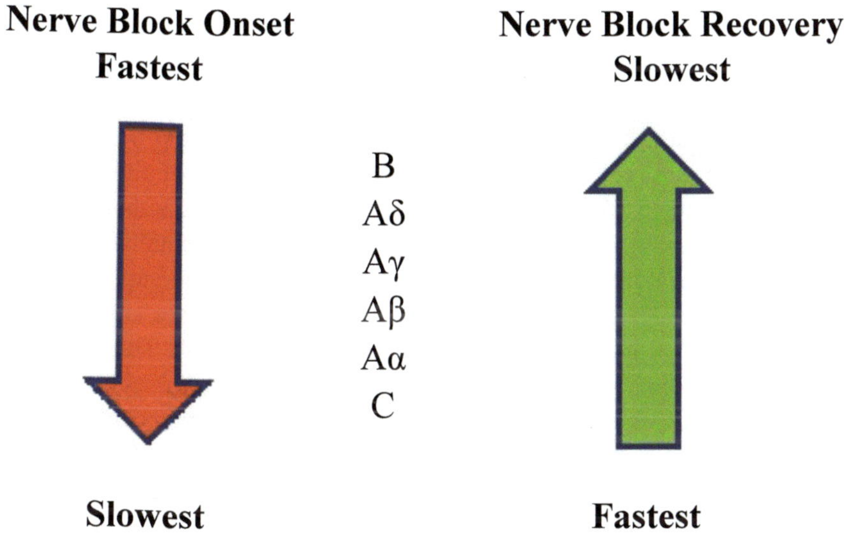

FIGURE 5.3 Differential rate of nerve blockade of receptors.

5.4.4 DURATION OF BLOCKADE

The duration of the nerve block (anesthesia) is influenced chiefly by the physical attributes (lipid solubility) of the LA and also over the presence or absence of vasoconstrictors. In general, LA can be classified into three categories: short-acting (e.g., 2-chloroprocaine, timing: 45–90 min), intermediate-acting (e.g., mepivacaine, lidocaine, timing: 90–180 min), and long-acting (e.g., bupivacaine, ropivacaine, levobupivacaine, timing: 4–18 h). The amount of blockade with the adding up of a vasoconstrictor results in mediating the intrinsic vasodilator properties of the LA. The more intrinsic vasodilator action the LA has, the more persistence is attained with the addition of a vasoconstrictor.

5.4.5 SENSITIVITY OF NERVE FIBERS TO LAS

Two important rules regarding the susceptibility of nerve fibers to LA are:

1. The minor nerve fibers are more vulnerable to the action of LA as compared to the large fibers. These small fibers are blocked superiorly because a shorter length of the axon is obligatory to be obstructed to arrest the conduction entirely.
2. Myelinated fibers are more simply be blocked as compared to nonmyelinated fibers because of LA pools nearer the axonal membrane. That is why C-fibers, which have a small diameter (but are nonmyelinated), are the most challenging fibers to LA.

The sensitivity of fiber to LAs is not estimated by whether it is sensory or motor. In fact, muscle proprioceptive afferent (A-β) and motor efferent fibers (A-α) are uniformly sensitive. These two types of fibers have an identical diameter, which is superior to that of the A-γ fibers that supply the muscle spindles. It is the more swift blockade of these smaller A-γ fibers, relatively than of the sensory fibers, which results in the superior loss of muscle reflexes. In the huge nerve trunks, motor fibers are situated in the outer segment of the bundle and are more reachable to LA. Thus, the motor fibers may be obstructed before sensory fibers in mostly mixed nerves. The discrepancy rate of blockade demonstrated by fibers of changeable sizes and firing rates is of substantial practical significance. The sensation of pain is typically the first modality to vanish pursued by the loss of sensations of warmth, cold, deep pressure, touch, and finally, loss of motor activity.

5.4.6 EFFECT OF pH

LAs pass through the nerve membrane in a nonionized lipid-soluble base form. When they are inside the axoplasm of the nerve, to be involved within the Na^+ channel, they must balance into an ionic form. Penetration of the LA through the nerve membrane is the rate-limiting step in this cascade. LAs are amines without protonation which appear to be relatively insoluble as such. They are produced as water-soluble salts, typically hydrochlorides, for this purpose. Though LAs are weak bases, their hydrochloride salts are slightly acidic (typical p*K*a values vary from 7.5 to 9). The stability of LA esters and any associated vasoconstrictor material is improved by this property.

Sodium bicarbonate ($NaHCO_3$) is frequently supplementary added to the LA to augment the quantity of drug in the base form, which somewhat curtail the commencement time. Noticeably, the limiting factor for the adjustment of pH is the solubility of the base form of the drug. Regrettably, only minor modifications in pH can be attained by the accumulation of bicarbonate because of the imperfect solubility of the base. Only diminutive reduction in the onset time is comprehended. For example, on alkalinization of bupivacaine, an augmentation in the quantity of base in the solution is inadequate by the negligible solubility of free base in the solution. For each LA, there is a pH at which the amount of the base in solution is maximal (a saturated solution). Further augmentation in pH result in precipitation of the drug and do not create an additional limitation of the onset time.

5.4.7 PROTEIN BINDING

LAs are bound to plasma and tissue proteins in large portions. They are, however, pharmacologically active in the open, unbound state only. Albumin and alpha 1 acid glycoprotein are the most important binding proteins for LAs in plasma (AAG). The binding to AAG is described as a high-affinity, but low capacity binding; thus, relative to albumin, LAs bind to AAG preferentially. Binding to AAG, however, is quickly saturated with LA blood levels that have been clinically achieved. When saturation of AAG occurs, all further binding is to albumin. At concentrations, several times greater than those clinically achieved, albumin will bind LA drugs in plasma.

The plasma protein binding associated with the period of activity is in the order: bupivacaine > etidocaine > ropivacaine > mepivacaine > lidocaine > procaine and 2-chloroprocaine. However, no straight connection subsisted between the LA plasma protein binding and binding to exact membrane-bound Na^+ channels. The more lipid soluble the drug, the more likely it will stay in the axon where the Na^+ channel exists in.

The degree of protein binding is concentration-dependent and influenced by the pH of the plasma. The percentage of the drug binding diminishes as the pH decreases. This is fairly necessary, as with the progression of acidosis, which may happen with cardiac arrest or/and LA-induced seizures, the concentration of the free drug elevates. The amount of this trend differs among LAs, and it is much more discrete with bupivacaine than with lidocaine. For occasion, as the binding reduces from 95 to 70% with acidosis, the concentration of nonbounded bupivacaine augments from 5 to 30% (a factor of 6), even though the total drug amount remains unaffected. Because of this augmentation in the amount of free drug, acidosis renders bupivacaine noticeably more toxic.

5.5 TYPES OF LOCAL ANASTHETICS

The LAs were classified into three categories based on the basic structure of the nucleus. They are (1) ester-linked, (2) amide-linked, and (3) miscellaneous.

5.5.1 ESTER-LINKED LAS

Ester-linked LAs are hydrolyzed in plasma by pseudocholinesterase at the ester linkage. The rate of ester-linked LAs hydrolysis depends on the form and position of the substitute in the aromatic ring. For example, 2-chloroprocaine is hydrolyzed about fourfold faster than procaine, which in turn is hydrolyzed about fourfold faster than tetracaine. However, in patients with uncharacteristic plasma pseudocholinesterase, the rate of hydrolysis of all ester-linked LAs is greatly decreased, and an extended epidural block in a patient with irregular pseudocholinesterase has been documented. Ester-linked LA metabolism by hydrolysis contributes to the growth of para-amino benzoic acid (PABA). PABA and its derivatives bear a minute risk potential for allergic reactions. However, allergic reactions can also extend from the application of multiple doses of amide-linked LA that contain PABA as a preservative. The following are the drugs belonging to this category (Figure 5.4).

5.5.1.1 COCAINE

Cocaine is derived from the coca shrub's leaves. That is a benzoic acid ester. The acts of cocaine are nerve impulse blockade and central vasoconstriction secondary to local norepinephrine reuptake inhibition. In modern medicine, its toxicity and the propensity for misuse have precluded greater therapeutic application of cocaine. Its euphoric properties are due to catecholamine uptake inhibition at CNS synapses, particularly dopamine. Other LAs do not obstruct norepinephrine absorption and do not induce cocaine sensitization to the characteristics of catecholamines, vasoconstriction, or mydriasis. Usually, cocaine is mostly used to deliver topical upper respiratory tract anesthesia, where the combination vasoconstrictor and LA properties provide single agent anesthesia and mucosal shrinkage. Cocaine is up to 99% metabolized predominantly in the liver. The metabolism is by hydrolytic ester cleavage, and benzoylecgonine (major metabolite) and ecgonine methyl ester and ecgonine are the removed metabolites. Minor metabolites of cocaine include norcocaine, *para*-hydroxycocaine, m-hydroxycocaine,

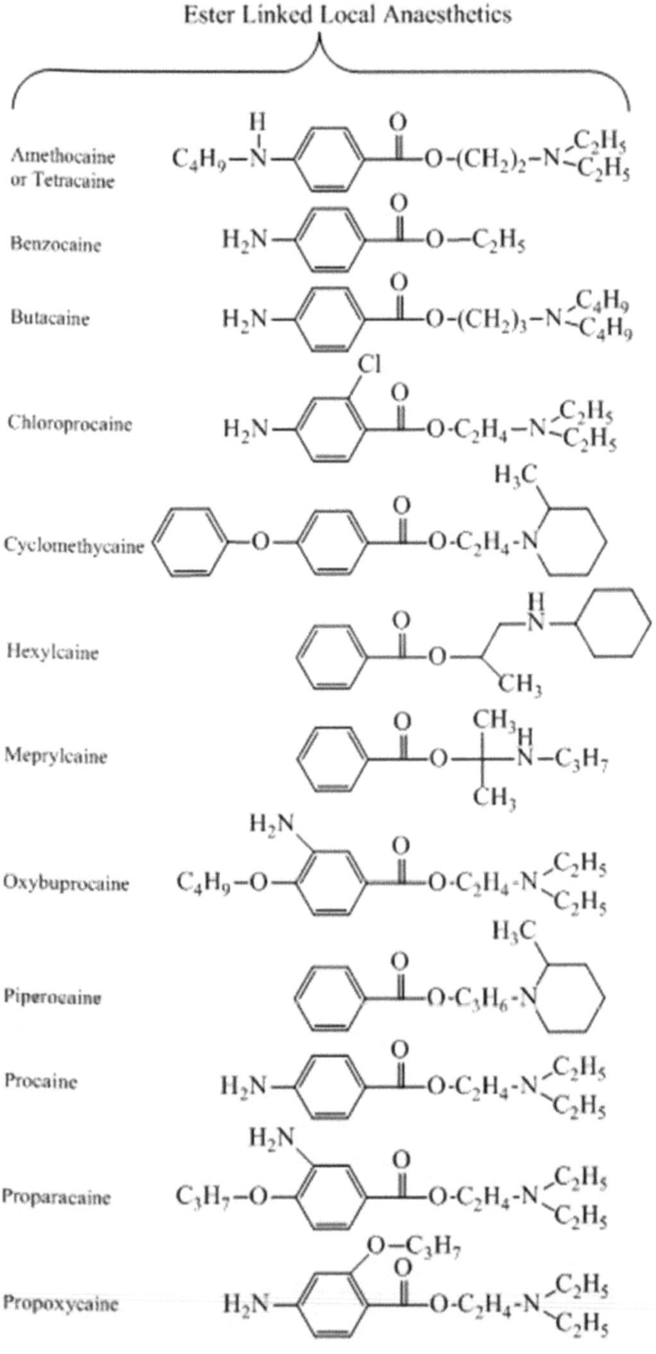

FIGURE 5.4 Examples of ester-linked LAs.

para-hydroxybenzoylecgonine, and *meta*-hydroxy benzoylecgonine. Within 4 h after cocaine consumption, benzoylecgonine can be found in the urine and is detectable at concentrations higher than 150 ng/mL for up to 8 days after cocaine is used. It is possible to detect an aggregation of cocaine metabolites in the hair of daily users. Cocaine binds to form cocaethylene with alcohol in the liver. Studies showed that cocaethylene is also more euphoric and has greater coronary toxicity than cocaine alone.

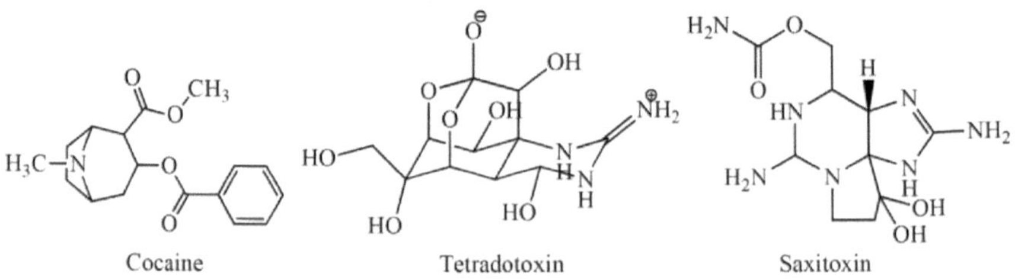

5.5.1.2 PROCAINE

This amino ester molecule (2-(diethylamino)ethyl 4-aminobenzoate) was the first synthesized LA. Procaine has low potency, slow onset, and short duration of action. Although for numerous decades, it was employed extensively, on the other hand, its applied is mainly restricted to infiltration anesthesia and possibly diagnostic nerve blocks.[8]

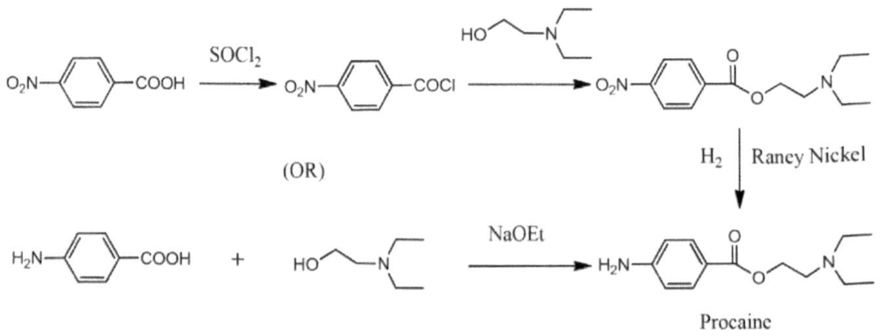

5.5.1.3 2-CHLOROPROCAINE

2-Chloroprocaine, an ester LA that was introduced in 1952, is a chlorinated procaine derivative. The most readily metabolized LA used at present is chloroprocaine.

5.5.1.4 TETRACAINE

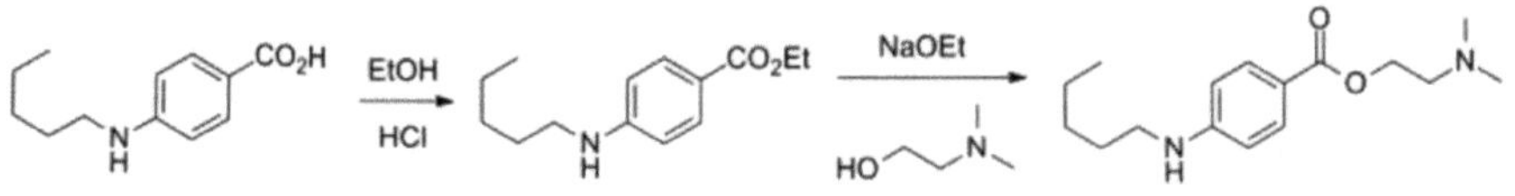

Tetracaine, 2-(dimethylamino) ethyl 4-(butylamino) benzoate, was introduced in 1932, and it is a long-acting amino ester. It is considerably more potent with a longer period of action than procaine and 2-chloroprocaine. Tetracaine is more gradually metabolized than the other commonly used ester LAs, and it is considerably more toxic. At the current situation, it is used in spinal anesthesia, where a drug of long duration is required, as well as in diverse topical anesthetic preparations. Because of its slow onset and potential for toxicity, tetracaine is employed infrequently for peripheral nerve blocks. It is used in combination with lidocaine and is often known as "supercaine."[9]

5.5.1.5 BENZOCAINE ETHYL 4-AMINOBENZOATE

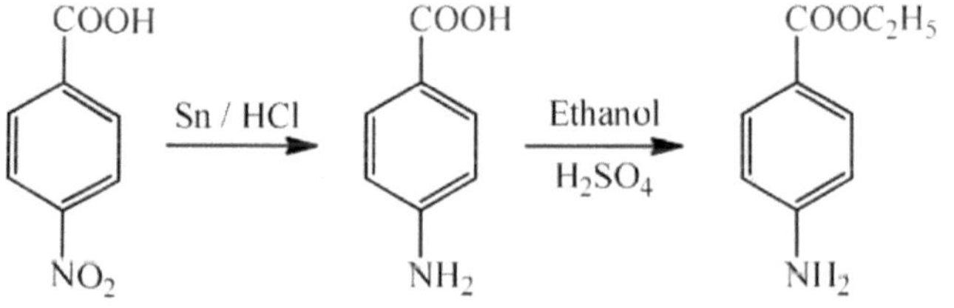

5.5.1.6 BUTCAINE

Butcaine, 4-aminobenzoic acid 3-(dibutylamino)propyl ester, is a LA used for minor operative purposes. Its synthesis is comparable to procaine, but it uses *n*-dibutylamine instead of diethylamine.

5.5.1.7 CHLORPROCAINE

IUPAC name: 2-diethylaminoethyl-4-amino-2-chloro-benzoate.[10]

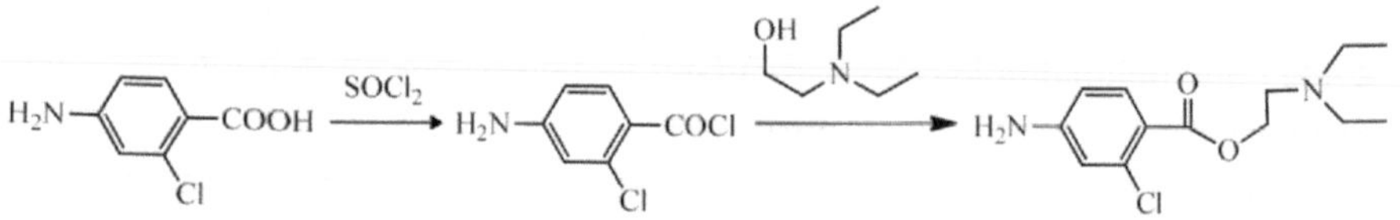

5.5.1.8 CYCLOMETHICAINE

IUPAC name: 4-(cyclohexoxy)benzoic acid 3-(2-methyl-1-piperidinyl) propyl ester.[11]

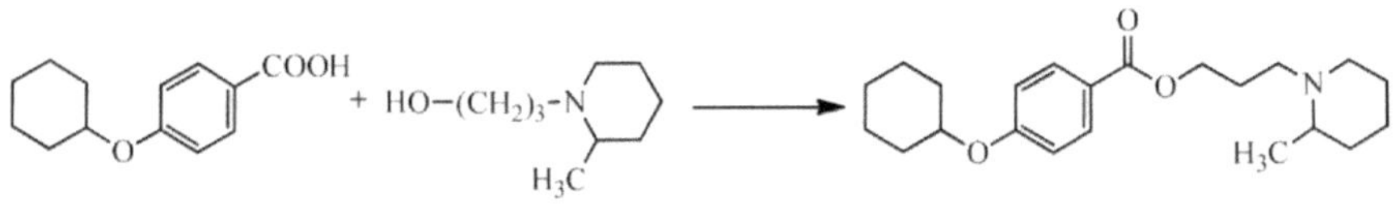

5.5.1.9 HEXYLCAINE

IUPAC name: 1-cyclohexylaminopropan-2-yl benzoate.[12]

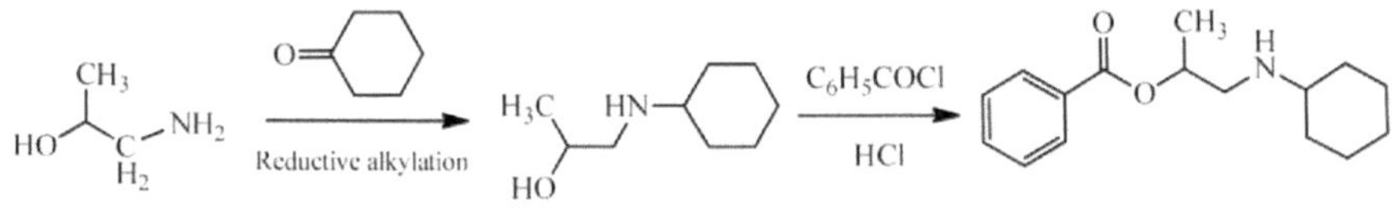

5.5.1.10 MEPRYLCAINE

IUPAC name: benzoic acid (2-methyl-2-propylaminopropyl) ester.[12]

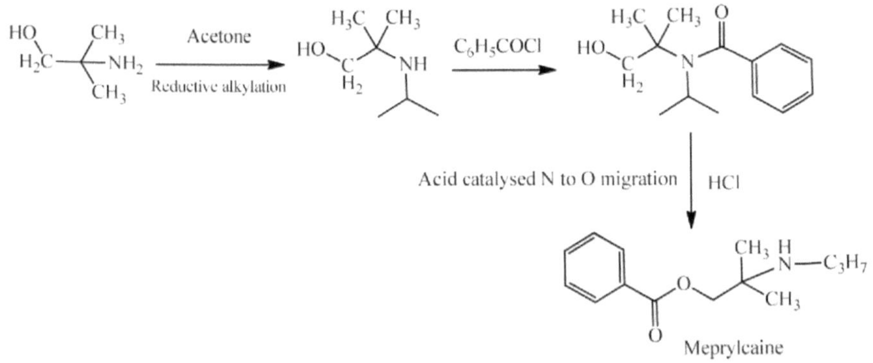

5.5.1.11 OXYBUPROCAINE

Oxybuprocaine is sometimes referred to as benoxinate and is used in otolaryngology and ophthalmology in particular. Oxybuprocaine (2-diethylaminoethyl-4-amino-3-butoxy-benzoate) is mismatched with silver and mercury salts, as well as basic substances. It also decreased the antimicrobial action of sulfonamides.[13]

It can cause hypersensitivity, irritation, irreversible corneal damage, and anaphylaxis in excess use.

5.5.1.12 PIPEROCAINE

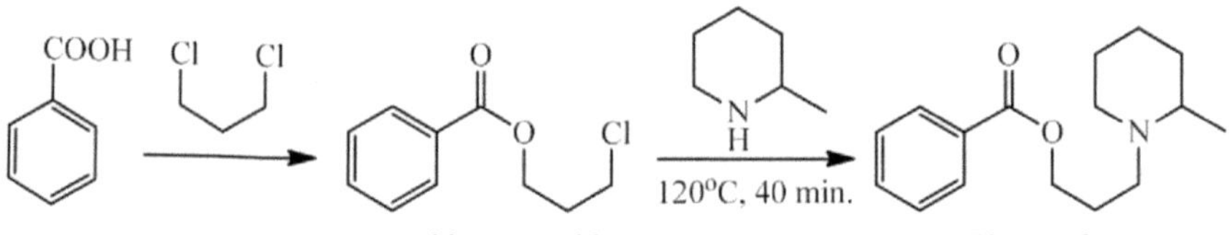

Piperocaine (3-(2-methylpiperidin-1-yl)propyl benzoate) is synthesized by employing the γ-chloro-propyl benzoate on reacting with 2-methyl piperidine at 120–140°C reflux for 40 min yields piperocaine.[14]

5.5.1.13 PROPARACAINE OR PROXYMETACAINE

Proxymetacaine (2-(diethylamino) ethyl 3-amino-4-propoxybenzoate) is for topical ophthalmic application only. It is specifically not deliberate for injection. Permanent corneal opacification with subsequent vision distortion may be caused by repeated application of this or some other topical ocular anesthetic. Proxymetacaine is available as its hydrochloride salt in ophthalmic solutions at a concentration of 0.5%.

5.5.1.14 PROPOXYCAINE

This ethyl 4-amino-2-propoxybenzoate LA (2-(diethylamino)) has a fast onset of action and a longer period of action relative to procaine hydrochloride.

5.5.2 AMIDE-LINKED LAS

These are metabolized by a dealkalization reaction in the liver in which the tertiary amine cleaves an ethyl group. The hepatic blood supply and function of the liver decide these anesthetics' hepatic clearance. Factors that reduce hepatic drug extraction or/and blood flow, ensuing an amplified elimination half-life in both cases. A slight quantity of unchanged drug (3–5%) of the total drug injected gets eliminated by urinary excretion. The following are the drugs that go into this category (Figure 5.5).

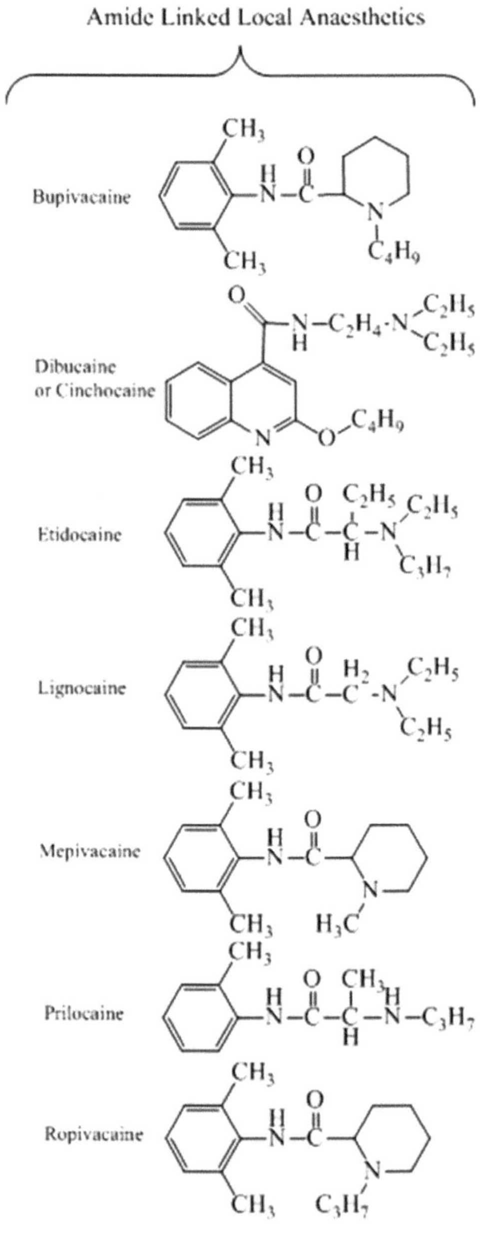

FIGURE 5.5 Examples of amide-linked LAs.

5.5.2.1 LIDOCAINE

Lidocaine (2-(diethylamino)-*N*-(2,6-dimethylphenyl)acetamide) was initiated in 1948 by Astra, and it stays one of the most flexible and widely used LAs. It is the model of the amide class of LAs. Lidocaine is absorbed speedily after parenteral administration and from the gastrointestinal and respiratory tracts. Lidocaine can be used in roughly any peripheral nerve block where an intermediate-length LA is needed. For surgical anesthesia, a dosage of 1.5–2% with or without the use of epinephrine is used more commonly. The 2,6-xylidine reaction with chloroacetyl chloride, followed by the reaction with diethylamine, is prepared in two steps.[15]

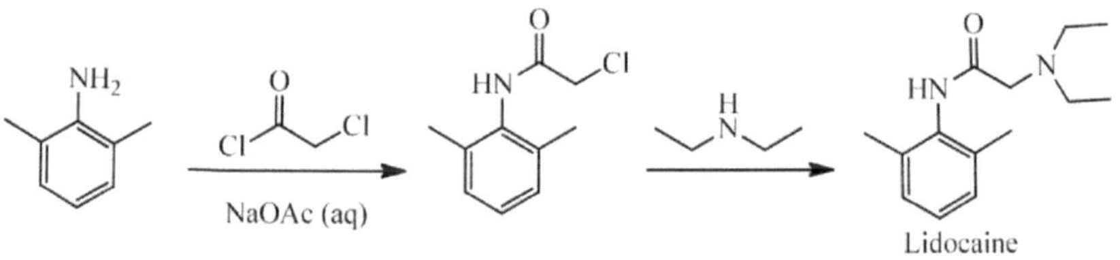

5.5.2.2 MEPIVACAINE

Mepivacaine, (*RS*)-*N*-(2,6-dimethylphenyl)-1-methyl-piperidine-2-carboxamide, is an intermediate-duration amino amide LA introduced in 1957. Its pharmacological properties are comparable to those of lidocaine. It is more toxic to neonates (and as such is not applied in obstetric anesthesia), and it appears to have a slightly superior therapeutic index in adults than lidocaine. Its initiation of action is similar to that of lidocaine, but its duration of action is significantly longer than that of lidocaine. In any peripheral nerve block technique, the initial option is 1.5% mepivacaine when an intermediate-duration blockade is needed (3–6 h, depending on the type of nerve block and the addition of a vasoconstrictor). It is synthesized by reacting with 2,6-dimethylanilinomagnesium bromide, which is synthesized from 2,6-xylidine and ethyl magnesium bromide, the ethyl ester of 1-methylpiperidine-2-carboxylic acid.[16]

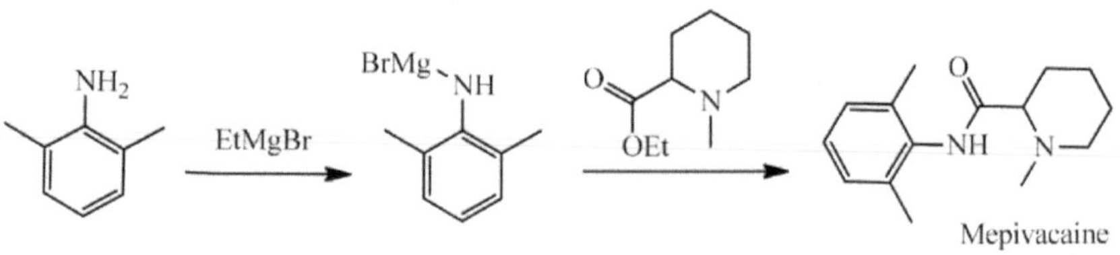

5.5.2.3 PRILOCAINE

Prilocaine, (*RS*)-*N*-(2-methylphenyl)-*N*,*N*-propylalaninamide, is an intermediate-duration amino amide LA with a pharmacological profile similar to that of lidocaine. The principal disparity is a lack of vasodilatation and an augment volume of distribution, which decreases its CNS toxicity. Though, it is exclusive among amide LAs for its tendency to cause methemoglobinemia, an effect of metabolism of the aromatic ring to o-toluidine. By IV administration of methylene blue (1–2 mg/kg), the toxicity may be treated. Occasionally, prilocaine is used for peripheral nerve blockade. It is synthesized by the addition of 2-bromopropionyl bromide to o-toluidine followed by the reaction of n-butylamine.

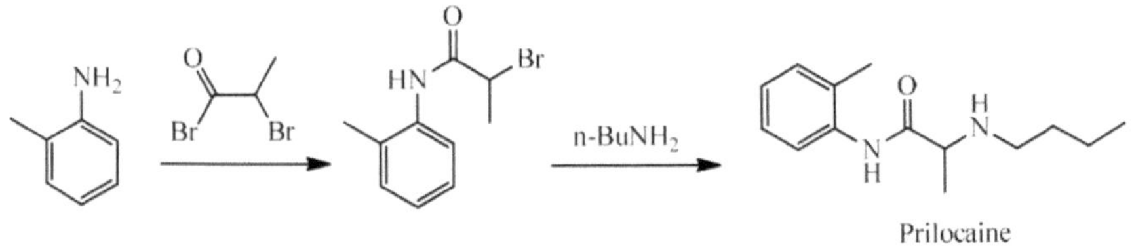

5.5.2.4 ETIDOCAINE

Etidocaine (*N*-(2,6-dimethylphenyl)-2-(ethyl(propyl)amino)butanamide) is a long-acting amino amide LA introduced in 1972. Etidocaine is structurally similar to lidocaine, with alkyl substitution between the hydrophilic amine and the amide linkage on the aliphatic binding group. This function increases the drug's lipid solubility and demonstrates greater potency than lidocaine, which has a very fast onset of action and an extended anesthesia time. Its initiation of action is comparable to that of lidocaine, and its duration is comparable to that of bupivacaine. The chief drawback of etidocaine treatment is its powerful motor blockade at therapeutic dosages, which frequently last longer than the sensory blockade. Therefore, etidocaine is in no way related to peripheral nerve blockade.

5.5.2.5 BUPIVACAINE

Bupivacaine, (*RS*)-1-Butyl-*N*-(2,6-dimethylphenyl)piperidine-2-carboxamide, is one of the most generally utilized LAs in regional and infiltration anesthesia.

It is introduced in 1963. It shares a widespread structural characteristic with lidocaine, except that the amine-containing group is a butylpiperidine. A long-acting agent that induces prolonged anesthesia is bupivacaine. By the incorporation of epinephrine, the analgesic effect can be extended much further. It has a higher degree of cardiotoxicity than lidocaine. Bupivacaine's cardiotoxicity is cumulative and considerably higher than can be expected from its LA potency. Since direct injection of diminutive volumes of bupivacaine into the medulla may produce malignant ventricular arrhythmias, at least part of the cardiotoxicity of bupivacaine can be centrally controlled. Cardiotoxicity caused by bupivacaine can be difficult to treat. In both neuraxial and peripheral nerve blockades, bupivacaine is widely used. A slower onset and a long (unpredictable) duration of the blockade define the blocking property. Significant doses of bupivacaine should be avoided due to its toxicity profile. In the presence of a base, it is synthesized from picolinic acid and ethyl chloroformate. The intermediate reaction with 2,6-xylidine gives amide which on treating with *n*-butyl iodide gives the pyridinium salt. The salt on reduction yields bupivacaine.

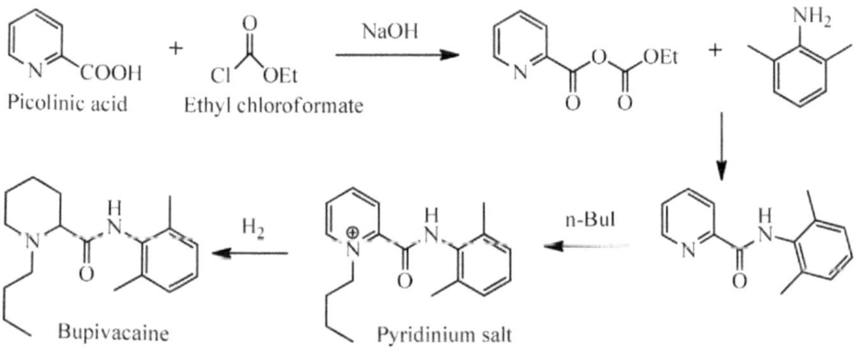

5.5.2.6 *ROPIVACAINE*

The cardiotoxicity of bupivacaine stimulated concern in creating a less toxic, long-lasting LA. This led to the development of ropivacaine ((*S*)-*N*-(2,6-dimethylphenyl)-1-propylpiperidine-2-carboxamide), the *S*-enantiomer of 1-propyl-2′,6′-pipecolocylidide. The S-enantiomer was selected because it has lower toxicity than the R-enantiomer, like most LAs with a chiral core. Apparently, this is due to slower penetration, which for a given dose results in lower blood pressures. During IV administration, Ropivacaine undergoes significant hepatic metabolism, with just 1% of the drug unchanged in the urine being removed. Ropivacaine is much less active than bupivacaine when used at lower doses to produce anesthesia. The onset of blockade is about as swift as 1.5% mepivacaine or 3% 2-chloroprocaine at concentrations of 0.75%, with abbreviated CNS toxicity and cardiotoxic potential and a smaller tendency for motor blockade than bupivacaine. For these factors, ropivacaine has been one of the long-acting LAs in peripheral nerve blockade that is most commonly used. Its synthesis is the same as bupivacaine, but n-propyl iodide is used instead of n-butyl iodide.

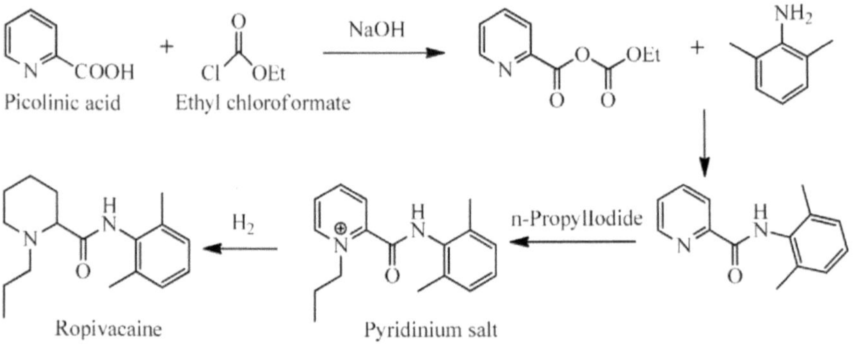

5.5.2.7 *LEVOBUPIVACAINE*

A single enantiomer of bupivacaine hydrochloride is found in levobupivacaine and is less cardiotoxic than bupivacaine. With no unchanged substance found in the urine, it is broadly metabolized. For peripheral nerve blockade, the properties of levobupivacaine are less well studied than those of ropivacaine. Levobupivacaine is an important and, at the same time, less toxic form of bupivacaine.

5.5.3 MISCELLANEOUS

The agents that go to this category are structurally dissimilar from each other. These agents, though having no association in conditions of basic structural similarity, still have LA activity. Hence, these drugs are categorized as miscellaneous (Figure 5.6), for example, diperidon, dimethisoquin, and so on.

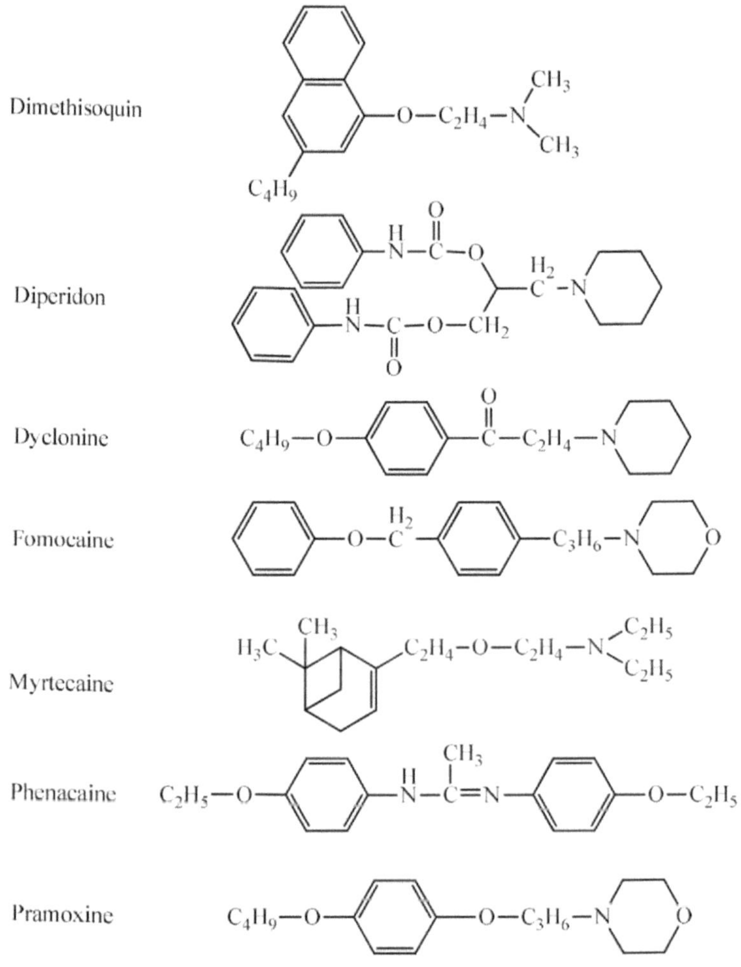

FIGURE 5.6 Examples of miscellaneous LAs.

5.6 ADDITIVES TO LAS

5.6.1 VASOCONSTRICTORS

The vascular absorption of a vasoconstrictor, which increases the degree of drug interaction with nerve tissues, is what adds up to a LA holdup. The net effect is the continuation of the blockade by as much as 50% and a decline in the systemic absorption of LA. These consequences differ considerably among diverse types of LAs and entity nerve blocks. For example, lidocaine is a normal vasodilator as compared with blocks that use ropivacaine having its own minor outcome of vasoconstriction, in which the effect is distinct for such blocks. In peripheral nerve blockade, epinephrine is the most commonly used vasoconstrictor. When paired with LAs, a reduction in nerve blood flow was associated with epinephrine. Though, this outcome was not distinguished when concentrations of epinephrine were preserved at 1:400,000 (2.5 μg/mL). As such, this is the suggested concentration when employed as an adjuvant.

5.6.2 OPIOIDS

In order to manage acute or chronic pain, the delivery of opioids into the epidural or subarachnoid space is focused on the evidence that opioid receptors are found in the spinal cord substantia gelatinosa. Thus, the combinations of a LA and an opiate are regularly fruitfully employed in the neuraxial blockade to prolong analgesia and augment the blockade. However, in peripheral nerves, analogous receptors are absent or the consequences of opiates are negligibly feeble. For this motive, opiates do not have a noteworthy clinical role in tangential nerve blockade.

5.6.3 CLONIDINE

Clonidine is a potent selective alpha2-adrenergic agonist that acts centrally. Because of its capability to reduce sympathetic nervous system output from the CNS, clonidine acts as an antihypertensive drug. It does not lead to depression of ventilation, nausea, pruritus, vomiting, and so on. By activating postsynaptic alpha2 receptors in the substantia gelatinosa of the spinal cord, clonidine induces analgesia. A new meta-analysis showed that clonidine extends the period of both the motor block and sensory block by approximately 1.5–2 h. The prominent side effects include orthostatic hypotension, fainting, and sedation may occur.

5.7 SELECTING LAS FOR PERIPHERAL NERVE BLOCKS

Choosing a correct LA is very significant and the subsequent positions have to be supposed while choosing an agent for nerve blockade. In general, the duration of a LA and onset of LA are analogous in nature. For example,

2-chloroprocaine has a short onset but also a comparatively concise duration. In contrast, bupivacaine and ropivacaine have longer durations of action, but take to some extent longer to apply their outcome (Table 5.1). The LA should be adjusted for the duration of the surgical procedure and the anticipated degree of pain. For example, the formation of an arteriovenous fistula is a reasonably diminutive operation with nominal postoperative pain. So, the use of a short-acting agent (e.g., mepivacaine) offers brilliant intraoperative circumstances, devoid of the trouble of an anesthetized limb for 10–18 h, postoperatively. A rotator cuff repair engrossed a superior amount of postoperative pain, and so utilizing a long-acting LA (e.g., ropivacaine) is appropriate. The duration and onset of a given LA vary according to the blocked nerve. For instance, 0.5% of ropivacaine can be required to provide 10–12 h of analgesia at the brachial plexus; up to 24 h of analgesia may provide the same concentration at the sciatic nerve. This is likely due to disparities in the local vascularity, which persuades the uptake of LA. Ropivacaine 0.2% is typically adequate to offer outstanding sensory analgesia but do without any motor blockade. The toxicity of the agent should be measured while administering. Bupivacaine affords the longest period of the frequently used LAs but also has the nastiest cardiotoxic profile. However, a lessening in the dose employed, seen more and more recurrently with ultrasound-guided nerve blocks, may afford a corresponding or superior safety profile than ropivacaine higher dose, the drug to which it is frequently contrasted.[15]

TABLE 5.1 LAs for Peripheral Nerve Blockade.

LA (Formulation)	**Onset of action (min)**	**Duration of anesthesia: analgesia (h)**
3% 2-Chlorprocaine in a	10	1: 2
3% 2-Chlorprocaine in a + b	10	1.5: 3
2% Lidocaine in a + b	10	3: 5
1.5% Mepivacaine in a	10	2.5: 4
1.5% Mepivacaine in a + b	10	4: 6
0.5% Ropivacaine	15	6: 10
0.75% Ropivacaine	10	8: 20
0.5% Bupivacaine or levobupivacaine + b	15	10: 25

A, bicarbonate; b, epinephrine. These values do not apply to all scenarios, all nerves, and all plexuses but can be used as a rough comparative guide to aid in decision-making.

Ambulatory patients who are experiencing lower limb surgery should be administered long-acting LAs with concern and with appropriate education concerning ambulation with support. Short procedures with minimal postoperative pain (most ambulatory procedures) only require a short- to intermediate-acting agent, and substantiation of block decree is often a necessity before discharge home.

A perineural catheter must be considered in case of anticipated pain lasting longer than 15–20 h. The admiration that patients demonstrate for some effectual block will lighten rapidly when it wears off in the middle of the night, parting them unwary, in pain, and unable to contact substitute pain therapies.

5.8 MIXING OF LAS

Mixing of LAs (bupivacaine and lidocaine) is frequently performed in clinical practice with the objective of attaining the earlier beginning of the shorter acting LAs and the longer duration of the longer acting LA. Regrettably, when LAs are mixed, their onset, duration, and potency become much less predictable, and the end result is distant from anticipated. For example, work carried out at our institution has demonstrated that combining bupivacaine 0.5% and mepivacaine 1.5% versus either drug unaccompanied for interscalene brachial plexus block results in modest clinical benefits. Onset times for all three solutions were identical, and the duration of the combined solution was considerably shorter than bupivacaine only. Furthermore, a part experiment looking at the progression of LAs (i.e., bupivacaine first, mepivacaine second versus mepivacaine first, bupivacaine second) showed no disparity in the onset and duration between groups.

Therefore, if a long duration is preferred, a long-acting drug only will offer the finest circumstances. In totaling, mixing LAs also bears a danger of a drug error. For this reason, we infrequently mix LAs; instead, we prefer drugs and concentrations of solitary agents to attain the preferred consequences. The bulk of nerve blockade can be met using one short/intermediate and one long-acting LA. At our organization, more than 90% of peripheral nerve blockades for surgery are carried out with one of two LAs: 1.5% mepivacaine or 0.5–0.75% ropivacaine. Certainly, there are times when other LAs are used. Chloroprocaine is injected for quick knee arthroscopies, where the postoperative pain is negligible and requisite a quick arrival to the normal state.[16]

5.9 SYSTEMIC TOXICITY OF LAS

In totaling to disrupting the peripheral nerve conduction, LA obstructed with the utility of all organs in which the transference or transmission of the nerve impulses results. The hazard of such adverse reactions is directly comparative with the LA concentration attained in the human circulation.[17]

The subsequent aspects establish the LA concentration in the plasma:

1. The rate of absorption of the drug.
2. The dose of the drug administered.
3. Site injected and use of vasoconstrictors.
4. Biotransformation and elimination of the drug from the circulation.

Even though the peak level of an LA is unswervingly correlated to the dose administered, administration of the identical amount at diverse positions results in noticeable disparities in the peak blood concentrations. This explains why huge amounts of LA can be applied with peripheral nerve blocks without the toxicity that would be perceived with intravenous (IV) injection or an intramuscular injection. It is also for this reason that the observance to sternly distinct maximum doses of LAs is irrational. Careful deliberation of patient factors and the necessities of the block must herald an assortment of LA type and dose for peripheral nerve block.

Short-acting ester LAs are intrinsically safer with deference to systemic toxicity owing to their clearance by pseudocholinesterase. In the case of 2-chloroprocaine, peak blood levels attained are pretentious by the rate at which the LA drug undertakes biotransformation process and elimination process (plasma half-life of about 1 min). In disparity, the peak blood levels of amide-linked LA drugs are chiefly the result of absorption.

5.9.1 CENTRAL NERVOUS SYSTEM TOXICITY

The indications of CNS toxicity linked to LAs are a role of their level in plasma. Characteristically, toxicity is initially articulated as producing restlessness, stimulation of the CNS, tremor, and disorientation. As the concentration in the human plasma of the LA enhances, tonic–clonic seizures results (Figure 5.7). The higher the potency of the LA, the higher the convulsions may be commenced. On higher administrations of LA, depression occurs and respiratory failure results which finishes in the stage of coma.[18]

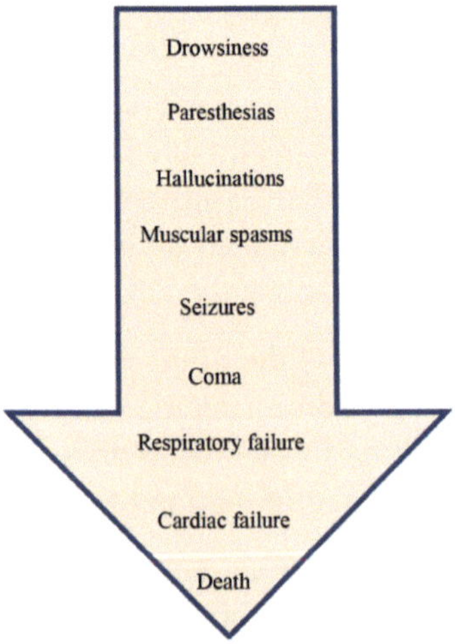

FIGURE 5.7 Progression of LA toxicity.

Noticeable stimulation and succeeding depression result by pertaining LAs to the CNS most likely due to dejection of the neuronal action. Discerning depression of the neurons of inhibitory types is considered to be related to in vivo excitatory phase. Nevertheless, quick systemic injection of any LA may result in death with no, or only ephemeral, signs of CNS stimulation. Beneath these circumstances, the drug concentration perhaps is ascended in such a speed that all the neurons are depressed concurrently. A diminutive dose of propofol (e.g., 0.5–1.0 mg/kg) or small doses benzodiazepines injected via IV route are the chosen drugs for aborting the convulsions. The depression of respiration with unwarranted sedation may lead to acidosis with a consequential higher level of the free drug in the human serum.

5.9.2 CARDIOVASCULAR TOXICITY

The chief area of activity in the cardiovascular system of any LAs is the myocardium, where they reduce the rate of conduction, the strength of myocardial contraction, and electrical excitability. The majority of the LAs cause arteriolar dilatation which contributes to hypotension. Cardiovascular effects characteristically happen at elevated systemic concentrations than those therapeutic dosages that produce effects on the CNS. Animal studies on cardiotoxic potentials of LA revealed that all caused dose-dependent depression of the cardiac muscle contraction. This depressant outcome on the cardiac contractility is analogous to the anesthetic effectiveness of the LA in neural blocking. Therefore, bupivacaine, which is four times more potent than lidocaine in blocking nerves, is also fourfold highly effective in dismaling the cardiac contractility. The deaths resulted from bupivacaine overdose have been connected with progressive persistence of ventricular conduction and widening of the QRS complex, followed by the unexpected commencement of arrhythmia like ventricular fibrillation.

5.9.3 PREGNANCY AND LA TOXICITY

Plasma concentration of the AAG is less in pregnant women and in newborns. This reduced amount successfully augments the free concentration of drug in the human plasma. For intermediate duration, in LAs like mepivacaine and lidocaine, slighter alterations in the protein binding may result during the gestation period, and the applications of these LAs are not linked with amplified danger of cardiac toxicity during the gestation period.

5.10 PHARMACODYNAMICS AND TREATMENT OF LA TOXICITY

The levels of lidocaine in blood are connected with the commencement of seizures emerge to be in the range of 10–12 μg/mL. The inhibitory pathways in the brain are disabled in a discerning way at these concentrations and the excitatory neurons can be in utility unrestricted. A further augmented concentration (20–25 μg/mL) lead to considerable cardiotoxicity and respiratory depression. In case of dissimilarity, for bupivacaine, the concentration in blood is 4 and 6 μg/mL, which eventually leads to seizures and cardiac toxicity. The long-acting LAs like etidocaine and bupivacaine carry an appreciably superior risk of cardiac arrest and hard resuscitation. The complications do not always occur immediately on injection of a LA and may be delayed up to 30 min. Numerous toxic responses result subsequent to unintentional intravascular administration of huge quantities of therapeutically active agents from tourniquet failure during IV administration or via an epidural catheter of a RA. The seizure can be arrested in patients with systemic neurotoxicity by injecting a small quantity of propofol (0.5–1.0 mg/kg) or benzodiazepine (specifically, midazolam 0.05–0.1 mg/kg) or though averting the detrimental effects of hypercarbia or hypoxia by aerating with 100% oxygen. The utilization of some vasopressor agents like vasopressin or epinephrine has extensively been an average fraction in managing the cardiac arrest, and at this moment, the instructions have not altered. Conversely, experimental substantiation involving animals has shown aggravate results when united with the lipid emulsion, maybe as a result of enlarged myocardial intracellular acidosis. The enclosure of these compounds as typical resuscitative drugs is necessitated. At last, in events of stern toxicity with huge doses of long-acting LAs, opportune organization of the cardiopulmonary avoid could establish lifesaving function.[19]

5.11 TECHNIQUES USED FOR LOCAL ANESTHESIA

LAs can block each nerve between the central nervous system and the peripheral nerve endings. The main peripheral technique is called *topical anesthesia* to the skin or additional body surfaces. The large peripheral nerves and small peripheral nerves can be anesthetized in anatomical nerve bundles (plexus anesthesia) or independently (peripheral nerve block).[20]

5.11.1 SURFACE ANESTHESIA

Application of LA as a spray, solution, or cream to the skin, the effect is short and is limited to the area of contact.

5.11.2 INFILTRATION ANESTHESIA

The infiltration anesthesia and surface anesthesia are jointly called as *topical anesthesia* that can be administered into the area to be anesthetized.

5.11.3 FIELD BLOCK

Subcutaneous administration of a LA in an area neighboring on the meadow to be anesthetized is called field block.

5.11.4 PERIPHERAL NERVE BLOCK

Administering the LA in the neighborhood of a peripheral nerve to anesthetize the area of interest is called peripheral nerve block.

5.11.5 PLEXUS ANESTHESIA

Plexus anesthesia is injecting a LA in the neighborhood of a nerve plexus, frequently inside a tissue compartment that prohibits the diffusion of the drug away from the projected site of activity.

5.11.6 EPIDURAL ANESTHESIA

Epidural anesthesia is administered into the epidural space where it acts essentially on the spinal nerve roots. Depending on the amount injected and the place of injection, the anesthetized area fluctuates.

5.11.7 SPINAL ANESTHESIA

Spinal anesthesia is administered into the cerebrospinal fluid, typically in the lumbar spine (particularly, in the lower back), where it proceeds to part of the spinal cord and on spinal nerve roots. The resulting anesthesia habitually broadens from the legs to the abdomen or even the chest.

5.11.8 INTRAVENOUS REGIONAL ANESTHESIA (BIER'S BLOCK)

The blood circulation of a limb is break off employing a tourniquet (a device analogous to a blood pressure cuff), then a great amount of LA is introduced into the peripheral vein. The drug fills into the venous system of the limb and diffuses into tissues where the nerve endings and peripheral nerves are anesthetized. The anesthetic effect is inadequate to the area that is prohibited from the blood circulation and resolution occurs swiftly once the circulation is reinstated.

5.11.9 LOCAL ANESTHESIA OF BODY CAVITIES

LA of body cavities include intrapleural anesthesia and intra-articular anesthesia.

5.11.10 TRANSINCISION (OR TRANSWOUND) CATHETER ANESTHESIA

A multilumen catheter is inserted through an incision or wound and aligned across it on the inside as the incision or wound is closed, providing continuous administration of LA along the incision or wound is called transincision (or transwound) catheter anesthesia.

5.12 CONCLUSION

The LAs of three categories, ester-amide, amide-linked, and miscellaneous, are specialized medicaments for carrying out the surgical and operative procedures. The physicochemical, pharmacodynamics, and pharmacokinetics of the drugs play an immense role in mediating the effects through diverse mechanisms. However, at this moment, further research, exploration, and development of the new generation are required for obtaining better analogs.

KEYWORDS

- **LAs**
- **mechanism of actions**
- **systemic toxicity**
- **structure–activity relationship**
- **nerve conduction**
- **essential chemical features**
- **administration techniques**

REFERENCES

1. Jeff, G. et al. Text Book on LAs: Clinical Pharmacology and Rational Selection, 2013.
2. Nerve Damage Associated with Peripheral Nerve Block (PDF). *Risks Associated with Your Anaesthetic* (The Royal College of Anaesthetists). Section 12. Jan 2006. Retrieved Oct 10, 2007
3. Picard, J.; Meek, T. Lipid Emulsion to Treat Overdose of Local Anaesthetic: The Gift of the Glob. *Anesthesia* **2006,** *61* (2), 107–109. DOI: 10.1111/j.1365-2044.2005.04494.x. PMID: 16430560.
4. Stern, A. *Pharmacology: Pre Test Self-Assessment and Review*; McGraw-Hill, Medical Pub: New York, 2002. Division. ISBN 0-07-136704-7.
5. Yagiela, J. A. Vasoconstrictor Agents for Local Anesthesia. *Anesth. Prog.* **1995,** *42* (3–4), 116–20. PMC: 2148913. PMID 8934977.

6. Nielsen, L. J.; Lumholt, P.; Hlmich, L. R. Local Anesthesia with Vasoconstrictor Is Safe to Use in Areas with End-Arteries in Fingers, Toes, Noses and Ears. *Ugeskr Laeger*. Oct 27, **2014,** *176* (44), pii: V04140238. Danish. PubMed PMID: 25354008.
7. Cocaine's use: From the Incas to the U.S. *Boca Raton News*. Apr 4, 1985. Retrieved Feb 2, 2014.
8. Einhorn, A.; Fiedler, K.; Ladisch, C.; Uhlfelder, E. (1909). Ueber p-Aminobenzoësäurealkaminester. *Justus Liebig's Annalen der Chemie* 1909, *371* (2), 142. DOI:10.1002/jlac.19093710204
9. Winthrop Chemical Company, Inc. U.S. Patent 1,889,645, 1932.
10. Marks, H. C.; Rubin, H. I. U.S. Patent 2,460,139, 1949.
11. McElvain, S. M.; Carney, Thomas P. Piperidine Derivatives. XVII. LAs Derived from Substituted Piperidinoalcohols. *J. Am. Chem. Soc.* **1946,** *68* (12), 2592–2600. doi:10.1021/ja01216a051.
12. Sriram, D. et al. Text Book of Medicinal Chemistry, 2000.
13. McElvain, S. M. U.S. Patent 1,784,903.
14. McGee, H. T.; Fraunfelder, F. W. Toxicities of Topical Ophthalmic Anesthetics. *Expert Opin. Drug Safet.* **2007,** *6* (6), 637–640.
15. Reilly, T. J. The Preparation of Lidocaine. *J. Chem. Educ.* 1999, *76* (11), 1557. DOI: 10.1021/ed076p1557.
16. Ekenstam, B. LAs I. N-Alkyl Pyrrolidine and N-Alkyl Piperidine Carboxylic Acid Amides. *Acta Chem. Scandinavica* **1957,** *11*, 1183–1190. DOI: 10.3891/acta.chem.scand.11-1183.
17. Löfgren, N. Studies on LAs. XX. Synthesis of Some alpha-Monoalkylamino-2-Methylpropionanilides. A New Useful LA.. *Acta Chem. Scandinavica* **1960,** *14*, 486–490. DOI: 10.3891/acta.chem.scand.14-0486.
18. Federsel hansjuergen et al: An Efficient Synthesis Of A New, Chiral 2',6'-Pipecoloxylidide La Agent *Acta Chem. Scandinavica Ser. B—Org. Chem. Biochem.* Jan 1, 1987, *B41* (10), 757–761.
19. Wilson and Gisvold's Text Book of Pharmacetuical and Medicinal Chemistry, 11th ed.
20. Hansh's Text Book of Medicinal Chemistry, 1999.

CHAPTER 6

Pharmacotherapeutic Perspectives on Adrenergic Agonists

GINPREET KAUR[1*], ERICA SEQUEIRA[1], HARPAL S. BUTTAR[2], and UPENDRA NAGAICH[3]

[1]*Department of Pharmacology, Shobhaben Pratapbhai Patel School of Pharmacy and Technology Management, SVKM'S NMIMS, Mumbai 400056, Maharashtra, India*

[2]*Department of Pathology & Laboratory Medicine, Faculty of Medicine, University of Ottawa, Ontario, Canada*

[3]*Amity Institute of Pharmacy, Amity University, Noida 201301, Uttar Pradesh, India*

[*]*Corresponding author. E-mail: ginpreet.aneja@gmail.com*

ABSTRACT

Adrenergic drugs primarily act by activating the adrenergic receptor (adrenoreceptor) and are called sympathomimetics since the sympathetic nervous system also acts through the same receptors. This chapter involves thoughtful insights on the classification, distribution, side effects, and mechanism of action of α-adrenergic agonists and β-adrenergic agonists.

6.1 CHEMICAL CLASSIFICATION OF ADRENERGIC AGONISTS

6.1.1 CATECHOLAMINES

These include norepinephrine, epinephrine, dopamine, and isoproterenol are called catecholamines since they contain a catechol function. In addition,

Biologically Active Small Molecules: Modern Applications and Therapeutic Perspectives.
Debarshi Kar Mahapatra, & Sanjay Kumar Bharti (Eds.)

since they are water soluble and are 50% bound to the plasma proteins, hence found in the bloodstream. Since they are polar, thus have poor CNS penetration but many do have effects on the central nervous system (CNS). They are highly potent in activating the α receptors or β receptors. These drugs undergo a rapid inactivation by monoamine oxidase (MAO) and catechol *O*-methyl transferase (COMT).

Dopamine

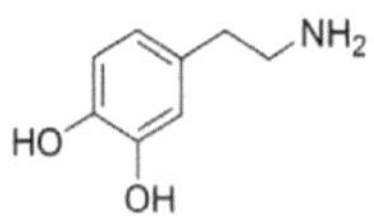

Norepinephrine

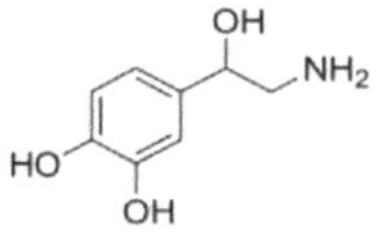

Isoproterenol

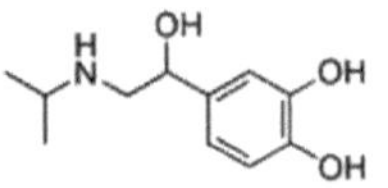

Epinephrine

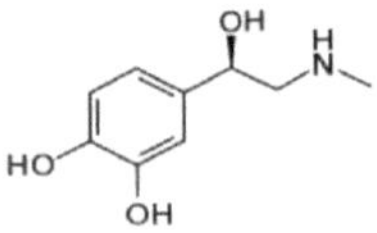

6.1.2 *NONCATECHOLAMINES*

These include amphetamine, ephedrine, and phenylephrine are called noncatecholamines since they lack a catechol hydroxy group. They are not inactivated by the enzyme COMT thus they have a longer half-life. Moreover, they are also poor substrates of MAO; thus exhibit an extended period of action. In contrast to catecholamines, they have an enhanced solubility in lipid, which allows them a larger entry to the CNS.

Phenylephrine

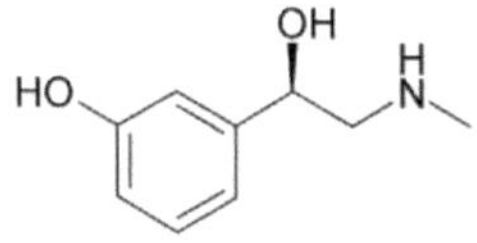

Pheylpropylamine

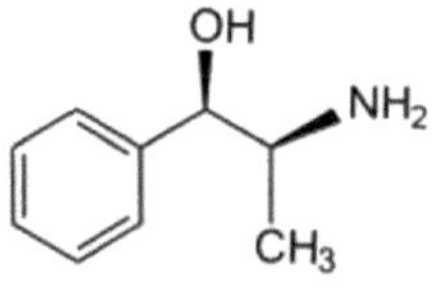

Amphetamine

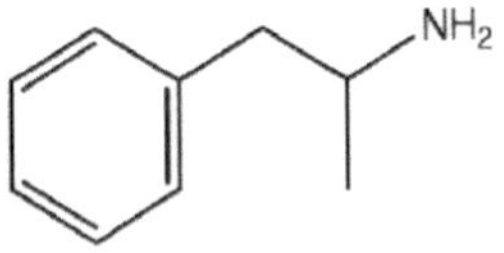

Ephedrine

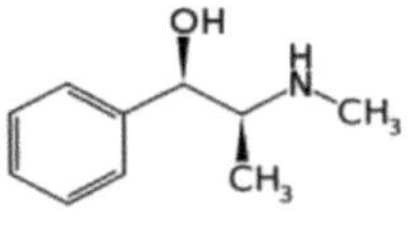

Albuterol

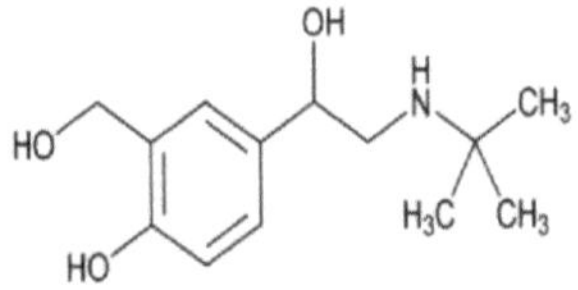

6.2 MECHANISM OF NEUROTRANSMISSION AT THE ADRENERGIC NEURON

Adrenergic drugs work exclusively on the adrenergic receptors and neurons that are located either postsynaptically on the effector organs or presynaptically on the neuron and these are accountable for norepinephrine release. The adrenergic neurotransmission involves the sympathetic division of the ANS.

6.2.1 *SYNTHESIS OF NOREPINEPHRINE*

The predecessor for the synthesis is phenylalanine, an amino acid. The rate-limiting enzyme, tyrosine hydroxylase (TH) and tyrosine, which is consequently produced, is transported via a sodium-linked carrier into the axoplasm of the neuron and undergoes a further hydroxylation by TH into dihydroxyphenylalanine (DOPA). The DOPA on decarboxylation yields dopamine (Figure 6.1).

6.2.2 *STORAGE OF NOREPINEPHRINE*

The final step where the dopamine formed gets hydroxylated by dopamine β-hydroxylase to form norepinephrine occurs inside the vesicle. Norepinephrine is stored as a complex with ATP inside the vesicle. In the adrenal medulla of kidney, the norepinephrine thus produced inside the chromaffin granules that slowly comes out into the cytoplasm through diffusion process, where it gets methylated by the enzyme *N*-methyltransferase and forms epinephrine.

6.2.3 *RELEASE OF NOREPINEPHRINE*

Once an action potential that is generated reaches the nerve junction, it triggers the entry of Ca^{2+}-ions which causes the vesicles present inside the neuron to combine with the cell membrane and expels its contents out into the synaptic cleft.

6.2.4 *BINDING TO THE RECEPTOR*

The released norepinephrine then binds to the postsynaptic receptors situated on the effector organ or to the presynaptic receptors on the neuron. This binding triggers a series of episodes that lead to the production of secondary messengers such as phosphoinositide cycle and cyclic adenosine monophosphate (cAMP).

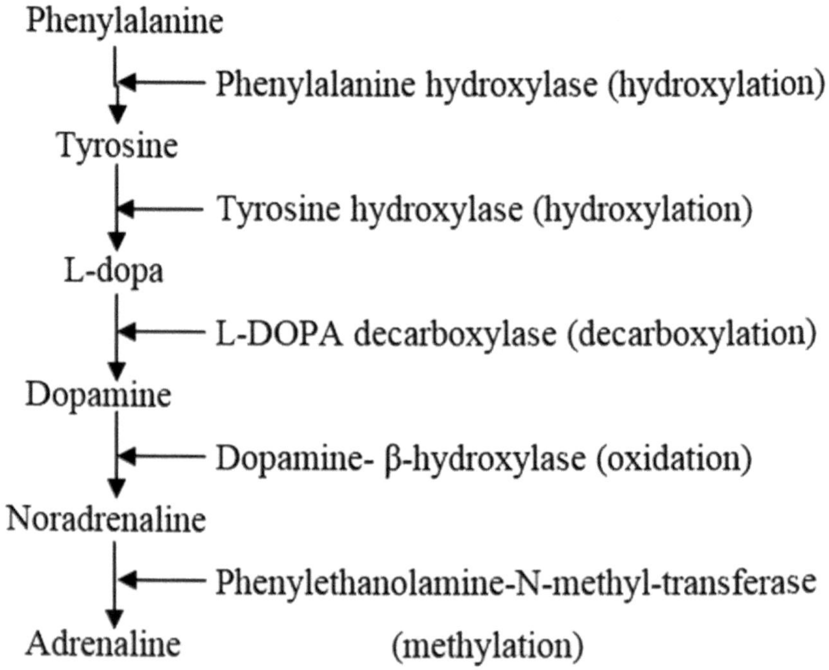

FIGURE 6.1 Synthesis of adrenaline (epinephrine).

6.2.5 UPTAKE OF NOREPINEPHRINE

Norepinephrine may either diffuse so as to enter general circulation or get metabolized by COMT to *O*-methyl derivatives or may undergo a reuptake back into the neuron. This is also important so as to maintain the norepinephrine content of the neuron.

6.2.6 METABOLISM OF THE NOREPINEPHRINE

Once the norepinephrine molecules re-enters into the adrenergic neuron, it is taken up further into the vesicles and remains sequestered for the release by another action potential. They are oxidized by monoamine oxidase and the major metabolites excreted in the urine are vanillylmandelic acid, normetanephrine, and metanephrine.

6.3 ADRENERGIC RECEPTORS

The adrenergic receptors are G-protein-coupled receptors which ultimately result in a decrease or increase in the formation of the secondary messengers. The two families of receptors can be designated as α and β receptors. Based on the relative differences in their organ specificities the β receptors were further subdivided into β_1 and β_2 subtypes. Later β_3 receptors were identified. Based on the pharmacological classification α receptors are also further classified as α1 and α_2 receptors. Sophisticated techniques such as molecular cloning further identified three subtypes of the α_1 receptors: α_{1A}, α_{1B}, and α_{1D} and three subtypes of α_2 receptors: α_{2A}, α_{2B}, and α_{2C}.

6.3.1 *α_1-RECEPTORS*

Activation of this receptor causes an increase in the production of diacylglycerol and inositol triphosphate leading to an increase in intracellular calcium ions. These receptors are primarily present on the postsynaptic membrane of the effector organs.

6.3.2 *α_2-RECEPTORS*

The norepinephrine that is produced moves through the synaptic cleft and related with the α_1 receptors. However, a section of it can respond with the α_2 receptors situated on the presynaptic membrane of the adrenoreceptor. This activates the feedback inhibition of the continuing discharge of norepinephrine from the already stimulated adrenergic neuron. In contrast to the α_1 receptor, the stimulation of the α_2 receptors results into the adenylyl cyclase inhibition and subsequently decrease in cAMP.

6.3.3 *β RECEPTORS*

- The β_1 receptors are primarily located in the juxtaglomerular (JG) cells in the kidney, heart, and has approximately equal affinities for both norepinephrine and epinephrine.

TABLE 6.1 Characteristic Properties of α and β-adrenoreceptors.

	α_1	α_2	β_1	β_2	β_3
Location	Postjunctionally on the effector organs	Prejunctionally on the nerve ending, postjunctionally in the brain, pancreatic β cells and extrajunctional in certain blood vessels, platelets	Heart, JG cells in kidney	Bronchi, blood vessels, liver, uterus, g.i.t, urinary tract	Adipose tissue
Major effects	Vasoconstriction, increased blood pressure, increased peripheral resistance, mydriasis and increased closure of the internal sphincter of urinary bladder	Inhibition of norepinephrine release, inhibition of insulin release, vasoconstriction, platelet aggregation.	Tachycardia, increased lipolysis, increased myocardial contractility	Vasodilation, slightly decreased peripheral resistance, bronchodilation, increased muscle and liver glycogenolysis, increased release of glucagon, relaxed uterine smooth muscle	
Selective agonist	Phenylephrine, Methoxamine	Clonidine	Dobutamine	Salbutamol, terbutaline	BRL 37344
Selective antagonist	Prazosin	Yohimbine, Rauwolscine	Metoprolol, Atenolol	ICI 118551, α-methyl propranolol	CGP 20712A, ICI 118551

- The β_2 receptors are located in the bronchi, blood vessels, uterus, liver, gastrointestinal tract, eye, and urinary tract and have a far-better affinity toward epinephrine as compared to the norepinephrine.
- Activation of the β_1 or β_2 ultimately results in the stimulation of adenylyl cyclase and thus an increased concentration of cAMP within the cell.

6.4 ADRENERGIC DRUGS (SYMPATHOMIMETICS)

These drugs have their mechanism of actions similar to that of epinephrine or of sympathetic stimulation.

6.4.1 DIRECT-ACTING SYMPATHOMIMETICS

They act directly as agonists on α and β receptors. Their responses are not reduced by a prior treatment with reserpine or guanethidine. After the transmitter depletion, the actions of these agents may increase because the loss of the neurotransmitter induces compensatory changes that upregulate receptors or enhance the signaling pathway. A few examples are isoproterenol, norepinephrine, phenylephrine, and epinephrine.

6.4.1.1 EPINEPHRINE

Epinephrine is one of the catecholamines and is a potent stimulant of both α receptors and β receptors and is generally used as a drug for therapies. It is produced from tyrosine and gets released together with norepinephrine (small quantities) in the blood. At lower doses, the β effects on the vascular system predominate and at higher quantity that α-effects prevail. It gets metabolized by COMT and MAO to vanillylmandelic acid and metanephrine, that are present in the urine.

6.4.1.1.1 Therapeutic Uses

- Used topically to reduce the intraocular pressure in open angular glaucoma.
- Used to treat type-I hypersensitivity reactions as drug of choice.

- Used to restore patient's cardiac rhythm with cardiac arrest.
- Used to treat respiratory tract infections when bronchoconstriction occurs in a reduced respiratory exchange, and thus can be used to treat acute anaphylactic shock and asthma.
- Used to amplify the length of local anesthesia action, when at the site of administration as they produce local vasoconstriction.

6.4.1.1.2 *Adverse Effects*

- Has effects on CNS such as anxiety, headache, fever, and tremors.
- Triggers cardiac arrhythmias, particularly if the patient is consuming digitalis.
- Induces pulmonary edema.
- Induces cerebral hemorrhage.
- Produces exacerbated cardiovascular actions in the presence of cocaine.
- Enhances cardiovascular actions are observed in patients with hyperthyroidism.
- Augments the discharge of endogenous glucose stores in case of diabetes.

6.4.1.2 NOREPINEPHRINE

Norepinephrine differs from adrenaline by lacking the methyl substitution on the aminoethanol group. Noradrenaline is a potent agonist for α receptors but has a little action on β_2 receptors. However, in comparison to epinephrine, noradrenaline is less potent at most α receptors.

6.4.1.2.1 *Therapeutic Use*

- Treats shock by enhancing the vascular resistance and subsequently increases blood pressure.
- Combats hypotension caused by the overdose of hypotensive drugs or removal of a pheochromocytoma.

6.4.1.2.2 *Adverse Effects*

- Adverse effects are similar to that of epinephrine. However, in addition, it can lead to blanching and sloughing of skin along with the injected vein.

- Since it is a potent vasoconstrictor, it can cause extravasation at the injected site.
- Cyclopropane or halothane anesthesia is contraindicated while receiving norepinephrine.
- Rapid treatment with the drug can produce cardiac arrhythmias. Extreme caution has to be taken in patients with enhanced blood pressure and cardiovascular diseases.

6.4.1.3 ISOPROTERENOL

The synthetic catecholamine, Isoproterenol has powerful effects on all β receptors and almost no action at α receptors. It augments the force and rate of heart contraction, in that way boosts the cardiac output. It causes the dilation of the arterioles of skeletal muscles, subsequently causing a dwindle in the peripheral resistance. It increases the systolic blood pressure, but decreases the diastolic blood pressure and mean arterial blood pressure. Since it produces a swift bronchodilation thus it alleviates the asthma attacks.

6.4.1.3.1 Therapeutic Uses

- Treatment for moderate-to-severe attack of bronchial asthma.
- Since it is a potent cardiac stimulant, it is useful for the immediate treatment for heart blockage.
- Can also be used in cardiogenic shock after a myocardial infection and in septicaemic shock.

6.4.1.3.2 Adverse Effects

- Commonly causes palpitation, headache, tachycardia, and flushing of the skin.
- Anginal pain, serious arrhythmias, tremors, dizziness, and sweating occur occasionally.

6.4.1.4 DOPAMINE

Dopamine is a naturally occurring catecholamine, the instantaneous metabolic predecessor of norepinephrine and is a dopamine (D_1 and D_2) as well as adrenergic (α and β_1) agonist. At higher doses, α effects predominate such as vasoconstriction and at lower doses, it stimulates the β_1 receptors present in cardiac regions. Peripherally at the proximal tubule, it exerts local diuretic and natriuretic effects. It is ineffective when taken orally since it is a substrate for both COMT and MAO.

6.4.1.4.1 Therapeutic Uses

- Dopamine is a drug of choice for treating the septic shock or cardiogenic wherein it raises the blood pressure by β_1 stimulation.
- Dopamine enhances the blood perfusion into the kidney, thus enhancing the glomerular filtration and thereby results in sodium excretion.

6.4.1.4.2 Adverse Effects

- Adverse effects are nausea, hypertension, arrhythmias, angina pain, tachycardia. Extravasation of large quantities of the drug during infusion can lead to ischemic necrosis and sloughing.
- Prior to the administration of dopamine for treating shock, hypovolemia should be corrected by transfusion of plasma, whole blood, or an appropriate fluid.
- Dopamine should be used at much-reduced levels or should be avoided if the patient utilizes a MAO inhibitor.

6.4.1.5 DOBUTAMINE

Dobutamine is a synthetic catecholamine and an analog of dopamine, but is not a D_1 and D_2 agonist. It interacts with both α and β receptors, although previously considered as a selective $\beta_1$receptor agonist, it is clear that the pharmacological effects are complex as it interacts with both α and β receptors.

6.4.1.5.1 Therapeutic Uses

- Used as an inotropic agent in congestive heart failure. The drug augments the cardiac output without a significant change in the heart rate, blood pressure, and peripheral resistance, which is a main benefit over other sympathomimetic drugs.

- Short-term treatment of cardiac decompensation after a cardiac surgery or in patients with acute myocardial failure, or congestive heart failure.

6.4.1.5.2 *Adverse Effects*

- Patients with atrial fibrillation must used this drug with caution, since it increases the atrioventricular conduction. In addition, tolerance of the drug on extended use has also been perceived.
- Some patients may show a rise in heart rate and blood pressure increase significantly during dobutamine administration.

6.4.1.6 *PHENYLEPHRINE*

Phenylephrine is a selective α_1 agonist and activates the β receptors at high concentrations only. It has no cardiac effect itself, but the reflex bradycardia is a characteristic effect of this drug. Phenylephrine differs from epinephrine only by lacking the 4-OH group on the benzene ring and thus it is resistant to COMT.

6.4.1.6.1 *Therapeutic Uses*

- Used as a nasal decongestant.
- Ophthalmic solution for treating mydriasis.
- Used to raise the blood pressure, chiefly to cease the episodes of supraventricular tachycardia.

6.4.1.6.2 *Adverse Effects*

High doses of phenylephrine may cause a hypertensive headache and irregularities in the cardiac function.

6.4.2 *INDIRECTLY ACTING SYMPATHOMIMETICS*

They act on the adrenergic neuron to release norepinephrine, which then acts on the adrenoreceptors. Their responses are abolished by earlier administration of reserpine or guanethidine. Examples: amphetamine, tyramine.

6.4.2.1 AMPHETAMINE

Amphetamine, racemic β-phenylisopropylamine is a synthetic drug and has a extensive period of action (4–6 h). The CNS activities are more pronounced which includes alertness, euphoria, increased work capacity. It can stimulate the respiratory center, especially if it has been depressed. It can also raise the blood pressure due to α and β effects. Its d-isomer shows a 3–4 times higher potency than the l-isomer. The other amphetamines that are used clinically are methamphetamine hydrochloride and hydroxyamphetamine hydrobromide.

6.4.2.1.1 Therapeutic Uses

- Therapy for depression.
- Treatment of attention-deficit/hyperactivity disorder, postencephalitic Parkinsonism, and epilepsy.
- Hyperactivity in pediatrics.
- Narcolepsy.
- Depression of appetite.

6.4.2.1.2 Disadvantages

- Contraindicated in the pregnant woman since it may cause unfavorable results on the development of the fetus.
- It is primarily a substance of abuse and can cause physical dependence.

6.4.2.1.3 Adverse Effects

- High doses can produce marked excitement, euphoria, and can progress to mental confusion, hallucinations, and psychotic conditions. Fatigue and depression usually follow CNS stimulation.
- Peripherally, it can lead to palpitations, arrhythmias, abdominal cramps, vomiting, and vascular collapse.
- Excessive sweating.
- Gastrointestinal symptoms include metallic taste, dry mouth, nausea, anorexia, diarrhea, vomiting, and abdominal cramps.
- Convulsions, coma, and cerebral hemorrhages which can be fatal.

6.4.2.2 TYRAMINE

Tyramine, a by-product of tyrosine metabolism, acts by entering the nerve terminal and displaces the stored norepinephrine. The released norepinephrine then acts on the adrenoreceptors. MOA causes the oxidation of tyramine but if the patient takes MAO inhibitors then it can results in vasopressor episodes.

6.4.3 MIXED ACTION SYMPATHOMIMETICS

They exhibit their action by acting directly as well as by acting indirectly. Their effects are not abolished but blunted, by a prior treatment with reserpine and guanethidine. These drugs stimulate the release of norepinephrine at the pre-synaptic end and activate the adrenergic receptors on the postsynaptic membrane. Examples are ephedrine, metaraminol, and methoxamine.

6.4.3.1 EPHEDRINE AND PSEUDOEPHEDRINE

These are naturally occurring plant alkaloids that stimulate both the α receptors and β receptors. Since these drugs are not catecholamines, thus are not metabolized by COMT and MAO and therefore they have a much longer period of action. Pseudoephedrine and Ephedrine have good oral absorption and penetrate the CNS.

6.4.3.1.1 Therapeutic Uses

- Ephedrine is principally used in the pharmacotherapy of asthma and aslo as a nasal decongestant due to its restricted vasoconstrictor activity and to elevate blood pressure.
- Ephedrine improves motor function and enhances contractility in myasthenia gravis.
- The mild stimulant effect of ephedrine on CNS decreases the fatigue, prevents sleep, and enhances the level of alertness. Moreover, it also improves athletic performance.
- Ephedrine is used as a prophylactic agent in the persistent asthmatic symptoms treatment.

- Pseudoephedrine is principally employed in treating the nasal congestion, eustachian tubes congestion, and sinus congestion.

However, ephedrine is banned by the US FDA since April 2004, due to severe life-threatening cardiovascular disorders. In addition, pseudoephedrine has been legally converted to methamphetamine.

6.4.3.2 METHOXAMINE

It is a parentral vasopressor which binds to α receptors and is also a selective α_1 agonist and has no action at β receptors. Since it is not metabolized by COMT thus its duration of action is comparatively longer than norepinephrine.

6.4.3.2.1 Therapeutic Uses

- Relieves the attacks of paroxysmal supraventricular tachycardia.
- To overcome hypotension in surgical cases which utilize halothane anesthetics
- Occasionally used as a pressor agent.

6.4.3.2.2 Adverse Effects

Adverse effects are hypertensive headache and vomiting.

6.5 CONCLUSION

Adrenergic agonist is an important class of drug with varied applications. The catecholamines, non- catecholamines, directly acting sympathomimetics (epinephrine, nor-epinephrine, dopamine, dobutamol, phenylephrine, isoproterenol), indirectly acting sympathomimetics (amphetamine, tyramine), and mixed action sympathomimetics (ephedrine, pseudoephedrine, methoxamine) have been found to express action through α (α_1, α_2) and β (β_1, β_2, β_3) adrenergic receptors present all across the human body.

KEYWORDS

- **adrenergic agonist**
- **sympathomimetics**
- **adverse effects**
- **neurotransmission**
- **receptors**
- **pharmacokinetics**

REFERENCES

1. *Goodman & Gilman's. The Pharmacological Basis of Therapeutics*, 12th ed.; Mcgraw-hill Medical Publishing Division: New York, 2011.
2. *Rang and Dale's Pharmacology*, 6th ed.; Churchill Livingstone: London, 2007.
3. *Lippincott's Illustrated Reviews: Pharmacology*, 4th ed., 2009.
4. Broadley, K. J. *Autonomic Pharmacology*; Taylor & Francis: London, 1996.
5. Cooper, J. R.; Bloom, F. E.; Roth, R. H. *The Biochemical Basis of Neuropharmacology*; Oxford University Press: New York, 1996.
6. Trendelenburg, U.; Weiner, N. *Catecholamines. Handbook of Experimental Pharmacology*, Vol 90, parts 1 and 2; Springer-Verlag: Berlin, 1988.
7. Guimares, S.; Moura, D. Vascular Adrenoreceptors: An Update. *Pharmacol. Rev.* **2001,** *53*, 319–356.
8. Insel, P. A. Adrenergic Receptors- Evolving Concepts and Clinical Implications. *New Engl. J. Med.* **1996,** *334*, 580–585.
9. Cunnane, T. C. The Mechanism of Neurotransmitter Release from Sympathetic Nerves. *Trends Neurosci.* **1984,** *7*, 248–253.

CHAPTER 7

Therapeutic Perspectives of Adrenergic Antagonists

ERICA SEQUEIRA[1], JAIMIN PATEL[1], HARPAL S. BUTTAR[3], GINPREET KAUR[1*], and UPENDRA NAGAICH[2]

[1]*Department of Pharmacology, Shobhaben Pratapbhai Patel School of Pharmacy and Technology Management, SVKM'S NMIMS, Mumbai 400056, Maharashtra, India*

[2]*Amity Institute of Pharmacy, Amity University, Noida 201301, Uttar Pradesh, India*

[3]*Department of Pathology & Laboratory Medicine, Faculty of Medicine, University of Ottawa, Ontario, Canada*

[*]*Corresponding author. E-mail: ginpreet.aneja@gmail.com*

ABSTRACT

Since catecholamines play a role in a variety of physiologic and pathophysiologic responses, the drugs that block adrenoceptors have important effects, some of which are of great clinical value. These effects vary dramatically according to the drug's selectivity for α and β receptors. This chapter deals with pharmacologically antagonistic drugs whose major effect is to occupy either α_1, α_2, or β receptors outside the central nervous system and avert their commencement by catecholamines and related agonists. The adrenergic antagonists are also called sympatholytic agents. These are the classes of drugs that completely antagonize the receptor action of adrenaline and related drugs. In addition, there is another class of drugs that affect the adrenergic function by interrupting the release or uptake of neurotransmitters into the synapse from the adrenergic neurons. This

Biologically Active Small Molecules: Modern Applications and Therapeutic Perspectives.
Debarshi Kar Mahapatra, & Sanjay Kumar Bharti (Eds.)

chapter involves the understanding of the classification, distribution, side effects, and mechanism of action of α and β adrenergic antagonists.

7.1 α-BLOCKERS

α-Adrenergic receptors intercede a lot of essential activity of endogenous catecholamines. Responses of particular clinical relevance include α_1 receptor mediated contraction of arterial and venous smooth muscle. The α_2 receptors are involved in increasing vagal tone, suppressing sympathetic output, inhibiting the release of norepinephrine and acetylcholine from nerve endings, contracting some arteries and veins, facilitating platelet aggregation, and controlling various metabolic effects. These results comprise repression of insulin secretion and lipolysis inhibition. α-Receptor antagonists possess multifarious pharmacological potentials and heterogeneous chemically. Various drugs of this class have dissimilar affinities for α_1 receptors and α_2 receptors. For example, prazosin has a higher potency as α_1-blocker as compared with α_2 receptors, phentolamine has equal potencies toward both the receptors, yohimbine is a α_2-selective inhibitor, and tamsulosin has much higher blocking potency at α_{1A} receptor than at α_{1B} receptor.

7.2 α-ADRENERGIC ANTAGONIST DRUGS

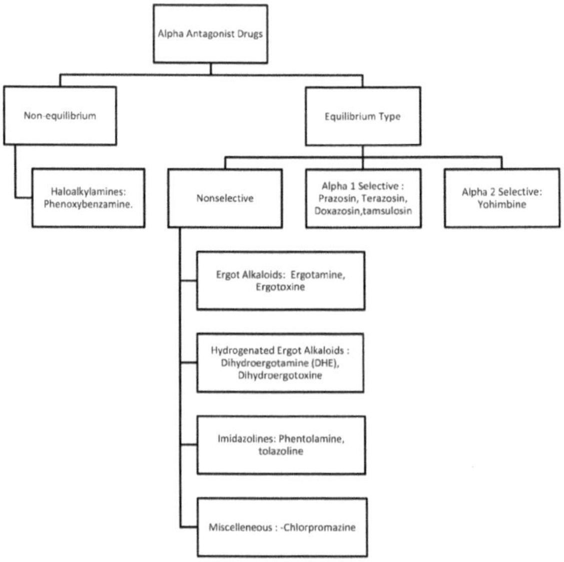

7.2.1 HALOALKYLAMINES

Phenoxybenzamine is a haloalkylamine that can block both the α_1 and α_2 receptors irreversibly. However, it shows more selectivity toward the α_1 receptors. The formation of the carbonium ion upon cleavage of the three-membered ring is directly accountable for the blockade. According to the presumption, the arylalkylamine moiety in the compound is accountable for the relative specificity of these molecules. Since the receptor blockade is irreversible, the restoration of the cellular responses toward the agonists requires the synthesis of new receptors.

7.2.1.1 PHARMACOKINETIC PROPERTIES

These properties are not completely understood. The half-life of phenoxybenzamine is approximately <24 h. However, since it causes the irreversible blockade of the adrenergic receptors, thus the duration of its action does not only reliant on the existence of the drug but also on the rate of synthesis of the adrenergic receptors.

7.2.1.2 THERAPEUTIC USES

- Used for the therapy of pheochromocytoma. In addition, it is also useful in the chronic management of these tumors.
- Effective in treating peripheral vasospatic conditions like Raynaud's disease.
- Management of autonomic hyperreflexia, which predisposes paraplegics to strokes.
- Used for treating the hypertension conditions, and particularly that are caused by pheochromocytoma.
- It is also the first α-blocker to be used for treating benign prostatic hyperplasia, although it is currently seldom used for that indication due to unfavorable side effects.
- It has been used in the treatment of hypoplastic left heart syndrome. It is also used in complex regional pain syndrome type-1 due to its antiadrenergic effects.

7.2.1.3 SIDE EFFECTS

- Common side effects include nasal congestion, bronchoconstriction, and miosis.
- Effects due to overdose include postural hypotension, reflex tachycardia, congestive heart failure, kidney failure, or cerebral stroke.

7.2.2 ERGOT ALKALOIDS

Ergot is obtained from *Claviceps purpurea* and contains 12 alkaloids, each derived from lysergic and isolysergic acids. The amino acids ergotamine and ergotoxine have partial agonistic or antagonistic effects at α-adrenergic, serotonin receptors, and dopamine receptors. Major actions attributed by the ergot alkaloids involve the stimulation of the smooth muscles and α-blocking activity. They produce a long-lasting vasoconstriction that predominates over their α blocking action. Ergotoxine is known to be a more potent α-blocker and less potent vasoconstrictor. They are principally used in the treatment of migraine. In addition, dihydroergotoxine has been used as a cognition enhancer. The amine alkaloid ergometrine has no α-blocking activity.

7.2.3 INDORAMINE

Indoramine is a selective competitive α_1-blocker and is primarily used in the treatment of hypertension. The drug is also known to reduce the incidence of attacks of Raynaud's phenomenon. In addition, it also acts as a competitive antagonist at the H_1 and 5-HT receptors.

7.2.3.1 PHARMACOKINETIC PROPERTIES

Since the drug undergoes an extensive first-pass metabolism, the bioavailability is less than 30%. Few of its metabolites may be active and little of the drug is eliminated unchanged in the urine. It has an elimination half-life of about 5 h.

7.2.3.2 ADVERSE EFFECTS

Adverse effects include dry mouth, sedation, and failure of ejaculation.

7.2.4 PRAZOSIN

Prazosin is a selective and a potent α_1 adrenergic receptor antagonist and is frequently used in the treatment of hypertension. It is a prototype of all the compounds that have a piperazinyl quinazoline nucleus. Due to its selectivity, it has found greater clinical applications and has largely replaced the classical non-selective haloalkylamines and imidazolines. It is a selective α_1 receptor blocker than α_2 receptor. The ratio of selectivity is 1000:1. In addition, prazosin is also a potent inhibitor of cyclic phosphodiesterases.

7.2.4.1 MECHANISM OF ACTION

Prazosin exerts its direct relaxant action over the vascular smooth muscle and causes a decline in the total peripheral resistance, whereas the vasodilatory effect of the drug is mainly exerted due to the blockade of postsynaptic α-adrenoceptors. The peripheral vasodilator effect of the drug is exerted mainly by the degree of arteriolar resistance. In contrast to the other α-blocker drugs, prazosin does not accompany reflex tachycardia. The α_1 receptor antagonists cause relaxation of bladder neck smooth muscles and prostate capsule and inhibit hypertrophy. The above helps in treating urinary retention associated with benign prostatic hypertrophy.

7.2.4.2 ADVERSE EFFECTS

- Less frequent adverse reactions which are reported to occur in 1–4% of patients are constipation, vomiting, edema, diarrhea, dyspnea, orthostatic hypotension, vertigo, syncope, nervousness, depression, urinary frequency, rash, reddened sclera, blurred vision, dry mouth, epistaxis, and nasal congestion.
- Rare side effects include fever, abdominal discomfort or abdominal pain, alopecia, pancreatitis, liver function abnormalities, positive ANA titer, hallucinations, pruritus, lichen planus, tachycardia, incontinence, impotence, paresthesia, priapism, tinnitus, arthralgia, and diaphoresis.

7.2.4.3 PHARMACOKINETICS

Prazosin is well absorbed following an oral administration and has a bioavailability of 50–70%. It reaches its peak concentrations 1–3 h after an oral dose.

It has a half-life of approximately 2–3 h and the effects last for up to 7–10 h. It shows a very high plasma protein binding and only 5% of the drug is available in its free form.

7.2.5 TERAZOSIN

It is a structural analog of prazosin with high selectivity for the α_1 receptors; however, it is less potent than prazosin. The major difference between these two drugs is their pharmacokinetic characteristics.

7.2.5.1 PHARMACOKINETIC PROPERTIES

Terazosin being more water-soluble has a bioavailability of more than 90%. It has a half-life of approximately 12 h and its duration of action extends beyond 18 h. Approximately, 10% of terazosin is excreted unchanged in the urine.

7.2.5.2 THERAPEUTIC USES

It is applied for treating hypertension and is more effectual than finasteride in treating benign prostatic hyperplasia (BPH).

7.2.6 CHLORPROMAZINE

Among the other neuroleptics, chlorpromazine also has α-adrenoreceptor-blocking activity.

7.2.7 YOHIMBINE

It is a naturally occurring plant alkaloid related to reserpine. It is found in the bark of *Pausinystalia yohimbine* and in *Rauwolfia* root. It is a competitive antagonist with selectivity toward the α_2 receptors. Its main actions involve the blocking of the α receptors and also block the 5-hydroxytryptamine receptors. It is also a strong stimulant of the brain and possesses aphrodisiac properties.

7.2.8 PHENTOLAMINE AND TOLAZOLINE

It is a selective α receptor antagonist and has affinities for both α_1 and α_2 receptors. It has similar cardiovascular effects as phenoxybenzamine. They are also known as the "classical" α-blockers so as to distinguish them from the recently developed drugs, such as prazosin and its derivatives. Other effects of phentolamine include blocking of the 5-HT receptors, the release of histamine from the mast cells, and blocking of the K^+ channels.

A related but less potent compound is tolazoline. Both tolazoline and phentolamine can stimulate the GI smooth muscle and enhance the gastric acid secretion. In addition, tolazoline can stimulate the secretion by salivary, lacrimal, and sweat glands as well.

7.2.8.1 PHARMACOKINETIC PROPERTIES

Pharmacokinetics of phentolamine is not well established, but the drug is extensively metabolized.

7.2.8.2 THERAPEUTIC USES

- Used in short-term management of hypertension in patients suffering from pheochromocytoma.
- Since it has an inhibitory effect on catecholamines in intestinal smooth muscle, it helps to relieve the pseudo-obstruction of the bowel in patients with pheochromocytoma.
- Used to prevent the dermal necrosis after the unintentional extravasation of α-adrenergic agonist.
- Treatment of hypertensive crisis that results following the sudden withdrawal of clonidine or due to the consumption of tyramine-containing foods during the use of monoamine oxidase (MAO) inhibitors.

7.2.8.3 ADVERSE EFFECTS

- Hypotension, reflex cardiac stimulation resulting in cardiac arrhythmias, tachycardia, and myocardial infarction.
- Since it causes GI stimulation, it can lead to nausea, exacerbation of peptic ulcer, and abdominal pain.

7.3 β-ADRENERGIC ANTAGONIST CLASSIFICATION

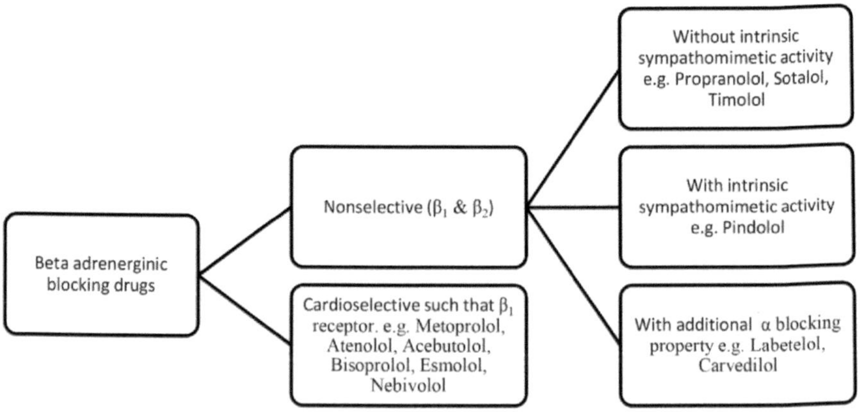

7.3.1 NONSELECTIVE B RECEPTOR ANTAGONISTS

Propranolol is a prototype to which all the β-antagonists are compared is a competitive nonselective β-adrenergic receptor antagonist, that is, it blocks both β_1 and β_2 receptors. Propranolol has an equal affinity for both β_1 and β_2 receptors, whereas the other agents such as metoprolol, acebutolol, bisoprolol, atenolol, and esmolol have a somewhat greater affinity for β_1 than for β_2. Several β-blockers such as pindolol and acebutol can activate the β receptors partially in the absence of catecholamines; however, propranolol is a pure antagonist and thus it cannot activate the β receptors. Such agents which have a partial agonistic effect are said to have an inherent sympathomimetic activity.

7.3.1.1 PROPRANOLOL

Propranolol is a prototype drug that interacts with the β_1 and β_2 receptors equally, lacks intrinsic sympathomimetic activity, and does not block α receptors.

7.3.1.1.1 Therapeutic Uses

- Used for the treatment of hypertension as well as for long-term management of angina pectoris.

- For the treatment of supraventricular arrhythmias/tachycardia, ventricular arrhythmias/tachycardia, premature ventricular contraction, myocardial infarction, and pheochromocytoma.
- Prophylaxis of migraine.
- Other uses include off-label indications including generalized anxiety disorder, parkinsonian tremors, antipsychotic drug-induced akathisia, and variceal bleeding in portal hypertension.

7.3.1.1.2 *Adverse Effects*

- Paradoxical hypertension, asthmatic attacks.
- Abrupt withdrawal can occasionally cause the recurrence of unstable angina, myocardial fibrillation, ventricular tachycardia, and a sudden death.
- Can cause or aggravate Raynaud's phenomenon.
- Propranolol can cross the placenta and can cause bradycardia, hypotension, and hypoglycemia in neonates.
- Sexual dysfunction, fever, rash, and alopecia.
- Fatigue and lethargy occur commonly.

7.3.1.1.3 *Pharmacokinetic Properties*

Since propranolol is a high lipophilic drug, it is nearly completely absorbed when given orally. However, since it undergoes an extensive first-pass metabolism, approximately only 25% reaches the systemic circulation. The bioavailability of the drugs improves when concomitantly with the ingestion of food and during long-term administration of the drug. The plasma protein binding is ~90%. The clearance of the drug can vary with the hepatic blood flow and liver diseases.

7.3.1.2 *NADOLOL*

Nadolol is a long-acting antagonist at both the β_1 and β_2 receptors and is devoid of intrinsic sympathomimetic activity.

7.3.1.2.1 *Therapeutic Uses*

- Effective in the treatment of both angina pectoris and hypertension.

- Other unlabeled uses include variceal bleeding in portal hypertension, Parkinsonian tremors, and migraine prophylaxis.

7.3.1.2.2 Pharmacokinetic Properties

Nadolol is highly water-soluble and is incompletely absorbed from the gut. It has a bioavailability of approximately 35% and shows a lesser interindividual variability than propranolol. It does not undergo an extensive liver metabolism and is thus excreted in its intact form in the urine. It has a long half-life of approximately 20 h.

7.3.1.3 TIMOLOL

Timolol is a nonsubtype selective β receptor antagonist and has no intrinsic sympathomimetic activity. Adverse effects can occur in susceptible patients with asthma or congestive heart failure.

7.3.1.3.1 Therapeutic Uses

- Treatment of congestive heart failure, hypertension, and prophylaxis of migraine.
- Widely used for the treatment of intraocular hypertension and open-angle glaucoma.

7.3.1.3.2 Pharmacokinetic Properties

Timolol is well absorbed and extensively metabolized by the liver, undergoing a first-pass metabolized.

7.3.1.4 PINDOLOL

Pindolol is a nonsubtype-selective β receptor antagonist and has an intrinsic sympathomimetic activity. The data on the clinical significance of partial agonists are limited, however, pindolol is used to block the exercise-induced increase in heart rate and cardiac output.

7.3.1.4.1 Pharmacokinetic Properties

Pindolol is completely absorbed following an oral administration and displays a moderately high bioavailability. Thus, this prevents the interindividual variation of the plasma drug concentration. In the liver, 50% of the drug is metabolized to hydroxylated derivatives that are subsequently conjugated with either glucoronide or sulfate. It has a half-life of ~4 h.

7.3.2 SELECTIVE β_1-ADRENERGIC RECEPTOR ANTAGONISTS

7.3.2.1 METOPROLOL

Metoprolol is a β_1-selective receptor antagonist and is devoid of the intrinsic sympathomimetic activity.

7.3.2.1.1 Pharmacokinetic Properties

Metoprolol is well absorbed following an oral administration and has a low bioavailability about 40% since it undergoes a first-pass metabolism. It undergoes an extensive liver metabolism with CYP2D6 and only 10% of the drug is excreted unchanged in the urine. It has a half-life of about 3–4 h.

7.3.2.1.2 Therapeutic Uses

Treatment of hypertension, stable angina, chronic heart failure, and acute myocardial infarction.

7.3.2.2 ATENOLOL

It is a β_1-selective antagonist and is devoid of the intrinsic sympathomimetic activity. Being hydrophilic, atenolol can penetrate the central nervous system (CNS) to a limited extent and has a longer half-life than that of metoprolol.

7.3.2.2.1 Pharmacokinetic Properties

Atenolol is not completely absorbed and shows a relatively little interindividual variation in the plasma concentrations. It has a half-life of 5–8 h and is mostly removed unchanged through the urine.

7.3.2.2.2 Therapeutic Uses

- Treatment of hypertension.
- Highly efficacious when used in combination with a diuretic in elderly patients with isolated systolic hypertension.

7.3.2.3 ESMOLOL

It is a β_1-selective antagonist with a very little intrinsic activity. It can be used when the β blockade is required for a short duration of time.

7.3.2.3.1 Pharmacokinetic Properties

Esmolol has a half-life of approximately 8 min. There is a rapid onset and cessation of the β receptor blockade. However, its carboxylic acid metabolite has a longer half-life of approximately 4 h. However, this metabolite has a very low potency as β-antagonist.

7.3.2.4 ACEBUTOLOL

Acebutolol is a β_1-selective antagonist with some intrinsic sympathomimetic activity.

7.3.2.4.1 Pharmacokinetic Properties

Acebutolol is well absorbed, however, it undergoes a significant first-pass metabolism to diacetolol, which is an active metabolite. Acebutolol has an elimination half-life of 3 h but the half-life of diacetolol is 8–12 h and is excreted in urine.

7.3.2.4.2 Therapeutic Uses

Effective for the treatment of ventricular arrhythmias and hypertension.

7.3.2.5 BISOPROLOL

Bisoprolol is a β_1-selective antagonist that does not have any inherent sympathomimetic activity. Primarily, it is rmployed for treating hypertension. It is generally used in combination with diuretics and ACE inhibitors, in patients with a chronic heart failure. It is generally being well tolerated; however, shows a few side effects including bradycardia, dizziness, fatigue, and hypotension.

7.3.2.5.1 Pharmacokinetic Properties

Bisoprolol, when given orally, is well absorbed and has a bioavailability of approximately 90%. It is 50% eliminated by renal excretion. The metabolism occurs in the liver and its metabolites are inactive. Bisoprol has a half-life of 11–17 h.

7.3.3 β RECEPTOR ANTAGONISTS WITH ADDITIONAL Α-BLOCKING PROPERTY

7.3.3.1 LABETALOL

Labetalol acts as a competitive antagonist at both the α_1 and β receptors. Its pharmacological activities are complex since it has optical isomers. However, the mixture displays a selective blockade at the α_1 receptors, blockade at the β_1 and β_2 receptors, and inhibits the neuronal uptake of norepinephrine. The potency to block the β receptors is 5–10 times higher than that for α1 receptors. It is used in the treatment of patients with a mild to moderate hypertension. It has an intrinsic sympathomimetic activity at the β_2 receptor, which contributes to the vasodilation.

7.3.3.1.1 Pharmacokinetic Properties

Labetolol is completely absorbed, but it undergoes a significant first-pass metabolism. It has an elimination half-life of 8 h.

7.3.3.2 RESERPINE

A naturally occurring plant alkaloid that is responsible for the depletion of biogenic amines by causing the blockade of the Mg^{2+}/adenosine triphosphate-dependent transport. It has a slow onset and a longer duration of action.

7.3.3.3 GUANETHIDE

It causes a gradual depletion of norepinephrine by blocking the release of the stored norepinephrine as well as by displacing the norepinephrine from storage vesicles.

Commonly occurring side effects include orthostatic hypotension, interference with male sexual function. In supersensitive patients, a hypertensive crisis may occur if the patient suffers from pheochromocytoma.

7.3.3.4 COCAINE

It is known to inhibit the norepinephrine uptake and is rarely used as a local anesthetic since it is a drug of abuse.

7.3.3.5 METHYLDOPA

Methyldopa is taken up by the noradrenergic neurons, following which it gets converted to the false neurotransmitter α-methylnoradrenaline. This transmitter does not undergo a deamination within the neuron by MAO, therefore it gets accumulated and displaces noradrenaline from the synaptic vesicles. α-methylnoradrenaline is released by the same mechanism as noradrenaline, but is less active than noradrenaline on α_1 receptors. By contrast, it is more active on α_2-receptors, so the autoinhibitory feedback mechanism operates more strongly than normal, thus reducing transmitter release below the normal levels.

7.4 CONCLUSION

Adrenergic antagonist is an important class of drug with varied applications. The α-adrenergic antagonists (haloalkylamines, ergot alkaloids, prazosine, indoramine, terazosine, chlorpromazine, yohimbine, phentolamine, tolazoline), β-adrenergic antagonists (propranolol, timolol, nadolol, pindolol, esmolol, metoprolol, atenolol), and mixed adrenergic antagonists (labetolol, reserpine, guanethide, methyldopa, cocaine) have been found to express through α (α_1, α_2) and β (β_1, β_2, β_3) adrenergic receptors present all across the human body.

KEYWORDS

- **adrenergic antagonist**
- **sympatholytics**
- **adverse effects**
- **neurotransmission**
- **receptors**
- **pharmacokinetics**

REFERENCES

1. *Goodman & Gilman's. The Pharmacological basis of Therapeutics*, 12th ed.; Mcgraw-hill Medical Publishing Division: New York, 2011.
2. *Rang and Dale's Pharmacology*, 6th ed.; Churchill Livingstone: London, 2007.
3. *Lippincott's Illustrated Reviews: Pharmacology*, 4th ed., 2009.
4. Cooper, J. R.; Bloom, F. E.; Roth, R. H. *The Biochemical Basis of Neuropharmacology*; Oxford University Press: New York, 1996.
5. Kuhlkamp, V.; Bosch, R.; Mewis, C.; Seipel, L. Use of β-blockers in atrial fibrillation. *Am. J. Cardiovasc. Drugs* **2002,** *2*, 37–42.
6. Kyprianu, N. Doxazosin and Terazosin Suppress Prostate Growth by Inducing Apoptosis. Clinical Significance. *J. Urol.* **2003,** *169*, 1520–1525.
7. McGavin, J. K.; Keating, G. M. Bisoprolol. A Review of Its Use in Chronic Heart Failure. *Drugs* **2002,** *62*, 2677–2696.
8. Tfelt-Hansen, P. Efficacy of β-Blockers in Migraine. A Critical Review. *Cephalgia* **1986,** *6* (Suppl 5), 15–24.
9. Toda, N. Vasodilating β-Adrenoceptor Blockers as Cardiovascular Therapeutics. *Pharmacol. Thers.* **2003,** *100*, 215–234.
10. Wadworth, A. N.; Murdoch, D.; Brogden, R. N. Atenolol. A Reappraisal of Its Pharmacological Properties and Therapeutic Use in Cardiovascular Disorders. *Drugs* **1991,** *42*, 468–510.
11. Hu, Z. W.; Shi, X. Y.; Hoffman, B. B. Doxazosin Inhibits Proliferation and Migration of Human Vascular Smooth-Muscle Cells Independent of α1- Adrenergic Receptor Antagonism. *J. Cardiovasc. Pharmacol.* **1998,** *31*, 833–839.

12. Babamoto, K. S.; Hirokawa, W. T. Doxazosin: A New α1- Adrenergic Antagonist. *Clin. Pharm.* **1992,** *11*, 415–427.
13. Kynel, J. J. Pharmacology of Terazosin: An α1- Selective Blocker. *J. Clin. Pharmacol.* **1993,** *33*, 878–883.
14. Rosendroff, C. β- Blocking Agents with Vasodilator Activity. *J. Hypertens. Suppl.* **1993,** *11*, S37–S40.
15. Westfall, D. P. Adrenoreceptor Antagonists. In *Modern Pharmacology*; Craig, C. R.; Stitzel, R. E., Eds.; Lippincott Williams & Williams: Baltimore, 2004; pp 109–120.
16. Brixius, K.; Bundkirchen, A.; Block, B.; Mehlhorn, U.; Schwinger, R. H. Nebivolol, Bucindolol, Metoprolol and Carvedilol Are Devoid of Intrinsic Sympathomimetic Activity in Human Myocardium. *Br. J. Pharmacol.* **2001,** *133*, 1330–1338.

CHAPTER 8

Skeletal Muscle Relaxants: Concepts and Developments

ERICA SEQUEIRA[1], GINPREET KAUR[1], and UPENDRA NAGAICH[2]

[1]*Department of Pharmacology, SPP School of Pharmacy & Technology Management, SVKM's NMIMS, V. L. Mehta Road, Mumbai 400056, Maharashtra, India*

[2]*Amity Institute of Pharmacy, Amity University, Noida 201301, Uttar Pradesh, India*

[*]*Corresponding author. E-mail: ginpreet.aneja@gmail.com*

ABSTRACT

Skeletal muscle relaxants find wide applications in the treatment of musculoskeletal conditions, including neck pain, lower back pain, fibromyalgia, tension headaches, and myofascial pain syndrome. These agents aim at managing the muscle pain and improvement of the functional status so that the patient can return to previous activities. The skeletal muscle relaxants may be classified as antispastic agents (baclofen, dantrolene, and tizanidine) and antispasmodic agents (cyclobenzaprine, metaxalone, and methocarbamol). Antispastic drugs are responsible for causing muscle tightness and cramping in cases of neurological disorders, spinal cord injury, whereas antispasm drugs prevent the use of related minor muscle spasms. Another way of classification includes neuromuscular blocking drugs, centrally acting drugs, and directly acting drugs. This chapter reviews the mechanism of neuromuscular transmission, types of skeletal muscle relaxants including their pharmacokinetics and therapeutic applications.

Biologically Active Small Molecules: Modern Applications and Therapeutic Perspectives.
Debarshi Kar Mahapatra, & Sanjay Kumar Bharti (Eds.)

8.1 INTRODUCTION

The skeletal muscle relaxants are commonly employed to treat two types of underlying conditions such as muscular pain or spasms from peripheral musculoskeletal conditions and spastic neurological conditions. These drugs act tangentially at the neuromuscular junction/at the muscle fiber itself (direct action) or act centrally at the cerebrospinal axis so as to reduce the muscle tone and/or the cause of paralysis. As per the reports from the Nutrition Examination (NHANES III) survey and Third National Health conducted in 1988–1994, it was found that 1% of American adults take muscle relaxants, often on a chronic basis.

In the early 1940s, the applications of the neuromuscular blockers marked the beginning of a new era in the area of surgical practice and anesthetics. At present, several surgeries have been performed with rapidity and safety owing to the ability of the drug to facilitate skeletal muscle paralysis. In 1942, Griffith and Johnson had suggested that *d*-tubocurarine (*d*Tc) is a safe drug to be used during surgery so as to ensure skeletal muscle relaxation. A year later, Cullen described that *d*Tc had been given to 131 patients under general anesthesia to produce additional skeletal muscle relaxation greater than that provided by the anesthetic alone. With the development of newer synthetic relaxants, there has been an augmentation in the clinician's options for providing skeletal muscle relaxation.

8.2 PRINCIPLES OF NEUROMUSCULAR TRANSMISSION

The neuromuscular junction is particular on the nerve side and on the muscle side so as to send out and obtain the chemical communication. From the ventral horn of the spinal cord, the motor neuron elongates to the neuromuscular junction via myelinated large axon. Once it moves toward the muscle, it repeatedly branches in order to form a motor unit. A typical motor nerve unit encompasses an anterior horn cell, its processes, and the skeletal muscle fibers it supplies to.

The neuromuscular transmission is arbitrated by the neurotransmitter acetylcholine (ACh), which is specifically released from the presynaptic vesicles. These molecules of ACh then diffuse across the synaptic cleft and attach to the acetylcholine nicotinic receptors at the postjunctional membrane, which in turn responds by unlocking the channels to facilitate the passage of calcium ions and following the depolarization of endplate, which leads to muscle contraction. The acetylcholine instantly separates from the receptor component and is rapidly hydrolyzed by the enzyme acetylcholinesterase enzyme.

The ACh part diffuses swiftly across the synaptic cleft to the nicotinic receptor situated on the motor end plate. The nicotinic receptor of the organ and skeletal vertebrate is a pentamer of four distinct subunits (2α, β, γ, and δ) in stoichiometric ratio of 2:1:1:1 and molecular mass of 40,000–60,000 daltons. The binding of two ACh molecules to receptors on α-β and γ-α subunits opens the channels.

8.3 CASCADE OF THE SEQUENTIAL EVENTS

1. Generation and arrival of the action potential at the nerve terminal.
2. Exocytosis of ACh into the synaptic cleft. Following its diffusion is its association with the nicotinic acetylcholine receptors.
3. Depolarization occurs at the motor end plate which gives rise to end-plate potential (epp).
4. This epp that is generated crosses the threshold potential and the action potential triggers the muscle cells, which causes the release of Ca^{2+}. This causes the subsequent shortening of the myofibrils and results in the development of muscle tension.
5. The ACh once dissociated from its receptor gets hydrolysis by AChE.
6. The choline returns back to the nerve terminal.
7. Resynthesis and storage of the ACh.

8.4 CLASSIFICATION

There are three distinct varieties of clinically useful skeletal muscle relaxants (Figure 8.1).

- Those that act peripherally at the neuromuscular junction, also known as *neuromuscular blocking drugs*.
- Those that act in the brain and spinal cord termed as *centrally acting muscle relaxants*.
- The *directly acting muscle relaxants*.

8.4.1 NEUROMUSCULARBLOCKING DRUGS

The neuromuscular blocking drugs obstruct the neurotransmission process by two ways: (1) by residing and triggering the nicotinic receptor for an extended duration which causes blockade. It is accomplished by *depolarizing agents* and (2) by *nondepolarizing agents* antagonizing the actions of ACh in a competitive manner at nicotinic acetylcholine receptors. Therefore, the chemical structures of the *depolarizing agents* closely bear a resemblance to ACh, in contrast, the *nondepolarizing agents* are quite bulky and bear rigid scaffold. They are broadly classified into two major categories (Figure 8.2).

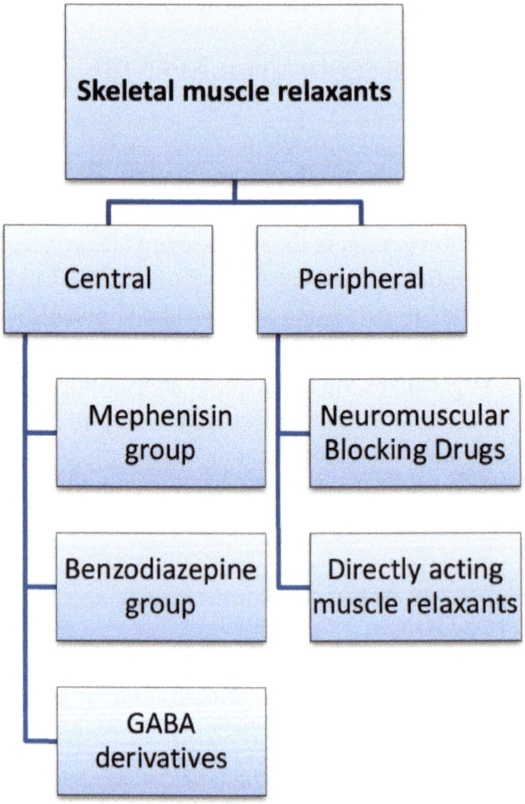

FIGURE 8.1 Broad classification of skeletal muscle relaxants.

8.4.1.1 NONDEPOLARIZING BLOCKERS

The nondepolarizing blockers exhibit their effects by acting as competitive antagonists at nicotinic receptors. They have slight or no agonist activity but can competitively reside in the receptor-binding site. In 1856, Claude Bernard showed that "curare" can cause paralysis by blocking neuromuscular transmission. Curare is a mixture of naturally occurring alkaloids found in various South American plants and is obtained from the natural sources of curare such as *Strychnos toxifera, Chondodendron tomentosum*, and related plants.

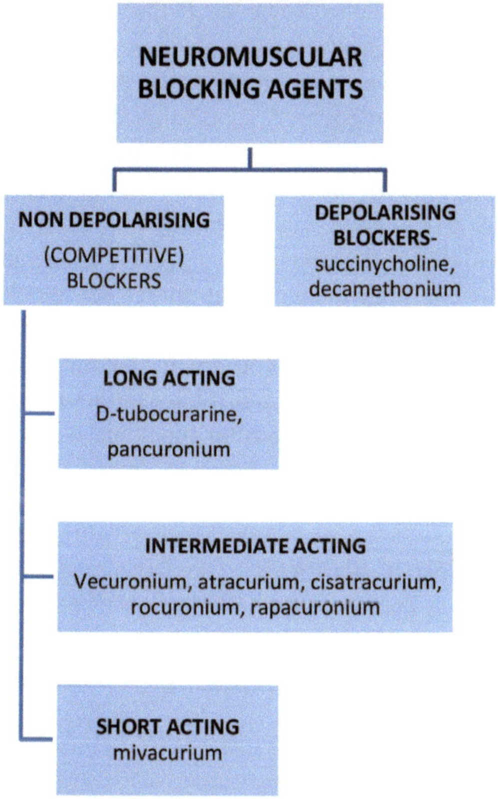

FIGURE 8.2 Detailed classification of neuromuscular blocking drugs.

Tubocurarine being one of the most important active principles was clinically used in the 1930s. However, now it is rarely used in medicine, being superseded by synthetic drugs with improved properties such as pancuronium, vecuronium, and atracurium which differ generally in their duration of action.

Nondepolarizing neuromuscular blocking drugs reduce the capacity of ACh to untie the ligand-gated cation channels in the skeletal muscle, thereby causing flaccid paralysis. Muscle contraction gets partly weakened when 75–80% of the receptors get dwelled in and gets inhibited completely when 90–95% receptors are fulfilled. Because nondepolarizing blockers contend with ACh, the blockade generated can also be reversed by escalating the quantity of ACh. This is done by inhibiting AChE, which hydrolyzes ACh. Pyridostigmine, neostigmine, and edrophonium are the AChE inhibitors employed clinically to overturn the neuromuscular blockade caused by the nondepolarizing blockers. However, if the quantity of the competitive blocking agent has been augmented to a degree that it is superior than that is required for 95% of the receptor blockade, AChE inhibitors will not be capable to enhance ACh adequately so as to overturn the block.

Reports of fatal ventilator depression are observed in anesthetized patients following the peritoneal or an immediate postoperative administration of antibiotics like gentamycin, streptomycin, neomycin, and polymyxin B, since these antibiotics themselves also possess a neuromuscular blocking activity. *d*Tc exhibits an enhanced neuromuscular blockade with general anesthetics, such as ether, halothane, cyclopropane, methoxyflurane; aminoglycoside antibiotics such as streptomycin, neomycin, gentamicin; also interacts with quinine, quinidine, and magnesium sulfate. *d*Tc, in contrast, shows a reduction in the neuromuscular blockade with anticholinesterases, such as neostigmine (Table 8.1).

8.4.1.2 DEPOLARIZING BLOCKERS

These drugs cause a relentless depolarization of the neuromuscular junction. There is only one depolarizing agent that is currently in clinical use, succinylcholine.

8.4.1.2.1 Mechanism of Action

This compound binds and causes the activation of the muscular nicotinic receptors in the similar way as ACh. However, succinylcholine is not metabolized by AChE, thereby resulting in the occupation of the receptor for an extended duration of time and thus this type of neuromuscular block cannot be antagonized by anticholinesterases. However, succinylcholine is hydrolyzed principally by butyrylcholinesterase, an enzyme situated in the plasma, but not in high amounts at the neuromuscular junction, resulting in a continuance of the muscle depolarization. The neuromuscular block that results from succinylcholine is typified by two cardinal phases. The first phase, also known as the phase-I block, is a result of the extended depolarization and renders the membrane unresponsive to any further stimuli. It is principally

TABLE 8.1 Characteristic Properties of Nondepolarizing Blockers.

Drug	Duration of action	Speed of onset	Characteristic side effects	Features and elimination
Tubocurarine	Long (1–2 h), competitive	Slow (> 5 min)	Hypotension, Bronchoconstriction (due to histamine release)	Plant alkaloid, now rarely used since replaced by synthetic drugs. Renal elimination; liver clearance
Pancuronium	Long, competitive	Intermediate (2–3 min)	Slight tachycardia, No hypotension	Approximately 5 times as potent as tubocurarine and better side effect profile. Renal elimination
Vercuronium	Intermediate (30–40 min), competitive	Intermediate	Muscle weakness, decreased respiratory reserve and apnea. Caution must be observed in patients of myasthenia gravis	Widely used. Occasionally causes prolonged paralysis probably owing to its active metabolite Rocuronium. Renal elimination
Atracurium	Intermediate (< 30 min), competitive	Intermediate	Transient hypotension due to histamine release, Bronchospasm, laryngospasm, prolonged muscle relaxation	Mechanism of elimination is unusual (Spontaneous nonenzymatic chemical degradation in plasma); degradation is slowed by acidosis
Mivacurium	Short (~15 min), competitive	Fast (~2 min)	Transient hypotension due to histamine release	A new drug that is chemically similar to atracurium, but rapidly inactivated by plasma cholinesterase

characterized by initial muscle fasciculations followed by a flaccid paralysis. The recovery of transmission marks the end of phase-I block.

With continued exposure to succinylcholine, the next phase, that is, phase-II block occurs. It is also known as the dual, biphasic, or antidepolarizing block. It differs markedly from phase-I block since it is slow in onset and the block appears similar to that produced by nondepolarizing agents, that is, it becomes receptive to high amounts of ACh and can be reversed by AChE inhibitors.

8.4.1.2.2 Pharmacokinetics

Succinylcholine has a rapid onset (1 min) and short duration of action (5 min) because it is rapidly hydrolyzed by the plasma cholinesterase (i.e., pseudocholinesterase) into the pharmacologically inactive succinylmonocholine and choline. Later succinylmonocholine is further hydrolyzed into succinic acid and choline.

8.4.1.2.3 Therapeutic Applications

Depolarizing blockers are used primarily as a short-acting muscle relaxant. It is used for brief procedure such as endoscopy, relief of laryngospasm, endotracheal intubation, orthopedic manipulations, and electroconvulsive therapy.

8.4.1.2.4 Toxic Effects and Contraindications

The initial stimulation of muscle fibers results in muscle fasciculations, which can cause considerable soreness of the muscles. Cardiac arrest and severe ventricular arrhythmias occur following succinylcholine administration to patients with massive trauma, severe burns, brain or spinal cord injuries, or tetanus. It can also precipitate malignant hyperthermia in genetically predisposed patients, especially when employed with inhalation anesthetics like halothane, cyclopropane, and diethylether.

8.4.2 CENTRALLY ACTING MUSCLE RELAXANTS

In addition to neuromuscular blocking agents, certain other compounds also decrease skeletal muscle tone by a discerning action in the cerebrospinal axis, without varying the perception. These agents specifically depress the supraspinal and spinal polysynaptic reflexes involved in the maintenance of muscle tone without having any considerable action on the monosynaptic reflexes. To a minor point, the polysynaptic pathways in the ascending reticular formation which are also accountable for restiveness are also depressed. They primarily are involved in reducing hyperreflexia, upper motor neuron spasticity, and decerebrate rigidity. In addition, it also finds its use in acute muscle spasms, neuralgia, anxiety, tension, spastic neurological diseases, tetanus, electroconvulsive therapy, and orthopedic procedures.

8.4.2.1 CLASSIFICATION

8.4.2.1.1 Mephenesin

Mephenesin is the first drug to be called as internuncial neuron blocking agent because its primary site of action is the spinal internuncial neuron that alters reflexes maintaining the muscle tone. However, it is not used clinically because it can cause gastric irritation and when given intravenously, it can cause thrombophlebitis, hemolysis, and a marked fall in blood pressure.

8.4.2.1.2 Carisoprodol

With a favorable muscle relaxant: sedative activity ratio, carisoprodol has weak antipyretic, anticholinergic, and analgesic actions in addition.

8.4.2.1.3 Chlorzoxazone

Chlorzoxazone is pharmacologically similar to mephenesin but exhibits a longer duration of action as well as it is better tolerated when given orally.

8.4.2.1.4 Chlormezanone

Chlormezanone has antianxiety and hypnotic actions as well. It is used for tension states associated with increased muscle tone.

8.4.2.1.5 Methocarbamol

Comparatively methocarbamol has a longer duration of action and a lesser sedative effect than mephenesin. When given orally, it has been used in chronic neurological diseases and reflex muscle spasms. It can be injected intravenously, without producing thrombophlebitis and hemolysis, and can be used for tetanus and orthopedic procedures.

8.4.2.1.6 Diazepam

Diazepam is a prototype of benzodiazepines that acts in the brain on particular receptors improving the GABAergic transmission. It reduces muscle tone by supraspinal action, is a muscle relaxant, and its sedative activity ratio is low. It has been found to be well tolerated with no gastric irritation, though sedation limits the dose which can be used for reducing muscle tone. Diazepam exerts both antispasm and antispasticity actions since it enhances the effects of GABA at $GABA_A$ receptors in the brain and spinal cord where it may have both presynaptic as well as postsynaptic effects. It is used in tetanus, spinal injuries, combined with analgesics, it is used for rheumatic disorders associated with muscle spasm.

8.4.2.1.7 Baclofen

Baclofen is a structural analog of γ-aminobutyric acid (GABA) and plays an essential role in decreasing spasticity by binding to $GABA_B$ receptors on presynaptic terminals of spinal interneurons. This results in hyperpolarization of the membrane, which results in the reduction of Ca^{2+} influx and thus decreases the release of the excitatory neurotransmitters, glutamate, and aspartate. The postsynaptic interactions with the sensory afferent terminals cause membrane hyperpolarization via a G-protein-coupled receptor which leads to an increase in potassium conductance and thereby enhances the inhibition. Baclofen may also be involved in inhibiting γ-motor neuron activity and therefore reduce the sensitivity of muscle spindle, leading to inhibition of mono-synaptic as well as poly-synaptic spinal reflexes. It is recommended as antispasticity drug.

8.4.2.1.8 Cyclobenzaprine

Cyclobenzaprine depresses the CNS and thereby causes sedation, which may be counterproductive for the physical pharmcotherapy, which is advocated in treating particular types of muscle spasms.

8.4.2.1.9 Tizanidine

Tizanidine is a derivative of clonidine which has short-acting α_2-adrenergic receptor (presynaptic) agonist actions. It is accountable for inhibiting the excitatory amino acids release and facilitating inhibitory transmitter glycine release in the spinal interneurons. Its side effects include drowsiness, dry mouth, hallucinations, and night-time insomnia (Table 8.2).

TABLE 8.2 Therapeutic Applications of the Centrally Acting Muscle Relaxants.

Compound	Therapeutic application
Diazepam	For skeletal muscle spasm and spasticity, control of seizures, convulsions in status epilepticus and tetanus, cerebral palsy and athetosis, stiff-man syndrome, and to facilitate orthopedic procedures
Carisoprodol	Central antispasmodic activity, analgesic activity
Methocarbamol	Inflammatory musculoskeletal conditions such as myositis, fibrositis, arthritis, and bursitis used to treat spasms and pain associated with muscle strain
Chlordiazepoxide	For delirium tremens that is caused due to the withdrawal of alcohol
Chlormezanone	Mild antispasmodic activity and antianxiety
Chlorzoxazone	Used in combination with analgesics
Baclofen	For spasticity in spinal cord injuries and in multiple sclerosis
Cyclobenaprine	For acute traumatic and inflammatory muscle spasms—not more than 2–3 weeks

8.4.3 DIRECTLY ACTING SKELETAL MUSCLE RELAXANTS

Dantrolene, a hydantoin derivative has direct effects on skeletal muscle by inhibiting the ryanodine receptor Ca^{2+} release channels present on the sarcoplasmic reticulum of the skeletal muscle, thus prevents their depolarization triggered opening. Dantrolene sodium has the advantage of having a long duration of time and also is orally active. It displays its effect in 1–2 h and the effect lasts for several hours. It facilitates the effects of GABA, thereby results in reduced efferent spinal motor neuron activity and depresses the brain stem reticular function.

8.4.3.1 THERAPEUTIC APPLICATION

When administered orally, dantrolene is responsible for reducing spasticity in upper motor neuron disorders, paraplegia, cerebral palsy, hemiplegia, and multiple sclerosis. Dantrolene also finds applications in patients with strokes, spinal cord injury, dystonia, and postoperative prophylaxis of malignant hyperthermia in high-risk patients. Dantrolene sodium is given in a dose of 25 mg twice a day orally initially, and the dosage may be increased gradually depending upon the effect required to a dose of 400 mg daily.

8.4.3.2 UNDESIRED EFFECTS

It causes malaise, sedation, lightheadedness, and other central effects occur but are less articulated. Troublesome diarrhea, long-term use-dependent liver toxicity also occur, and this restricts its use in chronic disorders.

8.4.4 MISCELLANEOUS AGENTS

8.4.4.1 MOTOR NERVE BLOCKER

Clostridium botulinum, an anaerobic bacterium, produces the botulinum neurotoxins. There are seven different types of the toxin such as the toxins (A-G), which are responsible for blocking the ACh release. Botulinum toxin is particularly involved in interfering with specific presynaptic proteins that are involved in exocytosis and thereby the release of synaptic vesicles which contain ACh. This results in a long duration (~3 months) flaccid paralysis of skeletal muscle and consequently diminishes the activity of sympathetic and parasympathetic cholinergic synapses. Continued use can result in immuno-resistance, the restoration of function requires nerve sprouting. Botulinum toxin has been primarily used to treat spasms and dystonias such as adductor spasmodic dysphonia, cervical dystonia, oromandibular dystonia, and spasms associated with the lower esophageal sphincter and anal fissures.

8.5 CONCLUSION

The application of neuromuscular-blocking agents is largely limited to anesthesia; the centrally acting muscle relaxants have clinical applications in musculoskeletal and neurological spastic disorders. In addition, the direct-acting type of skeletal muscle relaxants such as dantrolene offers a newer application to the treatment of spasticity.

KEYWORDS

- **acetylcholine**
- **γ-aminobutyric acid**
- **neuromuscular transmission**
- **neuromuscular blocking agents**
- **directly acting skeletal muscle relaxants**
- **centrally acting skeletal muscle relaxants**

REFERENCES

1. Bowman, W. C. Neuromuscular Block. *Br. J. Pharmacol.* **2006,** *147* (Suppl 1), S277–S286.
2. Lee, C. Conformation, Action, and Mechanism of Action of Neuromuscular Blocking Muscle Relaxants. *Pharmacol. Ther.* **2003,** *98*, 143–169.
3. Dillon, C.; Paulose-Ram, R.; Hirsch, R. et al. Skeletal Muscle Relaxant Use in the United States: Data from the Third National Health and Nutrition Examination Survey (NHANES III) Spine. **2004,** *29* (8), 892–896.
4. Montecucco, C.; Schiavo, G. Structure and Function of Botulinum Neurotoxins. *Q. Rev. Biophys.* **1995,** *28*, 423–472.
5. Prior, C.; Tian, L.; Dempster, J.; Marshall, I. G. Prejunctional Actions of Muscle Relaxants: Synaptic Vesicles and Transmitter Mobilization as Sites of Action. *Gen. Pharmacol.* **1995,** *26*, 659–666.
6. Tsui, J. K. Botulinum Toxin as a Therapeutic Agents. *Pharmacol. Ther.* **1996,** *72*, 13–24.
7. Papapetropoulos, S.; Singer, C. Botulinum Toxin in Movement Disorders. *Semin. Neurol.* **2007,** *27*, 183–194.
8. Kita, M.; Goodkin, D. E. Drugs Used to Treat Spasticity. *Drugs* **2000,** *59*, 487–485.
9. Choosing a Skeletal Muscle Relaxant. Sharon See and Regina Ginzburg. *Am. Fam. Phys.* Aug 1, **2008,** *78* (3), 365–370.
10. Bevan, D. R. Newer Neuromuscular Blocking Agents. *Pharmacol. Toxicol.* **1994,** *74*, 3–9.

11. Bowman, W. C.; Prior, C.; Marshall, I. G. Presynaptic Receptors in the Neuromuscular Junction. *Ann. NY Acad. Sci.* **1990,** *604*, 69–81.
12. Davis, E.; Barnes, M. P. Botulinum Toxin and Spasticity. *J. Neurol. Neurosurg. Psych.* **2000,** *68*, 141–147.
13. Pollard, B. J. Interactions Involving Relaxants. In *Applied Neuromuscular Pharmacology*; Pollard, B. J., Ed.; Oxford University Press: Oxford, 1994; pp 202–248.
14. Savarese, J. J.; Caldwell, J. E.; Lein, C. A.; Miller, R. D. Pharmacology of Muscle Relaxants and Their Antagonists. In *Anesthesia*; Miller, R. D., Ed., 5th ed.; Churchill-Livingstone: Philadelphia, 2000; pp 412–490.

CHAPTER 9

Nonsteroidal Anti-Inflammatory Drugs: Concepts and Innovations

UPENDRA NAGAICH[1*] and VANDANA CHAUDHARY[2]

[1]*Amity Institute of Pharmacy, Amity University, Noida 20130, Uttar Pradesh, India*

[2]*B. S. Anangpuria Institute of Pharmacy, Faridabad, Haryana, India*

[*]*Corresponding author. E-mail: upendra_nagaich@hotmail.com*

ABSTRACT

The non-selective, or classic, as well as the cyclooxygenase-2-specific ones, are all types of non-steroidal anti-inflammatory medications (NSAIDs). The most typical use of these medications is to treat musculoskeletal pain. Traditional or nonselective agents are still used extensively and are easily accessible as over-the-counter analgesics. However, especially in individuals who already have a high-risk profile, they come with a risk of significant cardiovascular, gastrointestinal, and renal side effects, such peptic ulcers and gastrointestinal hemorrhage. Physicians must be aware of these risk considerations in order to select drugs that have the best benefit-to-risk ratio while also taking the patient's unique demands and risk profile into account. Numerous anti-inflammatory compounds, their mechanisms of action, and their pharmacological features are the main topics of this chapter.

9.1 INTRODUCTION

Nonsteroidal anti-inflammatory agents (NSAIAs) or nonsteroidal anti-inflammatory drugs (NSAIDs) are a form of medication with antipyretic

Biologically Active Small Molecules: Modern Applications and Therapeutic Perspectives.
Debarshi Kar Mahapatra, & Sanjay Kumar Bharti (Eds.)

(fever-lowering), analgesic (pain-killing), and anti-inflammatory properties. The word “nonsteroidal” distinguishes these drugs from steroids, which have a similar eicosanoid-depressing, anti-inflammatory, and a host of other consequences. As analgesics, NSAIDs are scarce, and since they are nonnarcotic, they are used as a nonaddictive alternative to opioids. An OTC NSAID on the generic medication label is recognized as an NSAID. Both cyclooxygenase-1 (COX-1) and cyclooxygenase-2 (COX-2) activities, and thus prostaglandin and thromboxane syntheses are inhibited by NSAIDs. COX-2 inhibition is believed to have analgesic, antipyretic, and anti-inflammatory effects. COX-1 inhibition, especially aspirin, can also be inhibited by these NSAIDs, which may trigger gastrointestinal bleeding and ulcers.[1]

9.2 CLASSIFICATION OF NSAIDS[2]

9.2.1 *ON THE BASIS OF MECHANISM OF ACTION*

- **COX-1 selective inhibitors:** Acetylsalicylic acid.
- **Nonselective COX-1 inhibitors:** Acetylsalicylic acid at high dosage, Diclofenac sodium, Diclofenac potassium, Ibuprofen, Ketoprofen, Flurbiprofen, Indomethacin, Piroxicam, Naproxen.
- **COX-2 preferential inhibitors:** Nimesulide, Etodolac, Meloxicam, Nabumetone.
- **COX-2 selective inhibitors:** Celecoxib, Etorcoxib, Valdecoxib.

9.2.2 *ON THE BASIS OF CHEMICAL FORMULA*

- **Salicylates:** Aspirin (acetylsalicylic acid), Diflunisal (Dolobid), Salicylic acid, and other salicylates, Salsalate (Disalcid).
- **Propionic acid derivatives:** Ibuprofen, Dexibuprofen, Naproxen, Fenoprofen, Ketoprofen, Dexketoprofen, Flurbiprofen, Oxaprozin, Loxoprofen.
- **Acetic acid derivatives:** Indomethacin, Tolmetin, Sulindac, Etodolac, Ketorolac, Diclofenac (safety alert by FDA), Aceclofenac, Nabumetone (drug itself is nonacidic but the active, principal metabolite has a carboxylic acid group).
- **Enolic acid (Oxicam) derivatives:** Piroxicam, Meloxicam, Tenoxicam, Droxicam, Lornoxicam, Isoxicam.

- **Anthranilic acid derivatives (Fenamates):** Mefenamic acid, Meclofenamic acid, Flufenamic acid, Tolfenamic acid.
- **Sulfonanilides:** Nimesulide (systemic formulations for the substantial threat of hepatotoxicity are forbidden by some nations).
- **Others:** Clonixin. Licofelone acts by inhibiting the lipooxygenase (LOX) and COX, and hence is known as 5-LOX/COX inhibitor.

9.3 PHARMACOLOGICAL EFFECTS OF NSAIDS[3]

9.3.1 ANTI-INFLAMMATORY EFFECTS OF NSAIDS

NSAID identification on the generic drug logo is an OTC NSAID. NSAIDs suppress COX-1 and COX-2 activities, as well as the production of prostaglandin and thromboxane. COX-2 inhibition is believed to have antipyretic, anti-inflammatory, and analgesic qualities, and that these NSAIDs, which can induce GIT leakage and gastric ulcers, can also inhibit COX-1. Aspirin, in fact, is thought to inhibit COX-1.

9.3.2 ANALGESIC EFFECTS OF NSAIDS

Prostaglandins induce hyperalgesia by compromising the transducing property of free nerve ends, causing impulses that normally do not elicit pain to do so. NSAIDs do not impact tenderness triggered by direct prostaglandin therapy, although they do stop pain sensitization caused by bradykinin, TNF-alpha, interleukin, and other algesics. As a result, they are more successful against pain caused by inflammation.

9.3.3 ANTIPYRESIS EFFECTS OF NSAIDS

NSAIDs lower body temperature during a fever but may not cause hypothermia in people who are normally warm. Fever is caused by the production of pyrogens, such as ILs, TNF, and interferons, which cause the hypothalamus to produce PGE2, which raises the fixed temperature point. NSAIDs block the activity of pyrogens, but PGE2 administered into the hypothalamus does not. COX-2 tends to be the isoforms present at this site. Fever, on the other hand, may be caused by a mechanism other than PG.

9.3.4 ANTIPLATELET AGGREGATORY EFFECTS OF NSAIDS

Platelet aggregation is inhibited by therapeutic doses of most NSAIDs by inhibiting both proaggregatory (PGI_2) and antiaggregatory TXA_2 (COX-1 generated). The bleeding time is extended.

9.3.5 ANAPHYLACTOID ACTIONS OF NSAIDS

Aspirin triggers asthma, angioedema, and rhinitis urticaris in certain individuals who are prone. These subjects react in the same way as chemically diverse NSAIDs do, but without the immunological cause for the response. COX inhibition, with corresponding arachidonic acid conversion to LTs and other lipooxygenase pathway components, may be involved, but no proof exists.

9.4 DRUG INTERACTIONS[4]

9.4.1 PHARMACODYNAMIC INTERACTION OF NSAIDS WITH OTHER DRUGS

- A reduced hypotensive influence is achieved by hypotensive medications (β-blockers, angiotensin converting enzyme (ACE) inhibitors, diuretics).
- Ethanol results in an elevated risk of urinary tract bleeding.
- An elevated risk of bleeding results from ticlopidine or clopidogrel.
- Using lithium increases the toxicity of lithium.
- A reduced nephrotoxicity of medications results from cyclosporine or ACE inhibitors or tacrolimus.
- Fluoroquinolones have an elevated toxic effect on the central nervous system (CNS).
- An elevated risk of hypoglycemia results from oral antidiabetic prescriptions.
- Coumarin is associated with an elevated risk of gastrointestinal tract bleeding.

9.4.2 PHARMACOKINETIC INTERACTION OF NSAIDS WITH OTHER DRUGS

- A prominent hazard of hypoglycemia results from oral antidiabetic prescriptions.

- Coumarin leads to a superior hazard of bleeding.
- Corticosteroids contribute to a prominent risk of gastropathy and gastrointestinal tract bleeding.
- Aminoglycoside results in improved aminoglycoside ototoxicity and nephrotoxicity.
- Fenytoine or valproinic acid refers to an improved action of valproinic acid or fenytoine.
- The activity of methotrexate or digoxin increases and increases the toxicity of methotrexate or digoxin.
- Tricycles of neuroleptic antidepressant drugs or antiarrhythmic or selective inhibitors of serotonin reuptake (SSRI) result in enhanced drug action.

NSAIDs are often used to manage both acute and chronic illnesses that cause pain and inflammation. NSAIDs are widely used to treat rheumatoid arthritis, osteoarthritis, moderate to extreme pain caused by inflammation, fatigue, and tissue damage, low back pain, inflammatory arthropathy (e.g., psoriatic arthritis, ankylosing spondylitis, Reiter syndrome), tennis elbow, headache, nausea, acute gout, menstrual discomfort, metastatic bone pain, among other conditions.

9.5 CONTRAINDICATIONS[5]

Individuals with the following circumstances should use NSAIDs with caution:

- Irritable bowel syndrome.
- People >50 years who have a family history of GIT disorders.
- People who had GIT issues with the use of NSAID in the past.

Individuals with the subsequent circumstances should generally avoid NSAIDs:

- Third trimester of pregnancy
- Coronary artery disease
- Peptic ulcer or stomach bleeding
- Uncontrolled hypertension
- Past stroke
- Past transient ischemic attack
- Past myocardial infarction
- Undergoing the surgeries associated with coronary arteries
- Kidney disease
- Taking aspirin for heart

9.6 ADVERSE EFFECTS[6]

- NSAIDs elevate GIT complications. NSAIDs used following surgery for pain relief increases the risk of kidney problems, dyspepsia, extreme upper gastrointestinal bleeding.
- As other medicines, NSAIDs can interfere with other medicines, for example, NSAIDs and quinolones when coadministered can elevate the hazard of adverse effects of quinolones on the central nervous system including seizures.
- The combinational possibility of a conventional NSAID (prescription or over the counter) COX-2 inhibitor should not be taken concurrently.
- Individuals undergoing daily aspirin treatment (e.g., to minimize cardiovascular risk) should be vigilant with other NSAIDs, as they can hinder aspirin's cardioprotective impact.

The adverse effects of NSAIDs on various body systems are described:

9.6.1 CARDIOVASCULAR

- In comparison to aspirin, NSAIDs, both modern and conventional anti-inflammatory selective COX-2 antagonists, elevate the possibilities of both stroke and MI, and ultimately death.
- Naproxen tends to be less toxic.
- Aside from (low-dose) aspirin, NSAIDs in individuals devoid of heart disease history are linked with a double chance of heart failure.

9.6.2 GASTROINTESTINAL

- Adverse drug reactions (ADRs) coupled with NSAIDs administration contribute to overt and indirect GIT discomfort. A simultaneous attack on the GI tract is caused by NSAIDs: acidic molecules specifically irritate the mucosa, and COXs inhibition decreases the levels of antioxidant prostaglandins.
- Elevated gastric acid secretion, decreased bicarbonate secretion, decreased mucus secretion, and decreased trophic effects on epithelial mucosa are caused by prostaglandin synthesis inhibition in the GIT. Nausea/vomiting, dyspepsia, gastric ulceration/bleeding, diarrhea, and so on are normal gastrointestinal ADRs.

- The tendency of individual agents to induce gastrointestinal ADRs also varies considerably. The highest frequency of gastric ADRs tends to be ketoprofen, indomethacin, and piroxicam, whereas Diclofenac and Ibuprofen have inferior chances.
- Gastric (but not always intestinal) side effects can typically be minimized by jamming the acid production by coadministering a proton pump inhibitor, such as omeprazole, esomeprazole, or misoprostol. The high occurrence of GIT ADRs with misoprostol is diarrhea. Although they are effectual but are costly as a maintenance therapy.

9.6.3 *INFLAMMATORY BOWEL DISEASE*

- For inflammatory bowel disease (e.g., Crohn's disease or ulcerative colitis), NSAIDs have plausibilities to induce gastric bleeding and ulceration in the gastric lining.
- Paracetamol (also known as acetaminophen) or codeine-containing medications (which slow down the activity of the intestines) are safer medicines for IBD pain relief.

9.6.4 *RENAL*

- A comparatively poor rate of renal ADRs is also associated with NSAIDs. The cause of these renal ADRs is attributed to transformations in the renal hemodynamics (kidney blood flow) that are normally regulated by NSAID-affected prostaglandins.
- Prostaglandins usually allow the afferent arterioles of glomeruli to be vasodilated. This helps preserve natural glomerular perfusion, a measure of renal activity, and glomerular filtration rate (GFR). This is principally vital in renal failure, where the kidney is struggling to sustain an elevated angiotensin II level of the renal perfusion pressure. Angiotensin II often constricts the afferent arteriole into the glomerulus at these elevated levels in comparison to the efferent arteriole usually constricted by it.
- Prostaglandins help to expand the afferent arteriole. NSAIDs persuade unopposed constriction of the afferent arteriole and lower renal perfusion pressure by jamming this prostaglandin-mediated effect, particularly in the case of renal failure. Salt (sodium) and fluid

accretion, hypertension, and common ADRs connected with changed renal utility.

- Renal dysfunction can also be caused by these agents, especially in combination with other nephrotoxic agents. In fact, renal dysfunction is a concern if the patient is also simultaneously taking an ACE inhibitor (which removes the vasoconstriction of the efferent arteriole by angiotensin II) and a diuretic (which decreases plasma volume and therefore RPF)-the so-called "triple whammy" effect.

9.6.5 PREGNANCY

- During pregnancy, particularly during the third trimester, NSAIDs, are not recommended. While NSAIDs are not direct teratogens as a class, they may induce premature closure in the fetus of the fetal ductus arteriosus and renal ADRs.
- Furthermore, they are concerned with premature delivery and miscarriage. However, aspirin is employed for pregnant individuals with antiphospholipid antibodies, along with heparin. Additionally, in pregnancy, Indomethacin is employed for polyhydramnios by dropping the formation of fetal urine by restricting the flow of fetal renal blood.
- In comparison, Paracetamol (acetaminophen) during breastfeeding is known to be safe and well-tolerated. Doses should be taken as designated, with the hazard of hepatotoxicity being endorsed to overdoses. In France, the country's health group contraindicates the NSAIDs applications until the sixth month of pregnancy.

9.6.6 PHOTOSENSITIVITY

- Photosensitivity reactions are most likely to be produced by 2-arylpropionic acids, although other NSAIDs, including Diclofenac, Piroxicam, and Benzydamine, have also been involved. The main photoactive NSAID was Benoxaprofen, since removed due to its hepatotoxicity.
- The photosensitivity process, responsible for the high photoactivity of the 2-arylpropionic acids, is the ready decarboxylation of the carboxylic acid moiety.

- The basic absorption properties of the various 2-aryl chromophoric substituents influence the mechanism of decarboxylation. While Ibuprofen has poor absorption, it has been described as a weak photosensitizing agent.

9.6.7 OTHERS

Elevated hepatic enzymes, headache, and dizziness are common medication adverse reactions (ADRs). Swelling of the face and/or body, confusion, hyperkalemia, rash, and bronchospasm are all uncommon ADRs. NSAIDs are also implicated with certain cases of Stevens–Johnson syndrome. The bulk of NSAIDs have a tough time accessing the CNS. However, the COX enzymes are constitutively produced in some areas of the CNS, which ensures that even mild penetration may cause side effects including sleepiness and dizziness.

9.7 MECHANISM OF ACTION[7]

NSAIDs function as nonselective COX enzyme inhibitors, inhibiting both COX-1 and COX-2 isoenzymes. In contrast to the aspirin step, which is a permanent inhibition, this inhibition is competitively reversible. COX uses arachidonic acid to catalyze the synthesis of prostaglandins and thromboxane. Inflammation is regulated by prostaglandins, which function as messenger molecules.

COX-1 is a naturally occurring enzyme that plays a "housekeeping" function in the control of a variety of physiological processes. Prostaglandins have a defensive function in the stomach lining, preventing the stomach mucosa from being eroded with their own acid. COX-2 is a heterotrophic inflammatory protein, and COX-2 inhibition is what causes the possible benefits of NSAIDs. When stomach prostaglandin levels are reduced by nonselective COX-1/COX-2 inhibitors, stomach ulcers or internal bleeding from the duodenum may occur.

However, several aspects of NSAIDs' mode of operation are unclear, and other COX methods have been proposed for this reason. The COX-3 pathway was believed to fill some of this gap, but recent experiments have cast doubt about whether it plays a significant role in humans, and alternative explanations have been proposed. NSAIDs are also used to treat severe gout pain since they block urate crystal phagocytosis in addition to inhibiting prostaglandin synthase.

9.8 PHARMACOKINETICS[8]

The rest of NSAIDs are mild acids with a pKa of 3–5. They are well absorbed by the intestine and intestinal mucosa. They are highly protein-bound (often >95%) in plasma, generally to albumin, and their distribution rate normally approximates plasma amount. The majority of NSAIDs are metabolized in the liver, where they are oxidized and further conjugated to inactive metabolites that are excreted in the urine, though some drugs are partially excreted in the bile. Metabolism may be hampered by some illnesses, and aggregation can occur even with standard dosing. Ibuprofen and Diclofenac have short half-lives (2–3 h). Certain NSAIDs have extremely long half-lives (e.g., 20–60 h) (typically oxicams).

9.9 ASPIRIN[9]

Aspirin is a salicylate. In the liver, it is rapidly converted to salicylic acid, which is responsible for the primary action. This works by lowering the levels of compounds in the body that cause fever, indigestion, and inflammation. Aspirin is a pain reliever that also helps to reduce inflammation and fever. It may also be used to cure or prevent heart disease, chest discomfort, and strokes (angina). And with the supervision of a specialist will aspirin be used to treat respiratory disorders.

9.9.1 IMPORTANT INFORMATION

- Aspirin should not be used in the case of NSAIDs, such as Motrin, Advil, Orudis, Aleve, Lodine, Indocin, Toradol, Voltaren, Relafen, Mobic, Feldene, and so on in bleeding conditions, such as hemophilia, gastrointestinal, or intestinal bleeding.
- Do not offer this medicine to an infant with fever, signs of flu, or chicken pox, or to a teenager. Aspirin may produce Reye's syndrome, which in children is a dangerous and often a fatal disease.
- Asthma or seasonal allergies, liver disease, stomach ulcers, bleeding or blood clotting disorder, kidney disease, excessive blood pressure,

cardiac disease, and congestive heart failure, nasal polyps, gout that make aspirin not truely secure.
- Aspirin can be toxic to the womb of an unborn infant and can also lower birth weight or have other hazardous consequences. Aspirin can transfer into breast milk and a breastfeeding baby may be affected.

9.9.2 PHARMACOLOGICAL EFFECTS

9.9.2.1 ANTIPYRETIC, ANALGESIC, AND ANTI-INFLAMMATORY ACTIONS

In comparison to opioid medications, aspirin is a bad analgesic. It effectively relieves inflammatory, tissue-related trauma, connective tissue, and integumental pain in severe visceral and ischemic pain, but it is comparatively ineffective in other types of pain. Analgesic discomfort is caused by the acquisition of peripheral pain receptors and the halting of PG-mediated nerve ending sensitization. There is a central subcortical function that raises the pain level, but there is no morphine-like effect on the pain part of psychic processing or reaction. Aspirin resets the hypothalamic thermostat and lowers fever rapidly by promoting heat loss, although it does not minimize heat production. Anti-inflammatory activity is added at elevated doses (3–6 g/day or 100 mg/kg/day). Inflammatory symptoms, such as aversion to pain, swelling, vasodilation, and leukocyte invasion are decreased. In addition to COX inhibition, quenching free radicals may contribute to anti-inflammatory effects. These are only needed for anti-inflammatory doses. Cellular metabolism is increased, especially in skeletal muscles, as a result of the uncoupling of oxidative phosphorylation-increased heat production. Consumption of glucose rises blood sugar level, whereas liver glycogen loss falls. Hypoglycemia is seen at toxic doses due to the central sympathetic stimulation-release of ADR and corticosteroids. Chronic usage of elevated doses causes negative N_2 equilibrium by increasing protein carbohydrate conversion. Plasma-free fatty acid and cholesterol levels decrease.

9.9.2.1.1 *Cardiovascular*

- Platelets: COX-1-derived TXA2 platelet aggregation inhibition with the net result of increasing bleeding time.
- Endothelial COX-2-dependent PGI_2 can inhibit platelets aggregation (inhibition enhance the chances of aggregation through TXA_2).

- Aspirin (acetylsalicylic acid) alters, and irreversibly inhibits platelet COX covalently. For the lifespan of the platelet (~8–11 days), the enzyme is blocked. The result was achieved at very low dosage levels. Aspirin shows therapeutic effectiveness basis in stroke and MI (reduces mortality and prevents recurrent events).
- Blood vessels/smooth muscle: The PGI2 derived from COX-2 can antagonize vasoconstriction caused by catecholamine and angiotensin II (NSAIDs can elevate BP).
- Atherosclerosis:Atherosclerotic plaques can be destabilized by COX-2 inhibition (due to its anti-inflammatory actions).

9.9.2.1.2 *Renal*

- PGs produced by COX-1 and COX-2 (PGF_2, TXA_2, PGI_2 (glom), PGE_2 (medulla), powerful vasodilators) will retain both incr and decr Na^+ (predominantly natriuresis), typically in response to tubular Cl^-, extracellular tonicity or low BP shifts.
- NSAIDs aim to facilitate preservation of Na+ and can also improve BP. Many antihypertensives will counteract the effects of diuretics, ACE inhibitors, and -AR antagonists.
- PGs have a limited effect on the regular flow of renal blood, but in the damaged kidney, they become significant. Patients (particularly the elderly and those with volume depletion) are at risk of NSAID renal ischemia.

9.9.2.1.3 *Gastrointestinal*

- PGs (generated via COX-1): Inhibit the release of stomach acid, promote the secretion of mucus and HCO_3, vasodilate, and are also cytoprotective for the gastric mucosa. Therefore, NSAIDs with COX-1 inhibitory action will generate conflicting consequence which leads to sudden acute hemorrhage, gastric distress, and bleeding (effects are dose-dependent).

9.9.2.1.4 *Gestation*

- In the commencement and succession of labor and delivery, PGs are involved. Inhibition of their growth by NSAIDs will then prolong gestation.

9.9.2.1.5 Respiratory system

- Salicylates allow oxidative phosphorylation to be partially uncoupled with increased CO_2 output (COX-independent effects). Increase in plasma CO_2 leads to hyperventilation. Higher doses results in respiratory depression.

9.9.3 PHARMACOKINETICS

Since very little of aspirin (salicylic acid) is ionized in the stomach, it is a gentle acid. It is poorly soluble in acidic environments of the intestine and can prolong the absorption of large doses by eight to 24 h. Because of the higher pH and wider surface region of the small intestine, aspirin is quickly absorbed, causing further salicylate to dissolve. Protein binds between 50% and 80% of blood salicylate. The remainder resides in the living, ionized environment. Binding site saturation raises the volume of free salicylates which increases the total toxicity. The distribution rate is about 10% of body weight. Acidosis raises the delivery concentration owing to improvements in salicylate tissue infiltration. The liver is responsible for metabolizing up to 80% of all doses. Conjugation of salicylic acid with glycine produces salicyluric acid or D-glucuronic acid and further produces glucuronide. These metabolic processes have a rather limited capacity. With large doses, the kinetics shift from first-order to zero-order, as metabolism is slowed and renal emission becomes increasingly greater. Salicylates are mostly excreted into the kidneys as free salicylic acid (10%), salicyluric acid (75%), acyl glucuronides, and salicylic phenol (10%). (5%). Since the metabolism processes involved in salicyluric acid synthesis and salicyl phenolic glucuronide are overloaded when larger salicylate levels (more than 4 g) are consumed, the half-life becomes much longer (15–30 h). Both pathways proceed by first-order kinetics at small doses >250 mg, with a removal half-life of about 4 h. Since it is highly susceptible to urinary pH changes, renal excretion of salicylic acid becomes essential as metabolic pathways become overloaded. When urinary pH is elevated from 5 to 8, renal clearance improves 10- to 20-fold. This unique function of salicylate reduction is linked to the applications of urinary alkalinization.

9.9.4 DOSAGE AND ADMINISTRATION

- Analgesic and antipyretic: Adults: 1–2 tablets (325–650 mg) every 4 h on an oral basis. Children under 12 years of age should not surpass the average maximum dosage of 2.4 g every 6 h, 10–15 mg/kg every 6 h.

- Anti-inflammatory: For optimal anti-inflammatory benefit, three tablets (975 mg) 4–6 times a day, up to 30 tablets a day may be needed. A blood level within the desirable therapeutic range is between 15 and 30 mg/100 mL.
- Children: In four to six split doses, 60–125 mg/kg daily.
- To minimize the risk of morbidity and mortality in patients with unstable angina and in patients with past myocardial infarction: 80–325 mg daily, as determined by the surgeon, based on the patient's specific needs. For these applications, the Coated Aspirin Daily Low Dose is clearly suggested.
- To minimize the risk of transient ischemic attacks (TIA) and secondary avoidance of atherothrombotic cerebral infarction: 80–325 mg daily, as determined by the practitioner, based on the patient's specific needs. For these applications, the Coated Aspirin Daily Low Dose is clearly suggested.
- For venous thromboembolism prophylaxis after complete hip replacement: 650 mg twice daily (1300 mg daily), began 1 day prior to surgery and lasted for 14 days.
- For other inhibitory applications of platelet aggregation: 325–1300 mg daily according to human preferences and widely agreed treatment requirements for each sign.

9.9.5 PRECAUTIONS

- Patients of asthma and other allergic disorders, a history of stomach ulcerations, bleeding tendencies, extreme anemia or hypoprothrombinemia should be given salicylates with caution.
- Patients taking ASA on a daily basis have an elevated chance of experiencing gastrointestinal bleeding after alcohol consumption.
- Caution is important when salicylates and anticoagulants are administered simultaneously, as salicylates can reduce plasma prothrombin concentrations.
- Diabetics undergoing combined salicylate–hypoglycemic treatment should be carefully monitored: it may be appropriate to decrease the dosage of the sulfonylurea hypoglycemic drug, the requirements for insulin may alter.

- Pregnancy: During pregnancy, high doses (3 g daily) of ASA can prolong gestation and childbirth.
- In thyroid function tests, salicylates can cause improvements. By salicylate administration, sodium excretion created by spironolactone can be minimized.
- Big doses of salicylates are uricosuricagents, smaller quantities can suppress the clearance of uric acid and thus reduce the uricosuric effects of other medicines.
- Salicylates also retard the renal elimination of methotrexate.

9.9.6 *ADVERSE REACTIONS*

- Gastrointestinal: Vomiting, nausea, GIT bleeding and/or ulceration, diarrhea, heartburn, and dyspepsia.
- Ear: Vertigo, tinnitus, and hearing loss.
- Hematologic: Thrombocytopenia, leukopenia, anemia, and purpura.
- Dermatologic and hypersensitivity: Angioedema, urticaria, skin eruptions, pruritus, anaphylaxis, and asthma.
- Miscellaneous: Drowsiness, mental confusion, sweating, and thirst.

9.9.7 *SYMPTOMS AND TREATMENT OF OVERDOSE*

- Symptoms: This can involve quick and nausea, deep breathing, vertigo, vomiting, flushing, tinnitus, thirst, tachycardia, and sweating during moderate overdosage. Acid–base abnormalities may arise in more serious situations, including metabolic acidosis and respiratory alkalosis. Fever, hemorrhage, confusion, excitement, coma, respiratory collapse, and convulsions may be seen in serious cases.
- Treatment: Treatment consists of acid–base and fluid and electrolyte disruption reduction and maintenance. By urinary flow and alkaline diuresis, kidney clearance is improved, but caution must be taken in this strategy to not further aggravate metabolic acidosis and hypokalemia. Administration of sufficient sodium-containing fluids and sodium bicarbonate can avoid acidemia. Hypoglycemia, which may be controlled by glucose solutions, is an occasional accompaniment to salicylate poisoning. Provide vitamin K if a hemorrhagic diathesis is visible. In complex acid–base disruptions, hemodialysis may be beneficial, particularly in the presence of abnormal renal function.

9.9.8 INDICATIONS AND CLINICAL USES

- ASA is also suggested for the entire content based on its platelet aggregation inhibition consequences: for reducing the risk of morbidity and mortality in patients with dysfunctional angina and in patients with prior myocardial infarction. For the risk mitigation of acute ischemic attacks (TIA) and for the secondary avoidance of cerebral atherothrombotic infarction.
- Venous thromboembolism prophylaxis after complete replacement of the hip in men.
- Reduction of platelet adhesive properties in patients undergoing carotid endarterectomy to prevent TIA recurrence and in patients undergoing hemodialysis with a synthetic rubber arteriovenous cannula.
- Pain: Aspirin is an effective analgesic for treating the acute pain, but is normally deemed inadequate for pain management to ibuprofen as aspirin is more sufficient to make GIT bleeding. For certain pains caused by bloating, muscle cramps, acute skin inflammation, and gastric distension, aspirin is usually ineffective. Combinations of aspirin and caffeine, as with other NSAIDs give slightly better pain relief than aspirin itself. Effervescent aspirin formulas, such as Alka-Seltzer or Blowfish, alleviate discomfort more rapidly than aspirin in pills, rendering them beneficial for migraine therapy. Topical aspirin can be helpful in managing some forms of neuropathic pain.
- Headache: Aspirin successfully treats certain forms of headache, either by itself or in a mixed formulation, but its effectiveness can be uncertain for others. A care practitioner should handle secondary headaches, meaning those caused by another illness or accident, immediately. The International Classification of Headache Disorders distinguishes between cluster headache, stress headache, and migraine, among primary headaches. For the treatment of stress headache, aspirin or other over-the-counter analgesics are generally accepted as safe. For the treatment of stress headache, aspirin or other over-the-counter analgesics are generally accepted as safe. Aspirin, particularly as a component of the formulation of acetaminophen/aspirin/caffeine, for example, Excedrin Migraine, is known to be first-line migraine treatment, and is equivalent to lower sumatriptan doses. It is most effective in treating migraines.
- Fever: The power of aspirin to regulate fever, like its capacity to control pain, is attributed to its function on the prostaglandin system

by its permanent COX inhibition. While the application of aspirin as an antipyretic agent in the adult patients is quite well-known, numerous regulatory agencies and medical societies (including the American Academy of Pediatrics, United States Food and Drug Administration (FDA), and the American Academy of Family Doctors) strongly warn against the use of aspirin for children's fever care due to the possibility of Reye's syndrome, an uncommon but sometimes lethal disease. In 1986, owing to the risk of Reye's syndrome in infants, the FDA required labeling for all drugs containing aspirin to warn against its use in children and adolescents.

- Swelling and inflammation: Aspirin is used as an anti-inflammatory drug for both acute inflammation and long-term inflammation, as well as to combat inflammatory diseases such as rheumatoid arthritis.
- Heart attacks and strokes: Aspirin can help avoid heart attacks and strokes. In individuals at risk of cardiovascular disease, aspirin has also been known to inhibit the worsening of established coronary heart disease at reduced doses and reduce the occurrence of these events. Aspirin continues to give individuals at decreased risk of heart attack or stroke no protection, such as those with a background of such accidents or preexisting circumstances. Some reports prescribe aspirin on a case-by-case basis, although others have stated that the risks of other incidents, such as GIT bleeding, are too large to overshadow the possible benefit, and have suggested that aspirin should not be used entirely for primary protection.
- The phenomenon of aspirin tolerance is a complication of the use of aspirin for protection. The effectiveness of aspirin is decreased for patients who are allergic, which can cause an increased risk of stroke. Some studies have recommended test regimens to classify patients who are immune to aspirin or other antithrombotic medications (such asclopidogrel).
- Aspirin has also been recommended for cardiovascular disease prevention as a part of a polypill.
- After surgery: Since percutaneous coronary interventions (PCIs), such as coronary artery stent insertion, a US Guidelines from the Department for Healthcare Research and Quality request that aspirin be used permanently. Aspirin is often paired with an inhibitor of the ADP receptor, such as Clopidogrel, prasugrel, or ticagrelor to eliminate blood clots. This is called dual treatment for antiplatelets (DAPT). Guidelines from the United States and the European Union somewhat vary about the amount of time and the indications for which this

combination therapy can begin after surgery. US guidelines prescribe DAPT for a period of 12 months, while EU guidelines recommend DAPT for 6–12 months following insertion of a drug-eluting stent. They accept, however, that after DAPT is complete, aspirin can proceed forever.

- Cancer prevention: Aspirin decreases the overall risk of cancer as well as dying from cancer. For colorectal cancer, this effect is particularly helpful (CRC).
- Other uses: For the symptoms of acute rheumatic fever owing to fever and joint discomfort, aspirin is a first-line treatment. Typically, rehabilitation lasts 1 to 2 weeks, but it is seldom recommended for longer periods. After fever and pain have dissipated, aspirin is no longer needed as it does not reduce the incidence of cardiac problems and recurrent rheumatic heart disease. Naproxen has been found to be as effective as aspirin and less toxic, but because of insufficient scientific awareness, naproxen is only prescribed as a second-line treatment. Along with rheumatic fever, Kawasaki disease is among the few explanations for aspirin use in children, despite a lack of high-quality evidence for its use. Low-dose supplementation with aspirin has moderate effects while mostly used as preeclampsia control.

9.10 CELECOXIB[10]

Celecoxib works by suppressing inflammatory and pain-inducing hormones with low ulcerative potential. It is used to relieve the pain and irritation caused by a variety of conditions, including asthma, menstrual pain, and ankylosing spondylitis. Celecoxib is also used to treat inherited polyps in the colon. Celecoxib has a low COX-2 selectivity (6–20 fold).

9.10.1 PHARMACOLOGICAL EFFECTS

- Anti-inflammatory: Cecoxib, a highly selective COX-2 isoform inhibitor, blocks the transfer of arachidonic acid to prostaglandin precursors. It also has properties that are antipyretic, analgesic, and anti-inflammatory. Both COX-1 and COX-2 are inhibited by unselective NSAIDs (such as aspirin, naproxen, and ibuprofen). Inhibition of COX-1 decreases prostaglandin production and platelet activator formation of thromboxane A2. Historically, COX-1 is known as a "housekeeping" enzyme that is biologically derived and plays a significant role in the defence of platelet thrombogenesis, gastrointestinal mucosa, and hemodynamics of the kidneys. COX-2, on the other hand, is heavily expressed in inflammatory cells and is upregulated by bacterial lipopolysaccharides, cytokines, tumor promoters, and growth factors. Celecoxib inhibits COX-2 nearly 10–20 times more selectively than COX-1. It binds with its polar sulfonamide side chain to a hydrophilic side pocket region near the active binding site for COX-2. This selectivity, in principle, helps celecoxib and other COX-2 inhibitors to minimize inflammation (and pain) while reducing gastrointestinal drug adverse reactions that are typical with nonselective NSAIDs (e.g., stomach ulcers).

- Anticancer: Celecoxib influences genes and mechanisms involved in inflammation and malignant transformation in cancers, but not in normal tissues, for use in reducing colon polyps. Cadherin-11 binds celecoxib.
- Celecoxib is used in persons with hereditary adenomatous polyposis for rheumatoid arthritis, osteoarthritis, painful menstruation, acute discomfort, ankylosing spondylitis, and to decrease the amount of colon and rectal polyps. It can also be used in children with juvenile rheumatoid arthritis who are over 2 years of age and who weigh more than 10 kg (22 lb).
- It is more or less equivalent to Ibuprofen for postoperative pain.[5] It is similar to Paracetamol for pain relief (acetaminophen). The first line of therapy for OA in osteoarthritis is acetaminophen.
- Initially, it was meant to ease discomfort thereby minimizing the harmful gastrointestinal consequences seen in traditional NSAIDs. In fact, its main application is in patients who require routine and long-term pain relief; there is potentially little benefit of using celecoxib over traditional NSAIDs for short term or immediate pain relief, even in situations where nonselective NSAIDs or aspirin induce cutaneous reactions (urticaria or "hives").
- Mental illness: In the treatment of a variety of mental illnesses, including severe depression, bipolar disorder, and schizophrenia, preliminary data suggests its application.
- Familial adenomatous polyposis: It is suggested for patients with familial adenomatous polyposis to reduce colon and rectal polyps, although it is not clear whether it lowers cancer rates, so it is not a safe option for this reason.

9.10.2 PHARMACOKINETICS

- As celecoxib is granted fasting, maximal plasma concentrations are reached after around 2–3 h. Under fasting conditions, all peak plasma values (C_{max}) and the area under the curve (AUC) are nearly equivalent for a 200 mg BD dose; higher doses produce fewer than proportional improvements in Cmax and AUC. Absolute bioavailability studies have not been conducted owing to celecoxib's low solubility in aqueous media. Celecoxib capsules have a 99% relative oral solubility as compared with a suspension. Celecoxib is heavily protein-bound (97%) within the therapeutic dose range. It attaches to albumin, and to a lesser extent, alpha-1glycoprotein, according to in vitro studies. The obvious volume of steady-state transmission in healthy young adults is about 400 L, suggesting widespread tissue distribution.
- Celecoxib is fully metabolized in the liver. In vitro and in vivo studies indicate that cytochrome P450 2C9 plays a significant role in metabolism. Three metabolites were discovered in human plasma: the primary alcohol, the resulting carboxylic acid, and the conjugate glucuronide. Pharmacological activity is found in the parent drug. In the main metabolites found in human plasma, there is no evidence of COX-1 or COX-2 inhibitory action. Cytochrome P450 2C9 production is reduced in people who have genetic polymorphisms that cause decreased enzyme output, such as those that are homozygous for the CYP 2C9*3 polymorphism. Celecoxib may be used with care in people who have been diagnosed or accused of having reduced P450 2C9 metabolizers in the past, since they may experience abnormally high plasma concentrations owing to diminished mitochondrial clearance. Start the treatment with half of the lowest dosage recommended.
- Hepatic metabolism eliminates the bulk of celecoxib, with 3% of the unchanged drug present in urine or feces. After a single oral dose of radiolabeled substance, some 57% of the dosage was excreted in the feces and 27% was excreted in the urine. In both the urine and the feces, the carboxylic acid metabolite (73% of the dose) was the dominant metabolite, with minimal amounts of glucuronide also appearing in the urine. At steady-state, the removal half-life ($t_{1/2}$) was 4–15 h, and clearance was about 500 mL/min. Because of the drug's low solubility, absorption takes longer, resulting in a wide range of terminal half-lives ($t_{1/2}$).

9.10.3 PRECAUTIONS

- Celecoxib may lead to heart or circulation ailments, such as a heart attack or stroke, particularly if you are using it for a long period of time.
- Celecoxib should not be used either before or after cardiac bypass surgery. Celecoxib may also cause severe signs of gastrointestinal or intestinal bleeding or ulceration. These disorders, especially in elderly people, can be fatal and can develop without warning when taking celecoxib.
- Stop alcohol intake. It can increase the risk of bleeding in the stomach.
- Class D in FDA maternity. Taking celecoxib can damage the unborn baby during the last 3 months of pregnancy. Celecoxib flows into breast milk and a breastfeeding infant can be affected by it.

9.10.4 CONTRAINDICATIONS

- Celecoxib should never be used in people who have acquired asthma, hives, or allergic reactions after administering ASA or other NSAIDs, including most COX-2-specific antagonists.
- Due to the lack of any evidence showing synergistic effects and the risk for additive adverse reactions, celecoxib cannot be combined with other NSAIDs.
- Celecoxib is contraindicated in:
- Patients with known ischemic heart disease, peripheral arterial disease and/or cerebrovascular disease that is dysfunctional or severe.
- Patients with peptic ulceration or GIT bleeding.
- Patients with approximate creatinine clearance <30 mL/min.
- Patients with congestive heart failure.
- Patients with stern liver injury.

9.10.5 ADVERSE EFFECTS

- *Cardiovascular Adverse Effects:* An eminent hazard of severe coronary thrombotic adverse effects, stroke, and MI, which can be fatal, has been linked with COX-2 inhibitors, including celecoxib. Both selective COX-2 and NSAIDs (nonselective) can cause an enlarged hazard of severe thrombotic cardiovascular events. With a period of use, this risk can increase. There may be an increased risk for people with documented coronary disease, a medical history of cardiovascular disease, or risk factors. Patients at danger of cardiovascular disorder (including patients with asthma, ischemic cardiac disease, heart failure, hyperlipidemia, hypertension, or smokers) who are having major surgery can be believed to be at superior threat of experiencing a cardiovascular incident. Patients with substantial hazard events for cardiovascular events should be treated with celecoxib only if the patient on the whole has hazard and the possible complications and advantages of alternate analgesic treatments have been methodically considered. In order to minimize the elevated risk of adverse heart disease in patients treated with celecoxib, the lowest appropriate dose can be used for the shortest possible period. For patients at high risk of cardiovascular adverse effects, prescribers may alert the particular patient of any elevated risks when administering celecoxib. Physicians and patients, even in the absence of prior cardiovascular signs, should be vigilant for such cases. Patients must be conscious of the prominent symptoms and signs of extreme coronary toxicity and of the steps they should take if they are present. Because of its absence of action on platelets, celecoxib is not a replacement for coronary prophylaxis. Thus, concurrent antiplatelet therapies should not be stopped. There is no confirmation that simultaneous applications of aspirin lowers the risk of COX-2 inhibitor-related cardiovascular adverse effects, including celecoxib.

- *Gastrointestinal Effects:* In patients treated with celecoxib, severe GIT toxicity has been observed, such as ulceration, swelling, and GI perforation (both upper and lower). Celecoxib demonstrated a low occurrence in clinical trials of gastroduodenal ulceration and severe clinically relevant GI cases. In patients treated with NSAIDs, significant GIT toxicity, such as peptic ulceration, perforation, and bleeding, often extreme and rarely fatal, may occur at any time, with or without warning signs. Minor upper GIT complications, such as dyspepsia, are normal and can also happen during NSAID therapy at any time. Therefore, in patients treated with NSAIDs, even in the deficiency of prior GI tract signs, physicians should stay vigilant for ulceration and bleeding. Patients should be aware of the signs and/or symptoms of extreme GI toxicity and of the measures that should be taken if they arise. Elderly patients, patients with coronary disorder, patients who use concomitant aspirin (even at low doses), corticosteroids or other NSAIDs, patients who drink alcohol, or patients with a history of or active GIT disease (such as GIT bleeding, ulceration, and inflammatory conditions) are the patients mostly at risk of experiencing GI problems with NSAIDs. For these patients, CELEBREX should be prescribed with strict caution. Doctors and patients should be careful about ulceration and GI bleeding, especially in the absence of symptoms. People with a family history of peptic ulcer disease and/or GIT bleeding who are using NSAIDs are more than 10 times more likely to cause GIT bleeding than patients without any of these lifestyle factors. The way this result refers to celecoxib is ambiguous. There is no conclusive evidence that coadministration of histamine H2 receptor antagonists and/or antacids would either avoid the development of GI side effects, or if these adverse reactions occur, cause celecoxib to proceed.

- *Anaphylactoid Reactions:* Anaphylactoid reactions have happened in patients with no known previous exposure to celecoxib, as with NSAIDs, in general. Rare reports of anaphylactoid reactions and angioedema have been accounted for patients seeking celecoxib for postmarketing experience. Patients with the aspirin triad must not be treated with celecoxib. Typically, this symptom complex exists in asthmatic patients who develop rhinitis with or without nasal polyps or who, after consuming aspirin or other NSAIDs, show severe, possibly fatal bronchospasm. In situations where an anaphylactoid reaction arises, medical aid can be obtained.
- *Serious Skin Reactions:* In association with the applications of celecoxib, stern skin reactions, such as Stevens-Johnson syndrome, exfoliative dermatitis, and toxic epidermal necrolysis have been infrequently traced. Early in the process of rehabilitation, patients tend to be at the greatest risk for certain events; the commencement of the incidence happens in majority of cases during the first month. At the primary manifestation of the mucosal lesions, skin rash, or some other symptom of hypersensitivity, celecoxib should be discontinued.
- *Hypertension:* Celecoxib, as with all NSAIDs, can result in a novel hypertension or deterioration of the preexisting hypertension, and overall leads to augmented occurrence of cardiovascular events. In patients with hypertension, NSAIDs, including celecoxib, should be treated with caution. During the start of treatment for celecoxib and right through the therapy period, the BP should be controlled closely.
- *Renal Effects:* Long-term management of NSAIDs has induced renal papillary necrosis and other renal injuries. Kidney toxicity has

also been shown in patients in whom renal prostaglandins play a compensatory function in preserving renal perfusion. The application of NSAIDs in these patients may lead to a dose-dependent decrease in prostaglandin production, and secondly, in renal blood flow, which may trigger overt renal decompensation. These patients should be carefully watched when seeking medication with celecoxib. Those with compromised kidney function, cardiac disease, liver dysfunction, diuretics, and ACE inhibitors are the patients at higher risk of this reaction. Use for ACE inhibitors, angiotensin receptor blockers, thiazide diuretics, and anti-inflammatory drugs) and the elderly. NSAID therapy discontinuation is typically accompanied by rehabilitation to the pretreatment condition.

- Celecoxib decreases urinary excretion of PGE_2 and 6-keto-PGF1af (a metabolite of prostacycline) but does not affect serum thromboxane B_2 (TXB_2) and urinary excretion of 11-dehydro-TXB_2, a metabolite of thromboxane (both products of COX-1). In patients with considerable dehydration, care should be used before starting care with celecoxib. First of all, it is advisable to rehydrate patients and then to continue care.

9.10.6 INTERACTIONS

- General: The metabolism of celecoxib is predominantly regulated in the liver through cytochrome P450 2C9. Celecoxib should be administered with care to patients who are documented or suspected to be weak CYP 2C9 metabolizers based on the previous history with other CYP 2C9 substrates as these patients may have strangely elevated plasma levels due to decreased metabolic clearance. It should be performed with caution to coadminister celecoxib with medications proven to suppress 2C9. Consider beginning therapy at half of the lowest dose prescribed.
- Associated treatment of celecoxib with CYP 2C9 inhibitors contribute to an improvement in celecoxib plasma concentrations. Therefore, where celecoxib is coadministered with CYP2C9 inhibitors, a lessening in the dosage of celecoxib may be essential.
- Concomitant administration of celecoxib with CYP 2C9 inducers (such as rifampicin, carbamazepine, and barbiturates) can result in decreases in celecoxib plasma concentrations. Therefore, where celecoxib is coadministered with CYP 2C9 inducers, an improvement in the dosage of celecoxib could be essential.

- Clinical pharmacokinetics and in vitro experiments show that while not a substrate, celecoxib is a cytochrome P450 2D6 inhibitor. Therefore, an in vivo drug interaction potential occurs with medicines that are metabolized by P450 2D6. Antihypertensives include ACE inhibitors, angiotensin II antagonists, diuretics, and beta-blockers. Prostaglandin inhibitors can minimize the impact of antihypertensives, such as ACE inhibitors, diuretics, and beta-blockers of angiotensin II antagonists (also known as angiotensin receptor blockers or ARBs). In patients taking celecoxib simultaneously with these drugs, this relationship should be taken into account.
- Coadministration with ACE inhibitors, angiotensin II antagonists or diuretics of NSAIDs, even COX-2 inhibitors (selective), can lead in impaired renal function, including likely acute renal failure, in aged, volume-depleted (including diuretic treatment) or impaired renal function patients. Usually, such results are reversible. It is therefore important to perform the simultaneous administration of these medications with caution. Patients should be properly hydrated, and at the onset of concomitant therapy and regularly thereafter, the therapeutic requirement to control kidney function should be determined.
- In a clinical trial, nearly half of the patients who received an ACE inhibitor, lisinopril, in combination with celecoxib, did not react to lisinopril at the final clinical visit, compared with fewer than one-third of the patients who received lisinopril in combination with placebo, a statistically relevant gap.

9.10.7 DOSAGE AND ADMINISTRATION

Patients receiving celecoxib should both start treatment at the lowest prescribed dose and be titrated for the shortest practicable duration to the lowest dose associated with successful symptom management.

- Adults: Without regard to the timing of meals, the following doses may be given.
- Osteoarthritis: 200 mg taken once or in two separated doses is the minimum daily dosage. If required, a dosage of 200 mg twice daily can be used.
- Rheumatoid arthritis: The minimum daily dosage in two separated doses is 200–400 mg.

- Ankylosing Spondylitis: 200 mg taken once or in two separated doses is the minimum daily dosage. A cumulative daily dosage of 400 mg may be effective for certain patients.
- Management of acute pain and treatment of primary dysmenorrhea: The prescribed dosage on the first day is 400 mg as a daily dose, followed on following days by 200 mg once a day. If required, patients may be advised to take an additional dose of 200 mg on any given day. 400 mg a day is the highest prescribed dosage. Celecoxib can be given up to 2 h before surgery is performed.
- Elderly: In general, no dose change is required. However, it is advisable to start treatment at the lowest prescribed dosage for older adults with a lower than normal body weight (<50 kg).

9.11 DICLOFENAC[11]

Diclofenac sodium is an anti-inflammatory, antipyretic, and analgesic drug used to manage sprains, severe gout, rheumatoid arthritis, backache, tenosynovitis, osteoarthritis, soft tissue sports trauma, bursitis, ankylosing spondylitis, strains, dislocations, bruises, tendinitis, and frozen hip. They are also used to relieve discomfort and irritation induced by orthopedic (bone and joint) operations, dental surgery, and minimal surgery. It inhibits PG synthesis while being somewhat selective for COX-2.

9.11.1 PHARMACOLOGICAL EFFECTS

- There are pronounced anti-inflammatory, antirheumatic, antipyretic, and analgesic effects of Diclofenac sodium, a nonsteroidal drug.
- Prostaglandin biosynthesis inhibition, which has been shown experimentally, is known to have a significant effect on its mechanism of action. Prostaglandins play a significant role in inducing fever, pain, and inflammation.
- The analgesic effects and anti-inflammatory effects of DICLOFENAC-GA in rheumatic diseases give rise to a therapeutic reaction marked by relaxation of signs and symptoms such as rest discomfort, movement pain, morning stiffness and joint swelling, as well as improvement of operation.
- In addition, clinical trials have shown that Diclofenac preparations are able to alleviate discomfort and reduce the frequency of bleeding in

primary dysmenorrhea. Low sodium Diclofenac concentrations prevent the accumulation of platelets caused by collagen and adenosine diphosphate in vitro.

- Diclofenac sodium in vitro does not, in canine cartilage, inhibit proteoglycan biosynthesis at concentrations comparable to those achieved in humans. Diclofenac sodium is not known to affect the durability of human osteoarthritic cartilage or not.

9.11.2 PHARMACOKINETICS

Diclofenac is fully extracted from the enteric-coated capsules as they move into the liver. When one tablet is taken with or during a meal, the passage across the stomach is slower than when it is taken before a meal, but the volume of active content absorbed remains the same. In fasting subjects, the mean peak plasma concentration is reached 2 h after ingestion of one 50 mg tablet. As calculated by the area under the time–concentration curve, plasma concentrations are linearly proportional to dose scale. During oral delivery, nearly half of the active agent is metabolized during the first movement into the liver (first-pass effect). Diclofenac binds to serum proteins 99.7% of the time, often albumin (99.4%). Diclofenac's total systemic clearance in plasma is 263 56 mL/min (mean meaning SD). The terminal half-life in plasma is 1–2 h. Following a 15-day course of Diclofenac at a dose of 25 mg 3 times a day, there was no evidence of opioid accumulation in plasma. Diclofenac reaches the synovial blood, where maximum amounts have been measured 2–4 h following oral administration. Synovial fluid removal had an apparent half-life of 3–6 h. As a result, just 4–6 h after administration, the amounts of active agent in the synovial fluid were already higher than in plasma, and they stayed higher for up to 12 h. This influence could explain why the clinical advantage lasts longer than the plasma half-life might suggest. The biotransformation of Diclofenac involves some intact molecule glucuronidation, but mainly single and multiple hydroxylation preceded by glucuronidation. In the urine, approximately, 60% of the dose is excreted as metabolites from one of these two pathways; less than 1% is excreted as an unchanged substance. The bile in the feces extracts the remainder of the dosage as metabolites. There have been no specific age-related differences in opioid absorption metabolism or excretion. In patients with renal impairment, no concentration of the unchanged active compound can be inferred by single dose kinetics by using the standard dosage protocol. At a creatinine clearance of less than 10 mL/min, the potential steady-state plasma amounts of metabolites are four times greater than in typical subject matters.

9.11.3 PRECAUTIONS

- If an allergy to Diclofenac sodium or aspirin continues, symptoms of allergic reactions can occur, including swelling of the face and mouth (angioedema), difficulty breathing, runny nose, skin rash, and so on.
- Allergic symptoms are caused in the event of a stomach (gastric) or duodenal (peptic) ulcer, or bleeding in the digestive tract. In the following instances, allergic responses occur: issues with the kidneys or stomach, porphyria, asthma, allergic rhinitis (including hay fever), nasal polyps, respiratory problems, including long-term chest inflammation, cardiac problems, lupus (SLE), or some other related disease, pathological allergy of such carbohydrates, such as lactose.
- During the last 3 months of pregnancy, Diclofenac sodium should be avoided as it may affect the circulation of the infant. Taking Diclofenac sodium can make getting pregnant more difficult. Avoid taking this prescription when you are breastfeeding, although only trace concentrations of Diclofenac sodium have also been found in breast milk.
- The lowest dose of Diclofenac sodium, especially in underweight or elderly patients, should be taken as soon as possible. When taking some drug such as Diclofenac, it raises the chance of heart disease or stroke. The chance of taking elevated doses for a long time is higher. Still follow the orders of the doctor as to how many to take and how long to take. Diclofenac sodium can reduce the symptoms, such as headache and high temperatures, of infection.

9.11.4 CONTRAINDICATIONS

Peptic ulcer, gastrointestinal inflammation, reported hypersensitivity to the active ingredient and all excipients, breastfeeding in the third trimester, and severe hepatic, renal, or cardiac insufficiency are all contraindications. Diclofenac, aspirin, and other NSAIDs may trigger asthma, hives, and other allergic reactions, which can be severe and even catastrophic.

9.11.5 INTERACTIONS

ACE inhibitors, anti-inflammatory drugs, and thiazide diuretics are used in conjunction. The usage of an anti-inflammatory medication, an ACE inhibitor, and a thiazide diuretic at the same time enhances the possibility of kidney failure. This means the fixed-combination goods can use more than one drug type. Increased serum creatinine regulation, particularly at the start of the combination, should be accompanied by the use of both drugs at the same time. In elderly or preexisting renal disorder patients, the combination of pharmaceutical drugs in all three categories should be used with care.

- Lithium/Digoxin: When applied along with formulations containing lithium or digoxin, Diclofenac can raise its plasma concentrations, and these levels should be monitored during GN-DICLOFENAC treatment.
- Diuretics: Different NSAIDs are liable to inhibit diuretic activity. Like other NSAIDs, coadministration of Diclofenac with diuretics or antihypertensive agents can contribute to a reduction in antihypertensive activity (ACE). Therefore, the mixture should be approached with caution, and blood pressure must be regularly checked in patients, particularly aged sufferers. Patients should be adequately hydrated and attention should be given to evaluation of renal function following initiation of concomitant therapy and frequently afterwards, in particular for diuretics, ACE inhibitors, owing to the elevated risk of nephrotoxicity. Elevated blood potassium levels can be related to potassium-sparing diuretics with concomitant treatment, making it necessary to watch the latter.
- Potent CYP2C9 Inhibitors: Caution is recommended when coprescribing Diclofenac with active inhibitors of CYP2C9, which leads to inhibition of Diclofenac synthesis, can result in a significant rise in peak plasma concentrations and in susceptibility to diclofenac. Plasma Diclofenac levels can be raised with the concurrent treatment of voriconazole with diclofenac.
- Other NSAIDs: Some anti-inflammatory and corticosteroid nonsteroidal medicines. Coadministration of Diclofenac with systemic NSAIDs, like selective COX-2 inhibitors, should be evaded owing to data deficiency suggesting synergistic advantages and the potential for adverse additive effects. The prevalence of the side effects can increase with the attendant administration of systemic NSAIDs.

Concurrent aspirin therapy decreases plasma concentration, plasma peak levels, and Diclofenac AUC values. It is not advisable to use both medications simultaneously.

- Warfarin: Given the connected administration can augment the bleeding risk, caution is advised. Instantaneous NSAIDs administration and warfarin has been linked with serious, often fatal, bleeding. The exact cause of the relationship between NSAIDs and warfarin is unclear, but may include increased bleeding from gastrointestinal ulceration caused by NSAIDs or an additive effect of warfarin anti-coagulation and NSAID inhibition of platelet activity. In conjunction with warfarin, DICLOFENAC-GA should be used with caution, and certain patients should be watched closely.

9.11.6 ADVERSE EFFECTS

- Gastrointestinal: ~20%: epigastric pain, other gastrointestinal disorders, such as vomiting, nausea, abdominal cramps, diarrhea, flatulence, dyspepsia, and anorexia. <2%: hematemesis, gastrointestinal bleeding, peptic ulcer with bleeding, melaena, bloody diarrhea, and perforation. In isolated cases: lower gut disorders such as Crohn's proctocolitis or exacerbation of ulcerative colitis and nonspecific hemorrhagic colitis, glossitis, aphthous stomatitis, constipation, esophageal lesions, pancreatitis, and diaphragm-like intestinal strictures.
- Dermatological: > 1%: rashes or skin eruptions. < 1%: urticaria. In isolated cases: eczema, bullous eruptions, Stevens-Johnson syndrome, erythema multiform, erythroderma (exfoliative dermatitis), Lyell's syndrome (acute toxic epidermolysis), photosensitivity reaction, loss of hair, pruritus, and allergic purpura.
- Renal: <1%: edema. In isolated cases: acute renal failure, urinary abnormalities, such as hematuria, proteinuria, interstitial nephritis, nephrotic syndrome, papillary necrosis.
- Hepatic: ~2%: elevation of serum aminotransferase enzymes (ALT, AST). <1%: hepatitis with jaundice. In isolated cases: hepatic necrosis, fulminant hepatitis, and hepatic failure.
- CNS: >1%: dizziness, headache, and vertigo. <1%: drowsiness. In isolated cases: disturbances of sensation, including memory disturbance, paraesthesia, disturbances of vision (diplopia and blurred vision), disorientation, tinnitus, impaired hearing, irritability, insomnia, depression, convulsions, nightmares, anxiety, psychotic reactions,

tremor, cerebrovascular accident, taste alteration disorders, aseptic meningitis, and myoclonic encephalopathy.

- Hematological: In isolated cases: leucopenia, thrombocytopenia, agranulocytosis, anemia (aplastic anemia and hemolytic anemia), and positive Coombs' test.
- Hematological Isolated Cases: Thrombocytopenia, leucopenia, anemia (hemolytic anemia, aplastic anemia), agranulocytosis, positive Coombs' test.
- Hypersensitivity: Less than 1%: bronchospasm; anaphylactic/anaphylactoid systemic reactions, including hypotension. Other. In isolated cases: impotence (it is unlikely to be associated with diclofenac), palpitation, chest pain, hypertension, cardiac disease, myocardial infarction, congestive heart failure, asthma. In patients administered NSAIDs postoperatively, toxic shock syndrome has been identified.

9.11.7 DOSAGE AND ADMINISTRATION

The recommended starting dose is 75–150 mg/day. Long-term therapy usually needs 75–100 mg a day. As a general rule, the normal dosage can be split into two or three different doses. For primary dysmenorrhea, the regular dosage is 50–150 mg, which can be adjusted separately. A starting dose of 50–100 mg should be recommended, with the dose progressively raised to a maximum of 200 mg/day over many menstrual cycles if desired. Depending on the symptomatology, treatment begins as soon as the first signs appear and last for a few days.

9.12 DICLOFENAC POTASSIUM[12]

Diclofenac potassium is a derivative of benzeneacetic acid. Diclofenac potassium is available for oral administration as immediate release tablets of 50 mg (light brown). 2-[(2,6-dichlorophenyl)amino] benzeneacetic acid, monopotassium salt, is the chemical name. 334.25 is the molecular weight.

9.12.1 CLINICAL PHARMACOLOGY

Diclofenac is an NSAID used to explain analgesic, anti-inflammatory, and antipyretic actions in mouse models. The mechanism of diclofenac, including that of other NSAIDs, is not very well but may be linked to prostaglandin synthetase repression.

9.12.2 PHARMACOKINETICS

- After oral administration, Diclofenac is 100% consumed relative to IV administration as calculated by urinary recovery. Due to first-pass digestion, nevertheless, only about 50% of the absorbed dose is systemically accessible. In certain fasting participants, detectable plasma levels are discovered within 10 min of ingesting with Diclofenac potassium immediate release capsules. Max plasma levels were observed for around 1 h in normal fasting volunteers, ranging from 0.33 to 2 h. Food should not have a significant effect on Diclofenac absorption levels. However, there is typically a pause in the start of absorption and a drop in peak plasma levels of around 30%.
- The apparent delivery volume of Diclofenac potassium (V/F) is 1.3 L/kg. Diclofenac, mainly albumin, is more than 99% bound to human serum proteins. The binding of serum protein over the concentration range 0.15–105 μg/mL obtained at the prescribed doses is stable. In and out of the synovial fluid, Diclofenac diffuses. Diffusion through the joint occurs when serum concentrations are larger than those from the synovial fluid, after which the process reverses and synovial fluid concentrations are greater than the plasma concentrations. It is not known if entry into the joint plays a part in the effectiveness of diclofenac.
- Five metabolites of Diclofenac were identified in human plasma and urine which contains the metabolites of 4',5-dihydroxy-, 5-hydroxy-, 4'-hydroxy-, 3'-hydroxy-, and 3'-diclofenac hydroxy-4'-methoxy. Peak concentrations of 4'-hydroxy- and 5-hydroxy-diclofenac metabolites were 50% and 4% of the parent compound in patients with renal dysfunction after single oral delivery, respectively, compared with 27% and 1% in normal stable subjects. Diclofenac metabolites, though, undergo more glucuronidation, accompanied by biliary excretion and sulfation. One 4'-hydroxy-diclofenac metabolite has very poor pharmacologic activity.
- Through absorption and subsequent urinary and biliary excretion, Diclofenac is removed by sulfate conjugates and glucuronide metabolites. There is little or no excretion of free, unchanged Diclofenac in

the urine. Around 65% of the dosage is excreted as conjugates of unchanged Diclofenac and metabolites in the urine and about 35% in the bile. Because renal removal is not a major elimination mechanism for unchanged Diclofenac, no dosing improvement is needed in patients with mild-to-moderate renal dysfunction. Unchanged Diclofenac has a terminal half-life of about 2 h.

9.12.3 INDICATIONS AND USAGE

Before choosing to use Diclofenac potassium immediate-release tablets, thoroughly examine the possible advantages and disadvantages of Diclofenac potassium immediate-release (IR) tablets and other treatment choices. Using the lowest appropriate dosage for the shortest duration, make sure it is compatible with the treatment objectives of the particular patient.

9.12.4 CONTRAINDICATIONS

In patients with known Diclofenac hypersensitivity, IR tablets of Diclofenac potassium are contraindicated. Patients who have developed allergies, hives, or allergic responses after taking aspirin or other NSAIDs should not be given Diclofenac potassium IR pills. In the sense of coronary artery bypass graft (CABG) surgery, Diclofenac potassium IR tablets are contraindicated for the management of perioperative pain.

Hypertension: NSAIDs, including IR tablets of Diclofenac potassium, can lead to the onset of new hypertension or to a worsening of preexisting hypertension, all of which can relate to an increased occurrence of CV. When taking NSAIDs, patients taking thiazide or loop diuretics can have an impaired response to these therapies. In patients with hypertension, NSAIDs, including Diclofenac potassium IR capsules, should be taken with caution. During the start of NSAID care and throughout the duration of rehabilitation, blood pressure (BP) should be monitored closely.

Congestive Heart Failure and Edema: Any patients taking NSAIDs have been found to have fluid accumulation and edema. In patients with fluid accumulation or cardiac disease, Diclofenac potassium IR tablets should be used with caution.

Gastrointestinal Effects: Significant gastrointestinal (GI) adverse events, including inflammation, swelling, ulceration, and perforation of the liver, small intestine, or large intestine may be caused by NSAIDs, including Diclofenac potassium IR capsules, which may be lethal. These extreme adverse effects can occur in patients treated with NSAIDs at any time, with or without warning signs.

Renal Effects: Renal papillary necrosis and other renal injuries have been caused by long-term administration of NSAIDs. In patients in which renal prostaglandins play a compensatory role in maintaining renal perfusion, renal toxicity has also been seen. The administration of NSAIDs in these patients may induce a dose-dependent decrease in the formation of prostaglandin, and secondarily, in the flow of renal blood, which may precipitate overt renal de-compensation. Patients with reduced kidney function, cardiac disease, liver disease, people taking diuretics and ACE inhibitors, and the elderly are at highest risk for this reaction. Recovery to the pretreatment condition is typically accompanied by discontinuation of nonsteroidal anti-inflammatory drug therapy.

Hepatic Effects: Elevations of one or more liver tests during treatment with IR tablets of Diclofenac potassium can occur. Transaminase elevations were more commonly found in patients with osteoarthritis than in patients with rheumatoid arthritis. In patients undergoing long-term treatment with diclofenac, doctors can regularly evaluate transaminases because extreme hepatotoxicity can occur without distinguishing symptoms from a prodrome. The best times are not specified for making the first and subsequent tests of transaminase. However, during therapy with diclofenac, serious liver reactions may occur at any time.

Anaphylactoid Reactions: As with other NSAIDs, in patients without documented prior exposure to Diclofenac potassium IR tablets, anaphylactoid reactions can occur. Patients with aspirin triads should not be given Diclofenac potassium IR capsules. Typically, this symptom complex exists in asthmatic patients who develop rhinitis with or without nasal polyps or who, after consuming aspirin or other NSAIDs, show severe, possibly fatal bronchospasm.

Skin Reactions: Significant skin adverse events, such as exfoliative dermatitis, Stevens-Johnson Syndrome (SJS), and severe epidermal necrolysis (TEN), which can be lethal, can be caused by NSAIDs, including Diclofenac potassium IR pills. Without warning, these extreme events can occur. Patients should be advised of the signs and effects of severe skin symptoms, and the medication should be stopped when skin rash or any other signs of hypersensitivity first appear.

Pregnancy: As with other NSAIDs, Diclofenac potassium IR tablets should be prevented in late pregnancy because it may cause the ductus arteriosus to close prematurely.

9.12.5 PRECAUTIONS

It is not predicted that Diclofenac potassium IR tablets will replace corticosteroids or treat corticosteroid insufficiency. Abrupt corticosteroid discontinuation can contribute to an exacerbation of the disease. When a recommendation to discontinue corticosteroids is taken, patients on extended corticosteroid treatment should have their therapy progressively reduced. Diclofenac's pharmacological role in reducing fever and inflammation may decrease the effectiveness of these diagnostic signs in recognizing complications of presumed painful, noninfectious conditions.

9.12.6 DRUG INTERACTIONS

- When IR tablets of Diclofenac potassium are administered with aspirin, its protein binding is reduced. The clinical importance of this relationship is not clear, but coadministration of Diclofenac and aspirin is usually not advised, as with other NSAIDs, due to the potential for increased adverse effects.
- It has been documented that NSAIDs competitively inhibit the aggregation of methotrexate in rabbit kidney slices. This could mean that the toxicity of methotrexate may be increased. When NSAIDs are treated simultaneously with methotrexate, caution should be used.
- Like other NSAIDs, Diclofenac potassium IR tablets can affect renal prostaglandins and increase the toxicity of some medications. Concomitant treatment with Diclofenac Potassium IR tablets can also improve the nephrotoxicity of cyclosporine. Caution can be used when IR tablets of Diclofenac potassium are given simultaneously with cyclosporine.
- NSAIDs can decrease the antihypertensive effect of ACE inhibitors, studies say. In patients taking NSAIDs simultaneously with ACE inhibitors, this relationship should be taken into account.

- An elevation of plasma lithium levels and a decrease in renal lithium clearance have been produced by NSAIDs. The mean minimum concentration of lithium rose by 15% and renal clearance decreased by around 20%. These findings have been due to NSAID suppression of the synthesis of renal prostaglandin. Thus, subjects should be closely monitored for symptoms of lithium toxicity when NSAIDs and lithium are administered simultaneously.
- The effects of warfarin and NSAIDs on GI bleeding are synergistic, so that the risk of severe GI bleeding is greater for users of both medications than for users of either drug alone.

9.12.7 DOSAGE AND ADMINISTRATION

- The prescribed dosage for the treatment of pain or primary dysmenorrhea is 50 mg t.i.d. With experience, doctors can find that an initial 100 mg dose of Diclofenac potassium IR pills, supplemented by 50 mg doses, offers greater relief in certain patients.
- The suggested dosage for the relief of osteoarthritis is 100–150 mg/day in divided doses, 50 mg b.i.d. or t.i.d. The recommended dosage for the relief of rheumatoid arthritis is 150–200 mg/day in divided doses, 50 mg t.i.d. or q.i.d.
- Different Diclofenac formulations, such as Diclofenac sodium enteric-coated tablets, extended release tablets of Diclofenac sodium, or instant-release tablets of Diclofenac potassium, are not inherently bioequivalent, even though the strength of the milligram is the same.

9.13 DICLOFENAC SODIUM/MISOPROSTOL[13]

For inflammation and swelling, Diclofenac sodium—misoprostol is used to manage rheumatoid arthritis and osteoarthritis. This medication can be sold in a variety of forms and/or under a number of distinct brand names. Any specific brand name of this medication might not be accessible in any of the forms or approved for any of the conditions specified here. Certain forms of this medication are often ineffective for any of the diseases mentioned here. The doctor may have recommended this medication for conditions other than those mentioned in these drug awareness documents. If you have not told your doctor or are not positive if you are on this prescription, speak to your doctor. Do not stop taking this medication without first seeing the doctor, and if they have the same side effects as you have, do not give this prescription to someone else. Individuals do not take this medication until a doctor has advised them to do so.

9.13.1 PHARMACOLOGICAL EFFECTS

Diclofenac, an NSAID, decreases inflammation, discomfort, and fever, possibly by inhibiting COX production and prostaglandin synthesis. Misoprostol, a prostaglandin E1 analog, has antisecretory and mucosal protective properties in the stomach.

9.13.2 CONTRAINDICATIONS

Pregnancy, aspirin, or other NSAID allergy, susceptibility to misoprostol or some other prostaglandin, history of asthma, urticaria, or other allergic responses when taking aspirin or another NSAID, perioperative CABG pain control.

9.13.3 DOSAGE AND ADMINISTRATION

Osteoarthritis (OA)—Adults: Three times regular at 50 mg/200 mcg. 50/200 or 75 mg/200 mcg twice daily can be used by patients with intolerance, but it is less effective at reducing ulcers. Dosages more than 150 mg/day of Diclofenac and 800 mcg/day of misoprostol for osteoarthritis are not recommended.

Rheumatoid Arthritis (RA)—Adults: Three to four times per day, 50 mg/200 mcg. 50/200 or 75 mg/200 mcg twice daily can be used by patients with intolerance, but it is less effective at reducing ulcers. Diclofenac concentrations greater than 225 mg/day and 800 mcg/day misoprostol in RA are not prescribed.

9.13.3 DRUG INTERACTIONS

ACE Inhibitors: Diclofenac can reduce the antihypertensive effect. In addition, it could be possible to raise the chance of nephrotoxicity. Near monitoring of BP. Try halting diclofenac/misoprostol if BP regulation deteriorates. Periodic measurement is warranted for renal activity and potassium concentrations.

Aminoglycosides (e.g., amikacin, gentamicin, and tobramycin): In premature children, aminoglycoside plasma concentrations can be elevated because Diclofenac may decrease the rate of glomerular filtration. If this mixture cannot be prevented, decrease the dosage of aminoglycosides prior to beginning diclofenac/misoprostol or discontinue diclofenac/misoprostol prior to continuing treatment with aminoglycosides. Track the accumulation and renal activity of aminoglycosides.

Antacids: Misoprostol's bioavailability may be decreased, and the absorption of Diclofenac may be delayed. Antacids that contain magnesium raise the risk of diarrhea associated with misoprostol.

Aspirin: It could be necessary to raise the risk of severe GI complications including bleeding. Moreover, it could be possible to reduce the cardioprotective function of aspirin. Diclofenac and aspirin are not commonly recommended to be used together.

Azole Antifungals (e.g., fluconazole, voriconazole): Plasma amounts of Diclofenac can be increased, thus increasing the pharmacological consequences and chances of adverse reactions. Track the therapeutic reaction, and if necessary, change the diclofenac/misoprostol dosage.

Bile Acid Sequestrants (e.g., cholestyramine, colestipol): There may be a reduction in the symptoms of diclofenac. Consider raising the dosage of diclofenac/misoprostol during cholestyramine administration if an interaction is suspected.

Bisphosphonates (e.g., alendronate): It is possible to increase the chance of gastric ulceration. Apply with caution. Check for signs of adverse reactions to GI, in particular gastric ulceration.

Cyclosporine: The risk of cyclosporine toxicity can be increased, including the risk of nephrotoxicity. Track the renal activity and amounts of cyclosporin closely. Consider reducing the cyclosporine dose or halting diclofenac/misoprostol if an interaction is suspected.

Digoxin: Diclofenac may increase digoxin levels, resulting in an increased risk of toxicity. Control for potential toxicity of digoxins.

Diuretics (e.g., loop diuretics [e.g., furosemide], potassium-sparing diuretics [e.g., triamterene], thiazide diuretics [e.g., chlorothiazide]): There may be a decrease in the natriuretic activity of furosemide and thiazide diuretics. Serum potassium levels can be improved by the coadministration of potassium-sparing diuretics.

Heparins (e.g., dalteparin, enoxaparin, heparin): It could be necessary to increase the risk of hemorrhagic adverse reactions. Near clinical and laboratory control is warranted if combined use cannot be prevented.

Hypoglycemic Agents, Oral: There could be both hypo- and hyperglycemia, necessitating a shift in the dosage of the hypoglycemic agent. Blood glucose control and changing the oral hypoglycemic dosage as required.

Lithium: The levels of lithium could be elevated, thereby increasing the possibility of toxicity. Control for lithium toxicity symptoms: Additional control of lithium plasma concentration is justified. As required, change the lithium dosage.

9.13.4 ADVERSE REACTIONS

Cardiovascular: Arrhythmia, atrial fibrillation, CHF, hypertension, hypotension, MI, palpitations, phlebitis, premature ventricular contractions, syncope, tachycardia, vasculitis.

CNS: Coma, convulsions, depression, hallucinations, meningitis, psychotic reaction.

Dermatologic: Erythema multiforme, exfoliative dermatitis, pemphigoid reaction, photosensitivity, skin ulceration, Stevens-Johnson syndrome, TEN.

GI: Esophageal ulceration, GI bleeding, GI neoplasm benign, hematemesis, intestinal perforation, peptic ulcer, rectal bleeding.

Genitourinary: Hematuria, interstitial nephritis, nephrotic syndrome, papillary necrosis, proteinuria, renal failure, vaginal hemorrhage.

Hematologic–Lymphatic: Agranulocytosis, aplastic anemia, eosinophilia, hemolyticanemia, leukocytosis, leukopenia, lymphadenopathy, pancytopenia, thrombocythemia, thrombocytopenia.

Hepatic: Abnormal hepatic function, hepatitis, jaundice, liver failure, pancreatitis.

9.13.5 PRECAUTIONS

Pregnancy: In females who are pregnant, misoprostol can induce miscarriage, premature birth, and birth defects. Uterine rupture was recorded after the eighth week of pregnancy when misoprostol was offered to pregnant women to cause labor or abortion. Ensure that people with childbearing potential should not use misoprostol because the woman needs NSAID treatment and is at elevated risk of developing gastric ulcers associated with the use of NSAID. These patients must have a negative serum paternity test within 2 weeks prior to beginning therapy, be able to comply with successful contraceptive measures, have both oral and written reminders about the risk of misoprostol, the risk of contraception failure, and the risk of childbearing capacity for other women if they accidentally take the drug, and commence the medication only on the second or third day of pregnancy failure.

CV Risk: An elevated risk of severe, potentially lethal, CV thrombotic incidents, MI, and stroke may be caused by NSAIDs. The period of therapy will raise this risk. There may be a higher risk for patients with CV infection or risk factors for CV disease. Diclofenac/misoprostol is contraindicated in the form of CABG surgery for the treatment of perioperative pain.

GI Risk: NSAIDs cause an elevated risk of severe, possibly lethal, GI adverse reactions, including swelling, ulceration, and stomach or intestinal perforation. These reactions can occur at any moment, with or without warning signs, during use. Elderly patients have a higher chance of witnessing severe GI accidents.

9.13.6 PHARMACOLOGICAL USES

- Diclofenac is used to alleviate arthritis-caused discomfort, swelling (inflammation) and joint stiffness. It makes you perform more of your normal everyday tasks by reducing these symptoms. Misoprostol prevents the liver from the irritating symptoms of diclofenac. In patients at high risk of contracting stomach ulcers and severe risks such as bleeding, this combination drug is used to treat arthritis.
- Other Uses: To treat gout attacks, this drug can also be used. To avoid stomach pain and to decrease the chances of diarrhea, take this drug with meals.

9.14 DIFLUNISAL[14]

This NSAID Diflunisal has been connected to an elevated risk of significant and frequently lethal cardiac and blood vessel problems (e.g., heart attack, stroke). If cardiac problems remain or if diflunisal care is prolonged, the risk can increase. Diflunisal cannot be used before or during bypass heart surgery. Diflunisal is used to manage osteoarthritis, rheumatoid arthritis, and moderate-to-extreme discomfort.

9.14.1 IMPORTANT INFORMATION

- This medication displayed undesirable responses during breastfeeding, pregnancy, or breastfeeding, either pharmaceutical or nonprescription medicine, herbal products or nutritional supplements, allergies to medications, diets or other ingredients, history of kidney or liver disease, diabetes or complications with the stomach or intestines (e.g., inflammation, perforation, ulcers), history of swelling or fluid retention, asthma, and so on.
- Diflunisal can give rise to dizziness or sleepiness. With alcohol or similar medications, these symptoms can be greater. Serious stomach ulcers or bleeding can occur with diflunisal administration. Taking it in high doses, smoking, or consuming alcohol for a long period raises the risk of these side effects. The risk of these consequences will not be decreased when diflunisal is taken with food.
- Do not take more than the prescribed dose without consulting or using a doctor for longer than is required.
- Aspirin has been associated with Reye's syndrome, a severe disease. Do not send diflunisal to an infant or teenager who has a viral fever, chickenpox, or flu.
- Lab tests can be conducted when you use diflunisal, including kidney function, full blood cell counts, and blood pressure. It is possible to use these procedures to track the health or to scan for side effects.
- Pregnancy and breastfeeding: The fetus may be affected by diflunisal. During the last 3 months of pregnancy, do not use it. Do not breastfeed during the therapy of diflunisal.

9.14.2 PHARMACOLOGICAL EFFECTS

- Difflunisal works like other NSAIDs by inhibiting the development of prostaglandins, hormones that are implicated in pain and inflammation. Diflunisal has an antipyretic effect as well, but the medication is not recommended for use. While diflunisal has an onset time of 1 h and maximal analgesia is 2–3 h, once repeated doses are taken, the plasma levels of diflunisal will not be stable. A distinguishing characteristic of diflunisal is the long plasma half-life relative to related products. Usually, a loading dosage is used to maximize the rate at which the levels of diflunisal plasma become stable. It is used mainly for the treatment of arthritis symptoms and for acute pain after oral surgery, especially for the removal of wisdom teeth.

- The effectiveness of diflunisal is close to that of other NSAIDs, but has a duration of operation of 12 h or more. For chronic administration, this means less doses a day are needed. In acute use, where a single dose after oral surgery can sustain analgesia until the patient is asleep that night, it is common in dentistry.

9.14.3 DRUG INTERACTION

- Aspirin, Warfarin, Heparin, Prednisone, Fluoxetine may enhance the risk of stomach bleeding.
- Antacids, other NSAIDs (e.g., ibuprofen), or probenecid because they may increase the risk of diflunisal's side effects.
- Cyclosporine, acetaminophen, methotrexate, lithium, quinolones, sulfonylureas enhance the side effects possibilities which may be augmented by diflunisal.
- ACE inhibitors (e.g., enalapril), angiotensin receptor blockers (e.g., losartan), or diuretics (e.g., furosemide, hydrochlorothiazide) because their effectiveness may be decreased by diflunisal.

9.14.4 PHARMACOLOGICAL USES

- Pain, mild-to-moderate
- Osteoarthritis
- Rheumatoid arthritis
- Injury to tendons
- Inflammation
- ATTR amyloidosis

9.14.5 DOSAGE ADMINISTRATION

- An alternative medical report sheet called a Drug Guide comes with diflunisal.
- Diflunisal is taken by mouth. If it upsets the stomach, it can be used with food. Taking it with food does not minimize the possibility of complications with the stomach or bowels (e.g., bleeding, ulcers).

9.15 ETODOLAC[15]

Etodolac is a NSAID that works by reducing inflammation and pain-causing hormones in the body. NSAIDs are used to relieve pain and inflammatory effects. Etodolac is a drug that is used to relieve pain and irritation caused by arthritis or osteoarthritis.

9.15.1 IMPORTANT INFORMATION

- Etodolac may increase the risk of cardiac or circulation complications that are life-threatening like heart disease or stroke. This risk grows the more you use Etodolac. This medicine should not be taken immediately before or after heart bypass surgery (also called coronary artery bypass graft, or CABG).
- Etodolac can also increase the risk of major bleeding or perforation of the intestines or stomach (forming of an orifice). This disorders may be lethal, and without warning at any moment, gastrointestinal symptoms can arise. The risk of these severe gastrointestinal side effects can be much higher in older adults. Do not take any such over-the-counter drug for colds, asthma, or discomfort without first asking the doctor or pharmacist. Many over-the-counter drugs contain aspirin or other Etodolac-like medicines (such as ibuprofen, ketoprofen, or naproxen). Alcohol can increase the risk of Etodolac-induced stomach bleeding. Avoid sunlight radiation or artificial UV rays (sunlamps or tanning beds).

9.15.2 PHARMACOLOGICAL ACTIONS

- For the treatment of mild-to-severe discomfort, fever, and inflammation, NSAIDs are used. They function by reducing the levels of prostaglandins, which are chemicals responsible for pain and inflammatory fever and tenderness. Etodolac inhibits the prostanoid-forming COX enzymes, resulting in lower prostaglandin concentrations. Inflammation, inflammation and fever are decreased as a result.
- Postmarketing trials have shown that Etodolac COX inhibition is very COX-2 selective, comparable to celecoxib, and other "COX-2 inhibitors." In comparison to rofecoxib, Etodolac and celecoxib can fully inhibit COX-1 and are designated to have "preferential selectivity"

against COX-2. The Etodolac (inactive against COX) r-enantiomer reduces levels of beta-catenin in hepatoma cells.

9.15.3 INTERACTIONS

Etodolac can be avoided in patients who have had asthma episodes, hives, or any adverse reactions to aspirin or other NSAIDs. Rare, however, severe allergic reactions have been observed in these persons. Patients with peptic ulcer disorder or compromised kidney function should stop taking this treatment, since it will cause either of these problems worse. Etodolac is used with care in patients who are taking blood thinners (anticoagulants), such as warfarin (Coumadin), and it improves the possibility of bleeding. Patients who take both lithium and Etodolac can experience toxic blood lithium levels. In contrast, certain antidepressant medications, such as sertraline or fluoxetine have been found to interact with Etodolac, thereby increasing the likelihood of stroke, cardiac failure, and other cardiovascular diseases. Patients taking cyclosporine can suffer kidney toxicity (Sandimmune). It has not been thoroughly studied in children. Etodolac is a nonaddictive drug. NSAIDs should be stopped before elective surgery due to a minor clotting interaction that is common in this class of medications. It is best to avoid taking Etodolac at least 4 days before your appointment. Persons who consume more than three alcoholic beverages a day while taking Etodolac or other NSAIDs are at an increased risk of developing stomach ulcers.

9.15.4 SIDE EFFECTS

- Side effects or adverse drug reactions can result from Etodolac. If all of these conditions are serious or prolonged, it is advisable to call a physician: constipation, diarrhea, gas, or bloating, vomiting, fatigue, dizziness, ringing in the ears, runny nose, sore throat, distorted vision.
- It can be serious with certain side effects. It is recommended to call a doctor immediately if any of the above conditions arise and to discontinue use at that time.
- An unexplained rise of weight, swelling of the eyes, ears, mouth, neck, neck, hands, feet, ankles or lower legs, fever or chills, blisters, vomiting, scratching, diarrhea, flushing, trouble breathing or swallowing, yellowing of the skin or eyes, extreme tiredness, irregular orbruising, lack of energy, loss of appetite, discomfort in the upper

right part of the stomach, flu-like symptoms, pale skin, excessive tiredness, bruising, lack of nutrition, loss of appetite, flu-like symptoms.

9.16 FENOPROFEN[16]

Fenoprofen refers to a class of drugs known as NSAIDs. It acts by reducing inflammation and pain-causing hormones in the body. Fenoprofen is a drug that is used to treat arthritis-related discomfort and inflammation. This medication can also raise the risk of serious gastrointestinal or digestive symptoms, such as bleeding or perforation (forming of a hole). These conditions may be fatal, and gastrointestinal signs may appear at any time without notice. Taking an NSAID increases the likelihood of life-threatening cardiac or circulation risks, such as heart failure or stroke. This chance will rise if an NSAID is used more often. It decreases inflammation, discomfort, and fever by inhibiting COX and releasing prostaglandin.

9.16.1 DRUG INTERACTION

- Aminoglycosides (e.g., gentamicin): The levels of plasma aminoglycosides can be elevated.
- ACE inhibitors: The antihypertensive activity of inhibitors of ACE can be minimized.
- Anticoagulants: Coadministration can prolong the lifespan of prothrombin.
- Aspirin: Fenoprofen can be increased. it is not approved for coadministration.
- Diuretics: Fenoprofen-treated patients may be immune to the effects of thiazide and loop diuretics.
- Hydantoins, sulfonamides, sulfonylureas: These drugs can be displaced from their binding position by Fenoprofen.
- Lithium: Lithium renal Cl may be decreased and plasma concentrations may be increased, which may increase the likelihood of lithium toxicity.
- Methotrexate: The levels of methotrexate can increase.
- Phenobarbital: $T_{1/2}$ of Fenoprofen can decrease. Fenoprofen dose modifications may be needed if phenobarbital is inserted or omitted.
- SSRIs (e.g., fluoxetine, citalopram): It could be necessary to increase the likelihood of GI effects.

9.16.2 DOSAGE ADMINISTRATION

Do not take whether you have severe kidney disorder, a stomach ulcer or inflammatory bowel disease, or a history of allergic responses to aspirin or other NSAIDs, pulmonary disease, stomach ulcer or inflammatory bowel disease, history of stomach ulcers or bleeding, stomach problems, diverticulosis, liver or kidney disease, asthma, nasal polyps, or a history of heart arrest, stroke, or blood clotting, history of pregnancy is graded as division C. It is uncertain if Fenoprofen is detrimental to an unborn child. Fenoprofen may pass into breast milk, putting a breastfeeding kid at risk. Also, do not provide this drug to a kid under the age of 14 without consulting a specialist. The recommended regular dosage of Fenoprofen for adults is 3 g (3000 mg).

9.17 FLURBIPROFEN[17]

Flurbiprofen functions by lowering hormones that trigger inflammation and discomfort in the body. Flurbiprofen is a drug that is used to treat arthritis-related pain and inflammation. This medication may often induce severe effects in the stomach or intestines, such as bleeding or perforation (forming of a hole). These conditions, which may be lethal, can occur suddenly when you take flurbiprofen, particularly in older adults. Flurbiprofen is an anti-inflammatory NSAID that acts to alleviate fever, pain, and inflammation in the body. Related drugs include ibuprofen (Motrin or Advil), naproxen (Naprosyn), and others. NSAIDs are nonnarcotic pain relievers used to manage a range of moderate to extreme pain problems, such as injury, menstrual cramps, arthritis, and other musculoskeletal disorders. They work by lowering the levels of prostaglandins, which are the chemicals that cause pain, fever, and inflammation. The enzyme that produces prostaglandins (COX) is inhibited by flurbiprofen, resulting in lower prostaglandin concentrations. As a consequence, inflammation, inflammation, and fever are reduced. Flurbiprofen was authorized by the FDA in October 1988.

If you have severe kidney disease or a history of aspirin or any NSAID allergies, you may not take this medication. If either of these other factors necessitate a modification of flurbiprofen dose or the usage of special tests: Stroke, heart attack, heart cancer, blood clotting, excessive blood pressure, congestive heart failure, stomach ulcer or bleeding, liver or kidney disease, asthma, nasal polyps, blood clotting deficiency or bleeding, or smoking. During breastfeeding, it belongs to category C. Taking flurbiprofen during the last 3 months of pregnancy will harm the unborn child. Flurbiprofen may pass into breast milk, causing problems for a breastfeeding infant. Do not offer this medicine to a boy without first finding a specialist.

9.17.1 PHARMACOLOGICAL EFFECTS

9.17.1.1 DRUG INTERACTION

In those taking blood-thinning drugs (anticoagulants) such as warfarin (Coumadin), flurbiprofen is usually treated with caution because of the potential risk of bleeding. Patients taking lithium (Lithobid, Eskalith) may accumulate toxic amounts of lithium in their blood because flurbiprofen can prevent the kidney's removal of lithium from the body. Flurbiprofen can also improve the side effects caused by methotrexate (Trexall, Rheumatrex) and cyclosporine. Since it induces or worsens high blood pressure, flurbiprofen can decrease the efficacy of medications used to treat high blood pressure. The blood pressure reducing effects of ACE inhibitors can be reduced by NSAIDs. Combining angiotensin receptor blockers (e.g., valsartan [Diovan], losartan [Cozaar], irbesartan [Avapro]) or angiotensin converting enzyme inhibitors (e.g., enalapril [Vasotec], captopril [Captoen]) with NSAIDs can result in diminished kidney tasks, such as the kidney failure, in patients who are elderly, volume-depleted (including those on diuretic therapy), or with impaired kidney function. Such results are characteristically reversible.

9.17.1.2 SIDE EFFECTS

With minimal side effects, most patients benefit from Flurbiprofen and other NSAIDs. Significant side effects may occur, however, and usually appear to be linked to the dosage. Therefore, to avoid side effects, it is also advisable to choose the lowest possible dosage. The gastrointestinal tract is implicated in the most frequent side effects of Flurbiprofen. It may lead to ulceration, burning of the belly, inflammation, cramping, nausea, gastritis, and even extreme gastrointestinal bleeding and toxicity of the liver. Often, without any abdominal pain, bleeding and stomach ulceration may occur. The only signs of internal bleeding may be fatigue, black tarry stools, and dizziness upon standing. There may also be a rash, renal failure, ringing in the ears, and light headedness. Patients with a history of asthma attacks, hives, or other adverse responses to aspirin or other NSAIDs should stop Flurbiprofen. In such people, uncommon but serious allergic reactions have been documented. Patients with ulcer disease or impaired kidney function should also stop it, since this drug can aggravate all conditions. Flurbiprofen can cause heart attacks and strokes, fluid accumulation and exacerbate heart disease, cause or worsen hypertension and kidney failure, much like other NSAIDS.

9.17.2 DOSAGE ADMINISTRATION

For adults, the recommended dose of Flurbiprofen is 400 mg a day. Store it away from moisture, sun, and light at room temperature. Nausea, vomiting, stomach pain, dizziness, drowsiness, dark or bloody stools, coughing up blood, urinating less than normal or not at all, shallow breathing, and fainting can be opioid signs.

9.17.3 PHARMACOLOGICAL USES

Flurbiprofen is used to alleviate discomfort, swelling, tenderness, and weakness caused by rheumatoid arthritis and osteoarthritis. Flurbiprofen belongs to a family of drugs referred to as NSAIDs. It functions by preventing a material that induces pain, fever, and inflammation from being produced by the body.

9.18 IBUPROFEN[18]

Ibuprofen (isobutylphenylpropanoic acid) is a propionic acid derivative of a NSAID that is used to relieve pain, reduce fever, and reduce inflammation. Approximately, 60% of people build a tolerance to each NSAID, so if one does not fit, it is suggested that you try another. It is possible that ibuprofen is a less effective anti-inflammatory than other NSAIDs. Ibuprofen acts by lowering hormones that cause inflammation and pain in the body. Ibuprofen is used to treat a variety of ailments, including headaches, toothaches, back pain, arthritis, menstrual cramps, and minor fractures, as well as to reduce fever. Overdosing on ibuprofen will affect your stomach and intestines. Smallest dose of ibuprofen is required to alleviate the headache, swelling, or fever.

9.18.1 PHARMACOLOGICAL EFFECTS

The precise mechanism of action of Ibuprofen is unclear. Ibuprofen is a nonselective inhibitor of COX, an enzyme implicated in the arachidonic acid pathway's prostaglandin synthesis. Its pharmacological effects are thought to be attributed to COX-2 inhibition, which lowers the output of prostaglandins involved in inflammation, discomfort, fever, and swelling. Antipyretic symptoms may be attributed to hypothalamic action, which causes enhanced blood pressure, vasodilation, and heat dissipation from the periphery. Some Ibuprofen side effects, such as stomach ulceration, are thought to be caused by COX-1 inhibition. Ibuprofen is provided as a racemic mixture. The R-enantiomer is commonly interconverted to the S-enantiomer in vivo. The S-enantiomer is considered to be the most pharmacologically involved.

Ibuprofen and other nonsteroidal anti-inflammatory drugs work by inhibiting the COX enzyme, which converts prostaglandin H2 to arachidonic acid (PGH2). Other enzymes convert PGH2 to a number of other prostaglandins (pain, inflammation, and fever mediators) as well as thromboxane A2 (which stimulates platelet aggregation, leading to the formation of blood clots).

Ibuprofen, like aspirin and indometacin, is a nonselective COX receptor since it prevents all COX-1 and COX-2. COX-2 inhibition is largely responsible for NSAIDs' analgesic, antipyretic, and anti-inflammatory effects, while COX-1 inhibition is responsible for gastrointestinal side effects. Individual COX isoforms' functions in the actions of NSAIDs on analgesic, anti-inflammatory, and gastric damage are unclear, and a number of compounds cause different degrees of analgesia and gastric damage.

9.18.2 SIDE EFFECTS

The side effects include nausea, dyspepsia, stomach ulceration/bleeding, increased liver enzymes, diarrhea, constipation, nose bleeding, fever, dizziness, acne, salt and fluid accumulation, and hypertension. Infrequent side effects include esophageal ulceration, cardiac disease, hyperkalemia, renal dysfunction, confusion, and bronchospasm. Asthma can be exacerbated by Ibuprofen, which can be fatal. Ibuprofen may be measured in blood, plasma, or serum to show the effect of the medication in an individual that has had an anaphylactic reaction, affirm a diagnosis of overdose in hospitalized patients, or aid in a medicolegal death inquiry. A nomogram relating to ibuprofen plasma concentrations, absorption time, and the risk of renal toxicity in patients with overdose has been published.

Chance of cardiovascular disease: Chronic usage of Ibuprofen, as well as many other NSAIDs, has been linked to an increased risk of hypertension and myocardial infarction (heart attack), particularly in those who take large doses on a regular basis. At a large regular dosage, Ibuprofen was found to significantly increase systolic blood pressure in aged, hypertensive patients treated with hydrochlorothiazide.

Ibuprofen has been attributed to the onset of bullous pemphigoid or pemphigoid-like blistering while taken with other NSAIDs. Ibuprofen, like other NSAIDs, has been shown to be a photosensitizing agent, but it is deemed a poor photosensitizing agent as compared with other 2-arylpropionic acid family members. Ibuprofen, like other NSAIDs, is a very unlikely source of Stevens-autoimmune Johnson's disease (SJS).

9.18.3 DOSAGE ADMINISTRATION

Overdosing on Ibuprofen will affect your stomach and intestines. The effective dosage of ibuprofen for adults is 800 mg/dose, or 3200 mg/day (four maximum doses). Usethe least dose of ibuprofen possible to alleviate the discomfort, swelling, or fever. Take Ibuprofen with food or milk to ease abdominal pain.

9.18.4 DRUG INTERACTION

Drinking alcohol when consuming Ibuprofen raises the possibility of gastrointestinal bleeding. "Ibuprofen can interact with the antiplatelet activity of low-dose aspirin, possibly making aspirin less efficient as used for cardioprotection and stroke prevention," according to the US FDA. Leave enough time between ibuprofen doses and IR aspirin to prevent this problem. The recommended interval between the ibuprofen and aspirin doses is determined by how long the ibuprofen was taken first. It will take 30 min or more if ibuprofen was taken after IR aspirin, and 8 h or more if ibuprofen was taken before IR aspirin. This pacing, though, is not indicated for enteric-coated aspirin. However, if ibuprofen is used only regularly without the recommended timing, the daily aspirin regimen has a slight reduction in cardioprotection and stroke protection.

9.18.5 PHARMACOLOGICAL USE

Ibuprofen is used to ease discomfort in a variety of conditions, including headaches, toothaches, menstrual cramps, stomach aches, and arthritis. It is often used to manage fevers and to relieve minor aches and pains induced by the common cold or flu. It works by avoiding the formation of those natural substances in your body that cause inflammation. This effect tends to minimize pain and heat, and it can even be used to manage gout attacks in chronic diseases including arthritis.

9.19 INDOMETHACIN[19]

Indometacin (INN and BAN) or indomethacin (AAN, USAN, and former BAN) is a NSAID that is often used to treat fever, stiffness, and swelling. It acts by stopping the emergence of these symptoms by suppressing the development of prostaglandins, the molecules that are thought to trigger them.

9.19.1 IMPORTANT INFORMATION

Indomethacin is used to treat a number of diseases that cause discomfort or inflammation, including asthma, gout, ankylosing spondylitis, bursitis, and tendinitis. It is in the FDA's pregnancy group C. This medication has the ability to harm an unborn child. Taking indomethacin during the last 3 months of pregnancy will harm the unborn child.

9.19.2 PHARMACOLOGICAL EFFECTS

Indomethacin is a NSAIAs that reduces fever, pain, and inflammation. Ibuprofen (Motrin) and naproxen are also pain relievers (Naprosyn, Aleve). Indomethacin works by inhibiting the synthesis of prostaglandins. Prostaglandins are chemicals produced by the body that cause fever and pain associated with inflammation. Indomethacin decreases prostaglandin levels by inhibiting prostaglandin-making enzymes (COX-1 and COX-2). As a result, fever, discomfort, and inflammation are reduced. Indomethacin is used in an extended release formulation. The FDA first authorized indomethacin in January 1965.

9.19.3 ADVERSE EFFECTS

Since indomethacin inhibits both COX-1 and COX-2, it prevents the creation of prostaglandins, which are required to maintain the mucous lining of the gastrointestinal tract in the stomach and intestines safe. However, indomethacin, like other nonselective COX inhibitors, can cause peptic ulcers. These ulcers may cause serious bleeding and/or perforation, requiring the patient to be admitted to the hospital. To reduce the incidence of peptic ulcers, indomethacin should be provided at the lowest dose necessary to achieve a therapeutic effect, normally 50–200 mg per day. It can always be taken for food. A preventive antiulcer medication is useful to nearly all patients (e.g., highly dosed antacids, ranitidine 150 mg at bedtime, or omeprazole 20 mg at bedtime). Any gastrointestinal effects, such as dyspepsia, heartburn, and mild diarrhea are less severe, and indomethacin is seldom stopped. Many NSAIDs, especially indomethacin, induce lithium retention by reducing lithium excretion by the kidneys. As a result, indomethacin patients have a significant risk of lithium toxicity. Lithium users can take NSAIDs like sulindac or aspirin, which are less risky (e.g., for the treatment of depression or bipolar disorder). Indomethacin also lowers plasma renin production and aldosterone levels while improving sodium and potassium retention. It also enhances the effects of vasopressin. Together these may lead to:

- edema (swelling due to fluid retention)
- hyperkalemia (high potassium levels)
- hypernatremia (high sodium levels)
- hypertension

The procedure may also induce serum creatinine elevation and more serious renal damage, such as acute renal impairment, persistent nephritis, and nephrotic syndrome. Edema and hyperkalemia are common side effects of these conditions. Furthermore, indomethacin (10–20%) may induce headaches, which are often accompanied by vertigo and dizziness, hearing loss, tinnitus, and blurred vision (with or without retinal damage). There have been reports of Parkinson's disease, epilepsy, and psychiatric disorders deteriorating, but most events are unjustified. There have been reports of life-threatening shock (including angioedema, sweating, intense hypotension and tachycardia, and acute bronchospasm), serious or fatal hepatitis, and severe bone marrow harm. Skin reactions and photosensitivity are also potential side effects. Because of its powerful antipyretic influence, indomethacin can obscure the clinical direction of serious infections. Psychosis and long-term usage have also been recorded. Today, indomethacin is a second-choice drug due to the frequency and severity of side effects, as well as the existence of well-tolerated alternatives. The usage in acute gout attacks and dysmenorrhea is well established since the duration of treatment with these conditions is just a few days, so serious adverse effects are unlikely.

9.19.4 PHARMACOLOGICAL USES

Hemicrania continua, ankylosing spondylitis, arthritic gout, Bartter syndrome, bursitis, cryoglobulinemia, dysmenorrhea (menstrual cramps), exertion headache, fever and pain associated with malignant diseases (especially tumor fever associated with liver involvement, lymphogranulomatosis), hemicrania continua, hypnic headache, juvenile arthritis, migraine are all clinical indications for indomethacin-reactive arthritis, renal colic (kidney stone pain), retinopathy of prematurity, rheumatoid arthritis, tendinitis, trigeminal autonomic cephalgias, retinopathy of prematurity, retinopathy of prematurity, retinopathy of prematurity, retinopathy of prematurity, retinopathy of prematurity, retinopathy of prematurity, headaches exacerbated by or arising from the Valsalva maneuver. Indomethacin was often used clinically to prolong early childbirth, decrease amniotic fluid in polyhydramnios, and close the patent ductus arteriosus. Indomethacin is a powerful drug with many severe side effects, and it cannot be used as an analgesic for minor aches and pains or fever. Rather than being an analgesic, the medication is better characterized as an anti-inflammatory. Indomethacin can interact with warfarin, triggering an increase in international normalized ratio ().

9.19.5 DOSAGE ADMINISTRATION

Stop UV rays from the sun or artificial UV rays (sunlamps or tanning beds). Sunburn may be caused by indomethacin and it makes the skin more sensitive to sunlight. Carry sunscreen (minimum SPF 15) and protective footwear if you may be standing in the sun. 25 and 50 mg tablets or pills, 50 and 100 mg suppositories, 75 mg modified-release capsules, injectable concentrate 50 mg syrup (25 mg/5 mL) for i.m. Injection, mist or cream, 0.5% by weight, 1% topical material containing patches are typical dose and strength dosages.

9.20 KETOPROFEN[20]

Ketoprofen, also known as (R,S)2-(3-benzoylphenyl)-propionic acid ($C_{16}H_{14}O_3$ molecular formula), is an analgesic and antipyretic NSAID propionic acid. It acts by stopping the body from developing prostaglandins.

9.20.1 IMPORTANT INFORMATION

Ketoprofen functions by lowering hormones that trigger inflammation and pain in the body. Ketoprofen is a medication that is used to alleviate arthritis-related discomfort and inflammation. It is also known to ease the suffering of menstruation. It is in the FDA's pregnancy group C. You should not take the medication in higher concentrations or for longer than recommended. Ketoprofen cannot be used to treat minor aches and pains.

9.20.2 PHARMACOLOGICAL EFFECTS

Ketoprofen is an NSAID that is taken orally. Some participants of this group include ibuprofen (Motrin), indomethacin (Indocin), naproxen (Aleve), and others. These drugs are used to treat moderate to extreme discomfort, fever, and inflammation. They act by lowering the levels of prostaglandins, which are the chemicals that cause discomfort, fever, and inflammation in the body. Ketoprofen destroys prostaglandins by inhibiting the enzyme that makes them (COX). As a consequence, inflammation, inflammation, and fever are reduced.

9.20.3 SIDE EFFECTS

NSAIDs reduce the blood's ability to clot, causing bleeding to speed up after an injury. Ketoprofen can cause gastrointestinal and intestinal bleeding in people who have ulcers. Gastrointestinal ulceration and intestinal bleeding can occur without any abdominal pain. Dark tarry stools (due to blood in the stool), fatigue, and dizziness upon standing may be the only signs of leakage (orthostatic hypotension). People who are allergic to other NSAIDs should stop taking ketoprofen. NSAIDs affect kidney function by reducing blood supply to the kidneys. Impairment is most likely to develop in patients with preexisting renal function impairment or congestive heart failure, so NSAIDs can be used with care in these patients.

9.20.4 DRUG INTERACTION

- Ketoprofen can increase the amount of lithium in the blood (Eskalith, Lithobid) by decreasing lithium excretion in the kidneys, which can contribute to lithium toxicity.
- Ketoprofen can decrease the effects of blood pressure-lowering medications on blood pressure. This is because prostaglandins play a part in blood pressure control.
- The blood levels of methotrexate or aminoglycoside can increase when NSAIDs are mixed with methotrexate (Rheumatrex, Trexall) or aminoglycosides (e.g., gentamicin), since their removal is decreased. This can contribute to further side effects of methotrexate or aminoglycosides.
- Ketoprofen should be stopped by people taking blood thinners or anticoagulants, such as warfarin (Coumadin), since ketoprofen often thins the blood, and unnecessary blood thinning can contribute to bleeding.
- The combination of NSAIDs such as ketoprofen with angiotensin receptor blockers (ARBs) (e.g., valsartan [Diovan], losartan [Cozaar], irbesartan [Avapro]) or ACE inhibitors (e.g., enalapril [Vasotec], captopril [Capoten]) in aged, volume-depleted (including those undergoing diuretic therapy) patients or patients with poor kidney function may lead to decreased renal function. Such results are typically reversible.
- When taking ketoprofen or other NSAIDs, persons who have more than three alcoholic drinks a day are at elevated risk of developing stomach ulcers.

9.20.5 PHARMACOLOGICAL USES

Ketoprofen is often used to relieve inflammatory discomfort triggered by acne or extreme toothaches that cause gum inflammation. The usage of topical Ketoprofen plasters is popular in the management of musculoskeletal pain. Plasters have been shown to offer fast and continuous distribution to the underlying tissues without dramatically increasing opioid concentrations in the blood as opposed to typical oral administration. Conjugation with glucuronic acid, hydroxylation of the benzoyl ring with CYP3A4 and CYP2C9, and removal of the keto role occur in the liver. Ketoprofen is used to reversibly inhibit COX-1 and COX-2 enzymes thus reducing the production of pro-inflammatory prostaglandin precursors due to its antipyretic, analgesic, and anti-inflammatory impact. Ketoprofen, in the form of cream, ointment, liquid, mist, or gel may be used to treat some discomfort, including nerve pain such as sciatica, postherpetic neuralgia, which radiculopathy pain, and can also include ketamine and lidocaine, as well as other pharmaceutical drugs that may be helpful, such as cyclobenzaprine, amitriptyline, acyclovir, gabapentin, or other medicinal products.

9.20.6 DOSAGE AND ADMINISTRATION

- The normal starting dosage of Ketoprofen is 50 or 75 mg/6–8 h with immediate release capsules, or 200 mg once daily with extended release capsules.
- The daily dosage is 300 mg of quick release capsules per day, or 200 mg of extended release capsules per day.
- In order to prevent stomach pain, Ketoprofen should be taken with food.
- Menstrual cramps are treated every 6–8 h with 25–50 mg using quick release capsules.
- Rheumatoid or osteoarthritis is treated with either 75 mg 3 times daily or 50 mg 4 times daily with either an immediate release capsule or an extended release capsule of 200 mg daily.

9.21 MAGNESIUM SALICYLATE[21]

9.21.1 IMPORTANT INFORMATION

This drug acts by lowering the amounts of substances in the body that trigger discomfort, fever, and inflammation. Magnesium salicylate is used to ease the pressure, swelling, and stiffness associated with arthritis. This medication should not be provided to a child or teenager who has a fever, particularly if the child has symptoms of influenza or chicken pox. Salicylates may cause Reye's syndrome, a serious and sometimes fatal condition in infants. Magnesium salicylate, particularly if used for a long period, can trigger life-threatening heart or circulation complications, such as a heart attack or stroke. Magnesium salicylate can cause serious side effects in the stomach and intestines, such as bleeding or perforation. When you take this medication, you can acquire this disease, which can be deadly, particularly in older persons. It is in the FDA's pregnancy group C. Magnesium salicylate can be harmful to the unborn baby if the mother takes it during the last 3 months of pregnancy. Magnesium salicylate can be taken up to four times a day. To prevent a disturbed stomach, should not lie down for at least 30 min after taking the medication. Avoid taking antibiotics for at least 2 h after taking magnesium salicylate.

9.22 MECLOFENAMATE SODIUM[22]

Anthranilic acid derivative, that is, a structurally related to Diclofenac and Mefenamic acid.

9.22.1 *PHARMACOLOGICAL EFFECTS*

COX-1 and COX-2 inhibitions. Similar to other prototypical NSAIAs, its pharmacologic acts demonstrate anti-inflammatory, analgesic, and anti-pyretic operation.

9.22.2 *SIDE EFFECTS*

- Cardiovascular Risk: Possible elevated risk of severe coronary thrombotic events (sometimes fatal) (e.g., MI, stroke). With period of use, risk can increase. Individuals with coronary disease or cardiovascular disease risk factors can have an elevated risk.
- GI Risk: Increased risk of severe GI accidents (sometimes fatal) (e.g., bleeding, ulceration, perforation of the stomach or intestine). Severe GI accidents may occur at any moment and warning signs and symptoms do not precede them. For extreme GI cases, geriatric individuals are at greater risk.
- Renal Effects: Patients undergoing long-term NSAIA therapy have reported direct renal damage, including renal papillary necrosis. Overt renal decompensation potential. Increased risk of renal toxicity in patients with compromised renal or hepatic activity or cardiac failure, in patients with geriatric function, in patients with volume loss and in patients undergoing a diuretic blocker, an ACE inhibitor or an angiotensin II receptor.

- Hypersensitivity Reactions: Anaphylactoid-reported reactions. Immediate emergency attention for anaphylaxis and discontinuation. Stop in aspirin triad patients (aspirin allergy, asthma, nasal polyps); caution in asthma patients.
- Dermatologic Reactions: Severe documented skin reactions (e.g., exfoliative dermatitis, Stevens-Johnson syndrome, toxic necrolysis of the epidermis) can occur without warning. Discontinue rash or some other indication of hypersensitivity at first appearance (e.g., blisters, fever, and pruritus).

9.22.3 CONTRAINDICATIONS

- Well-known hypersensitivity to sodium meclofenamate or other component in the formulation.
- History of aspirin or other NSAIAs precipitated asthma, urticaria, or other sensitivity response.
- Management of perioperative pain in CABG surgery conditions.

9.22.4 DRUG INTERACTIONS

- ACE Inhibitors: Reduced potential BP reaction to an ACE inhibitor. Possible degradation of renal function in people with compromised renal function.
- Angiotensin II Receptor Antagonists: Reduced potential reaction of BP to angiotensin II receptor antagonist. Potential deterioration in renal function has been demonstrated in people with renal impairment.
- Antacids (aluminum- and magnesium-containing): No effect on the sodium absorption of meclofenamate.
- Aspirin: Plasma amounts of meclofenamic acid can be decreased. Increased incidence and other risks of GI ulceration. There is no clear indication that the elevated risk of major coronary problems associated with NSAIAs is mitigated by low-dose aspirin. Concomitant use is not advised.
- Diuretics (furosemide, thiazides): Reduced natriuretic effects, diuretic effectiveness, and renal insufficiency control.
- Lithium: Increased lithium plasma concentrations, lithium toxicity control.

- Methotrexate: Reduced natriuretic effects, diuretic effectiveness, and renal insufficiency control.

9.22.5 PHARMACOLOGICAL USES

Inflammatory Diseases: Symptomatic treatment with severe and recurrent arthritis and rheumatoid arthritis, ankylosing spondylitis or acute gouty arthritis, psoriatic arthritis, discomfort, mild-to-moderate pain reduction, dysmenorrhea and menorrhagia, primary dysmenorrhea treatment, idiopathic menorrhagia, inflammatory disorders. Heavy idiopathic menstrual discharge (i.e., it is possible to find no underlying pathophysiological cause); primary dysmenorrhea or idiopathic menorrhagia.

9.22.6 DOSAGE AND ADMINISTRATION

- Oral Administration: Food, milk, or antacid administration can mitigate adverse GI effects.
- Dosage: Accessible as sodium meclofenamate, dose represented in meclofenamic acid terminology. Using the lowest possible dose and shortest period of therapy in accordance with the patient's care objectives to reduce the future likelihood of adverse cardiovascular and/or GI events. Change the dose to the lowest appropriate dosage depending on individual criteria and response, attempt to titrate.
- Inflammatory Diseases: Osteoarthritis or rheumatoid arthritis: Oral dose 200–400 mg daily in three or four evenly divided doses. Initiate at a lower dose and change the dose and duration, depending on the severity of symptoms and therapeutic reaction if appropriate (maximum 400 mg daily).
- Pain: Per 4–6 h, oral-50 mg. For adequate pain relief, some patients may need doses of 100 mg to maximum 400 mg daily.
- Dysmenorrhea and Menorrhagia: Three times daily, oral dose 100 mg. Initiated at menstrual initiation and persisted for 6 days or before menstrual cessation.
- Renal Impairment: Recommended dose reduction in patients with renal impairment, control of renal activity. Not recommended in advanced kidney disease patients.
- Geriatric Patients: Take into account the decreased initial dosage, track carefully.

9.23 MEFENAMIC ACID[23]

As an NSAID, mefenamic acid is used to alleviate pain, especially menstrual pain. It is generally indicated for oral administration. Ponstel is the brand name for mefenamic acid in the United States, and Ponstan is the brand name in Canada and the United Kingdom. Ponstan is being sold by ERFA Canada 2012 Inc. Mefenamic acid decreases inflammation (swelling) and uterine contractions by an unexplained process. The inhibition of prostaglandin synthesis is assumed to be the cause. There is also proof that mefenamic acid may be used to avoid perimenstrual migraine headaches, with therapy occurring 2 days before the start of flow or 1 day before the predicted beginning of headache and continuing until the period of menstruation. Since hepatic metabolism is so critical in the reduction of mefenamic acid, patients with a possible liver defect should be given lower doses. Kidney failure will cause the medication and its metabolites to clump together in the excretory system. Mefenamic acid can also not be offered to those who have kidney disorders.

9.23.1 IMPORTANT INFORMATIONS

Mefenamic acid is listed as pregnancy category C by the FDA. At doses 10 times the human dosage, reduced pregnancy, delayed childbirth, and decreased weaning survival were observed. There is no controlled data of human reproduction. Mefenamic acid is only recommended for usage during breastfeeding where the benefit outweighs the harm. Mefenamic acid cannot be used during late pregnancy since NSAIDs are thought to trigger premature ductus arteriosus closure. Mefenamic acid is excreted in small amounts in human milk. According to the maker, it should be decided whether to stop breastfeeding or stop taking mefenamic acid, taking into account the medication's importance to the baby.

9.23.2 PHARMACOLOGICAL EFFECTS

Mefenamic acid is a competitive COX-1 and COX-2 inhibitor responsible for the first biosynthesis step of prostaglandin. Thus, reducing the activity of these enzymes decreases the development of prostaglandins that are involved in the processes of inflammation and pain.

9.23.3 ADVERSE REACTIONS

- Cardiovascular: hypertension, CHF, tachycardia, and syncope.
- CNS: Headache and dizziness.
- Dermatologic: Rashes and Pruritus.
- GI: Constipation, abdominal pain, dyspepsia, diarrhea, GI ulcers (duodenal/gastric), flatulence, heartburn, gross bleeding/perforation, vomiting, and nausea.
- Hematologic: Increased bleeding time and Anemia.
- Hepatic: Elevated liver enzymes.
- Miscellaneous: Abnormal renal function, edema, and tinnitus.

9.23.4 PHARMACOLOGICAL USES

Mefenamic acid is used to relieve mild-to-severe discomfort from different illnesses over the short term. It is also used to decrease suffering from menstrual cycles and blood loss.

9.24 MELOXICAM[24]

Mobic (meloxicam) is a NSAID. Meloxicam acts by lowering hormones that cause inflammation and pain in the body. Mobic is a prescription medication used to treat osteoarthritis and rheumatoid arthritis in adults and adolescents above the age of 2.

9.24.1 IMPORTANT INFORMATION

Avoid taking alcohol for the remainder of your treatment. It will raise the chance of abdominal bleeding. Mobics may trigger ovulation to be delayed (the release of an egg from an ovary). If pregnancy therapy is suggested, avoid using meloxicam, especially if you have no intentions to get pregnant. It is listed as pregnancy class D by the FDA. Taking Mobic in the last 3 months of pregnancy may injure the unborn child. Meloxicam may cross into breast milk, placing a breastfeeding kid at risk. Shake the oral suspension (liquid) well before measuring the dose. Instead of using a standard kitchen spoon to measure the liquid, use a specific dose-measuring spoon or medicine cup.

9.24.2 PHARMACOLOGICAL EFFECTS

Meloxicam inhibits the enzyme COX, which transforms prostaglandin H2 to arachidonic acid, the first stage in the production of inflammation-mediating prostaglandins. Meloxicam has been shown to selectively suppress COX-2 over COX-1, also at minimal clinical levels. Meloxicam concentrations of synovial fluid range from 40% to 50% of plasma concentrations. Since synovial fluid has a lower albumin amount than plasma, the free fraction in synovial fluid is 2.5 times higher than that in plasma. Although the significance of this penetration is unknown, it could explain why it works so effectively in animal models of arthritis care.

9.24.3 SIDE EFFECTS

Meloxicam usage may cause gastrointestinal toxicity and bleeding, headaches, rash, and very dark or black stools (a sign of intestinal bleeding). Its use, like that of other NSAIDs, is linked to an increased risk of medical complications, such as cardiac failure and stroke. It has less gastrointestinal adverse effects than Diclofenac, piroxicam, naproxen, and probably all non-COX-2-selective NSAIDs. Meloxicam does not tend to suppress thromboxane A at amounts that would interact with platelet function. Meloxicam was associated with a statistically marginally lower risk of thromboembolic symptoms than Diclofenac NSAID (0.2% vs. 0.8%, respectively) in a pooled study of randomized, supervised meloxicam treatment trials lasting up to 60 days, but with a comparable rate of thromboembolic events to naproxen and piroxicam.

9.24.4 PHARMACOLOGICAL USES

For the prevention of arthritis, Meloxicam is used. It decreases joint discomfort, swelling, and stiffness.

9.24.5 DOSAGE AND ADMINISTRATION

For adults, the recommended dose of meloxicam is 15 mg a day. Symptoms of overdose may include nausea, vomiting, dizziness, sleepiness, dark or bloody stools, blood coughing, fever, little or no urination, shallow breathing, fainting, seizures (convulsions), or coma.

9.25 NABUMETONE[25]

Nabumetone is a nonacidic NSAID that is rapidly converted in the liver to 6-methoxy-2-naphthyl acetic acid, a significant active metabolite. The active metabolite of nabumetone inhibits the enzyme COX and prevents the activity of COX-2 preferentially, as seen with previous NSAIDs (which is indirectly responsible for the production of inflammation and pain during arthritis). The active metabolite of nabumetone is considered to be the compound responsible for the medicinal gain. The parent drug, on the other hand, is a poor COX-2 inhibitor for drugs, particularly prostaglandins. It has the ability to be less nephrotoxic than indomethacin. Nabumetone belongs to a modern class of NSAIDs that have a lower chance of causing gastrointestinal mucosal inflammation and platelet function suppression, have little impact on renal prostaglandin release, and have less interactions with heart failure than older medications. Similarly to paracetamol, nabumetone has a good impact on the control of blood pressure in hypertensive patients taking ACE inhibitors.

9.25.1 IMPORTANT INFORMATION

Relafen belongs to a family of medications classified as NSAIDs. It acts by reducing inflammation and pain-causing hormones in the body. Relafen is a medicine that is used to alleviate arthritis-related pain and inflammation. Relafen may increase the likelihood of life-threatening cardiac or circulation risks, such as heart failure or stroke. This chance can rise if you use Relafen more often. This drug also raises the likelihood of serious gastrointestinal or intestinal complications, such as bleeding or perforation (forming of a hole). Both signs are potentially fatal, and gastrointestinal complications may strike at any time when taking Relafen. In older people, the probability of these serious gastrointestinal side effects is far greater. May not take some over-the-counter temperature, asthma, or pain medication without first seeing a doctor or pharmacist. Stop UV rays from the sun or artificial UV rays (sunlamps or tanning beds). Relafen can make the skin more susceptible to sunlight, raising the risk of sunburn. It is in the FDA's pregnancy group C. This medication has the ability to harm an unborn child. Inform the doctor whether you are pregnant or intend to become pregnant as a result of your procedure. Taking Relafen during the last 3 months of pregnancy will affect the unborn child.

9.25.2 PHARMACOLOGICAL ACTION

Nabumetone is part of a class of medications known as NSAIDs. Some members in this class include ibuprofen (Motrin), indomethacin (Indocin), naproxen (Aleve), among others. These drugs are used to treat moderate to extreme discomfort, fever, and inflammation. They work by lowering the levels of prostaglandins, which are chemicals produced by the body and are responsible for discomfort, fever, and inflammation. NSAIDs inhibit the enzyme that generates prostaglandins (COX), resulting in lower prostaglandin concentrations. As a consequence, inflammation, inflammation, and fever are reduced. Since the reaction to various NSAIDs varies from patient to patient, it is normal for a doctor to try a variety of NSAIDs for a specific illness. The FDA licensed nabumetone in December 1991.

9.25.3 SIDE EFFECTS

As it is a nonacidic prodrug which is then metabolized into its active 6MNA (6-methoxy-2-naphthylacetic acid) form, it has been shown to have a significantly lower chance of gastrointestinal side effects than most other nonselective NSAIDs.

9.25.4 DOSING

It can be taken with or without meals. For osteoarthritis or rheumatoid arthritis, the minimum starting dosage is a single dose of 1000 mg daily. Some patients may respond to 1500 or 2000 mg daily in a better way. It is also possible to separate and prescribe doses twice daily. It is best to use the lowest effective dose.

9.25.5 DRUG INTERACTIONS

Like other NSAIDs, nabumetone is related to a slew of probable or probable interactions that may change the way other medications function. Nabumetone can raise lithium levels in the blood by reducing lithium excretion by the kidneys (Eskalith, Lithobid). Lithium toxicity may occur as lithium concentrations grow too high. Nabumetone can reduce the effects of blood pressure-lowering drugs on blood pressure. This may happen because prostaglandins play a part in blood pressure regulation. As nabumetone is combined with aminoglycosides (e.g., gentamicin [Garamycin]), blood levels of aminoglycosides will rise, owing to a decrease in aminoglycoside elimination from the body. Which may contribute to more aminoglycoside-related side effects. Those who take oral blood thinners or anticoagulants (such as warfarin [Coumadin]) can stop nabumetone since it thins the blood as well. Unnecessary blood thinning will cause bleeding. The combination of methotrexate-containing NSAIDs (Rheumatrex, Trexall) decrease methotrexate removal from the bloodstream, lift methotrexate levels in the body, and increase methotrexate withdrawal symptoms. When taking nabumetone or other NSAIDs, people who drink more than three alcoholic drinks a day are more likely to experience stomach ulcers.

9.25.6 PHARMACOLOGICAL USES

It is used to relieve arthritis-induced discomfort or inflammation or other inflammatory disorders and illnesses such as synovitis. Nabumetone functions by lowering the effects of pain-causing enzymes and inflammation.

9.26 OXAPROZIN[26]

Oxaprozin, also known as Oxaprozinum (sold as Daypro, Dayrun, and Duraprox), is a NSAID used to treat osteoarthritis and rheumatoid arthritis-related inflammation, weakness, and joint pain. That is a propionic acid derivative chemically. It is available in 600 mg capsules. Adults can take 1200 mg a day, and no more than 1800 mg a day. Only adolescents over the age of 6 with adolescent rheumatoid arthritis have shown protection and efficacy, and the aged community remains at a higher risk of adverse reactions.

9.26.1 IMPORTANT INFORMATION

This drug can increase the risk of life-threatening cardiac or circulation complications, such as heart failure or stroke. This medication can never be used during or during heart bypass surgery (also called coronary artery bypass graft, or CABG). Taking an NSAID increases the likelihood of life-threatening cardiac or circulation risks, such as heart failure or stroke. This chance will rise if you use an NSAID more often. It is in the FDA's pregnancy group C. This medication has the ability to harm an unborn child. Taking oxaprozin during the last 3 months of pregnancy will harm the unborn child.

9.26.2 PHARMACOLOGICAL ACTIONS

Oxaprozin is part of a range of medications classified as NSAIDs. Some members in this group contain ibuprofen (Motrin), indomethacin (Indocin), naproxen (Aleve), and others. These drugs are used to treat moderate to extreme discomfort, fever, and inflammation. They work by lowering the levels of prostaglandins, which are the chemicals that cause pain, fever, and inflammation. The prostaglandin-making enzyme (COX) is inhibited by oxaprozin, resulting in lower prostaglandin concentrations. As a consequence, inflammation and fever are reduced. Oxaprozin was authorized by the FDA in October 1992.

9.26.3 DRUG INTERACTION

Oxaprozin can raise lithium levels by reducing lithium excretion in the kidneys (Eskalith, Lithobid). Lithium toxicity may result from high lithium levels. Oxaprozin can reduce the efficacy of blood pressure medications by reducing blood pressure. Since prostaglandins have a function in blood pressure management, this may happen. As oxaprozin is mixed with methotrexate (Rheumatrex, Trexall) or aminoglycoside antibiotics (e.g., gentamicin), serum levels of methotrexate or aminoglycoside antibiotics will grow, most likely due to decreased methotrexate or aminoglycoside removal. More methotrexate or aminoglycoside side effects can result as a result of this. People taking oral blood thinners or anticoagulants, such as warfarin (Coumadin), should avoid taking oxaprozin because it thins the blood, and excessive blood thinning causes bleeding. Oxaprozin improves the harmful impact of cyclosporine on kidney function. When aspirin is paired with oxaprozin, the likelihood of getting an ulcer rises. People who consume more than three alcoholic beverages a day while taking oxaprozin or other NSAIDs might be at an increased risk of developing stomach ulcers.

9.26.4 PHARMACOLOGICAL USES

For the prevention of arthritis, oxaprozin is used. It decreases joint discomfort, swelling, and stiffness.

9.26.5 DOSAGE AND ADMINISTRATION

For adults, the recommended volume of oxaprozin is 1200 mg a day. Symptoms of overdose can include nausea, vomiting, dizziness, drowsiness, dark or bloody stools, blood coughs, fever, little or no urination at all, shallow breathing, fainting, or coma.

9.27 PIROXICAM[27]

Piroxicam is an oxicam-type NSAID used to treat stressful, inflammatory conditions including arthritis. It is a nonselective COX inhibitor with analgesic and antipyretic properties. It acts by stopping the output of prostaglandins, a type of body chemical involved in the relaxation of pain, stiffness, tenderness, and swelling. The medication is sold in bottles, pills, and a 0.5% prescription-free gel (not in all countries). It is also available in a betadex type, which makes for faster piroxicam absorption from the gastrointestinal tract.

9.27.1 PRECAUTIONS

Get immediate medical treatment if you have symptoms of heart or circulation issues, such as chest pressure, exhaustion, shortness of breath, slurred expression, or vision or balance problems. When piroxicam is being eaten, do not consume alcohol. Alcohol may raise the likelihood of stomach bleeding caused by piroxicam. UV radiation from the sun or artificial UV rays can be stopped. Piroxicam can render the skin more sensitive to sunshine, which can lead to sunburn. It is in the FDA's pregnancy group C. This medication has the ability to harm an unborn child. During the last 3 months of pregnancy, piroxicam will affect the unborn child. Piroxicam may pass into breast milk, putting a breastfeeding kid at risk. Do not give this medication to any person without first seeing a doctor.

9.27.2 PHARMACOLOGICAL ACTIONS

9.27.2.1 ADVERSE EFFECTS

Intestinal symptoms such as nausea, discomfort, diarrhea, and gastrointestinal bleeding or ulceration, as well as fatigue, dizziness, nervousness, fatigue, sleepiness, anxiety, vertigo, hearing disturbance (such as tinnitus), elevated blood pressure, edema, resistance of light, and skin reactions (including, though seldom, Stevens-Johnson syndrome) are also common side effects. It is more likely than other NSAIDs to trigger GI disturbances and serious skin reactions.

9.27.3 PHARMACOLOGICAL USES

It is used to treat rheumatoid arthritis and osteoarthritis, as well as primary dysmenorrhea and postsurgical discomfort, and it serves as an analgesic, specifically when inflammation is present. In 2007, the European Medicines Agency published a report on its use, recommending that it be used only to combat chronic inflammatory disorders since the risk–benefit relationship is only favorable in these situations.

9.28 ROFECOXIB[28]

Rofecoxib is a NSAID that has been phased out owing to safety concerns. Merck & Co. promoted it to cure osteoarthritis, severe pain conditions, and dysmenorrhea. Rofecoxib was approved by the FDA on May 20, 1999). Merck withdrew rofecoxib off the market on September 30, 2004, leading to worries over an enhanced risk of heart failure and stroke linked with long-term, high-dose use. Merck abandoned rofecoxib after learning that it had hidden details regarding the drug's risks from physicians and patients for over 5 years, resulting in between 88,000 and 140,000 instances of serious cardiac failure. Rofecoxib was one of the most widely prescribed medications ever taken off the market.

9.28.1 IMPORTANT INFORMATION

Rofecoxib acts by reducing compounds in the body that induce pain, irritation, and fever. Rofecoxib is used to reduce swelling, inflammation, and stiffness induced by osteoarthritis, rheumatoid arthritis, and certain forms of childhood rheumatoid arthritis, to relieve severe pain in adults, to manage migraines, and to treat menstrual pain. If you have had an allergic reaction to ibuprofen (Motrin, Advil, Nuprin), naproxen (Aleve, Naprosyn, Anaprox), ketoprofen (Orudis KT, Orudis, Oruvail), Diclofenac (Voltaren, Cataflam), diflunisal (Dolobid), Etodolac (Lodine), Fenoprofen (Nalfon (Mobic). It is in the FDA's pregnancy Category C. Rofecoxib may not be taken until the third trimester of pregnancy since it may influence the infant's heart formation. It is unknown if rofecoxib moves into breast milk.

9.28.2 PHARMACOLOGICAL ACTIONS

COX-1 and COX-2 are two well-studied isoforms of the COX enzyme. COX-1 is involved in the production of prostaglandins that protect the intestinal lining, while COX-2 is involved in the production of pain and inflammation-related prostaglandins. The development of "selective" NSAIDs that inhibit COX-2 but not COX-1 provides the same pain relief as traditional NSAIDs, but with a significantly lower risk of lethal or debilitating peptic ulcers. Rofecoxib is a selective COX-2 inhibitor, commonly known as "coxib." Others contain Merck's etoricoxib (Arcoxia), Pfizer's celecoxib (Celebrex), and Valdecoxib (Bextra). At the time of its elimination, rofecoxib was the only coxib with experimental evidence of a better gastrointestinal adverse effect profile than conventional NSAIDs. This was mainly based on the VIGOR (Vioxx GI Results Research) review, which compared the efficacy and adverse impact profiles of rofecoxib and naproxen.

9.28.3 ADVERSE DRUG REACTION

Rofecoxib has a comparable side effect profile to other NSAIDs, with the exception of a lower risk of gastric ulceration. Prostaglandins are a lipid family with a wide number of members. Prostaglandin I2/PGI2/prostacyclin is the only member. Some prostaglandins (such as PGE2) play a significant role in vascular tone regulation. Since prostacycline/thromboxane releases both COX-1 although COX-2, and rofecoxib inhibits only the COX-2 enzyme, there is little reason to suggest that the medication substantially reduces prostacycline levels. There is also no reason to conclude that the equilibrium of prostacyclin and thromboxane concentrations defines vascular sound. In animal studies, Merck reported that there was no effect on prostacycline activity in blood vessels. Other reports also proposed that when rofecoxib becomes ionized under physiological circumstances, maleic anhydride metabolites can be related to cardiotoxicity.

9.29 SULINDAC[29]

9.29.1 IMPORTANT INFORMATION

Sulindac functions by decreasing hormones that trigger inflammation and pain in the body. Sulindac is used to treat arthritis, ankylosing spondylitis, tendinitis, bursitis, and gout-related pain and inflammation. This drug, particularly if held for a long period, may cause life-threatening heart or circulation complications, such as a heart attack or stroke. It is in the FDA's pregnancy group C. It is uncertain if sulindac will affect an unborn child. Sulindac use during the last 3 months of pregnancy may cause birth defects. It is unknown if sulindac moves into breast milk or whether it might cause any damage to a breastfeeding infant.

9.29.2 PHARMACOLOGICAL ACTIONS

Sulindac is a NSAID used to treat discomfort, diarrhea, and inflammation. Some NSAIDs include indomethacin, ibuprofen, nabumetone, among others. They act by lowering the levels of prostaglandins, which are chemicals created by the body and are responsible for inflammation, fever, and pain. Sulindac inhibits the enzyme that produces prostaglandins (COX), resulting in lower prostaglandin concentrations. As a consequence, inflammation, inflammation, and fever are reduced. Sulindac was authorized by the FDA in September 1978.

9.29.3 DOSAGE ADMINISTRATION

The normal adult dosage is 150–200 mg, taken with meals twice a day. There is a cumulative dosage of 400 mg daily.

9.29.4 DRUG INTERACTIONS

Sulindac can boost lithium levels in the blood by reducing lithium elimination by the kidneys (Eskalith, Lithobid). Lithium toxicity may be worsened by elevated lithium levels. Sulindac works by lowering blood pressure, which helps blood pressure drugs work better. Since prostaglandins have a function in blood pressure management, this may happen (reduction). In old, fluid-depleted (including diuretic therapy) or chronically performing kidney patients, combining NSAIDs like sulindac with angiotensin receptor blockers (e.g., valsartan[Diovan], losartan[Cozaar], irbesartan[Avapro]) or ACE antagonists (e.g., enalapril[Vasotec], cap This kind of consequences are normally reversible. When sulindac is combined with methotrexate (Rheumatrex, Trexall) or aminoglycoside antibiotics (such as gentamicin), blood levels of methotrexate or aminoglycosides can rise, likely due to a reduction in the removal of methotrexate or aminoglycosides from the body. This could lead to further methotrexate or aminoglycoside-related side effects. Sulindac can be used for patients who are taking oral blood thinners or anticoagulants like warfarin (Coumadin), since it thins the blood so much and may induce bleeding. People who eat more than three alcoholic beverages a day while taking sulindac or other NSAIDs are at an increased risk of developing stomach ulcers.

9.29.5 PHARMACOLOGICAL USES

It is useful in the treatment of acute and chronic inflammatory diseases, much like other NSAIDs. Sulindac is a sulfinylindene-dependent prodrug that the body transforms to an active NSAID. More specifically, the agent is transformed to a sulfide by liver enzymes, and is excreted in the bile and then reabsorbed by the stomach. This is thought to help control constant blood levels by reducing gastrointestinal side effects. Sulindac has been shown in several studies to be less irritating to the stomach than other NSAIDs, with the exception of COX-2 inhibitors. The exact function of its NSAID properties is unknown, but it is thought to work by inhibiting the synthesis of prostaglandin by COX-1 and COX-2 enzymes. The usual dosage is 150–200 mg twice a day, taken with food. People who have had serious allergic reactions to aspirin or other NSAIDs in the past (urticaria or anaphylaxis) should avoid it, and anyone who have a history of peptic ulcer disease should use it with care. Sulindac is much more likely than other NSAIDs to damage the liver or pancreas. Sulindac tends to minimize the aggregation of polyps and precancerous lesions in the colon, regardless of COX inhibition, particularly when used in accordance with familial adenomatous polyposis, and may have other anticancer properties. Sulindac is a tocolytic agent that may be used in the management of preterm birth. Sulindac, like other NSAIDs, is currently being researched for its function in the management of Alzheimer's disease.

9.30 TOLMETIN SODIUM[30]

Tolmetin is an anti-inflammatory arylalkanoic acid that is nonsteroidal. It is commonly used to suppress hormones that cause inflammation, swelling, tenderness, and stiffness in disorders, such as osteoarthritis and rheumatoid arthritis, particularly juvenile rheumatoid arthritis.

9.30.1 PHARMACOLOGICAL ACTIONS

Although the mechanism of action of tolmetin is unknown, experiments on humans and animals have shown that it does not induce an anti-inflammatory response by activating the adrenal or pituitary glands. Instead, tolmetin has been shown to inhibit in vitro prostaglandin synthetase and lower plasma levels of prostaglandin E, implying that it induces an anti-inflammatory response.

When tested in rats, tolmetin prevented experimentally induced polyarthritis and reduced inflammation. In patients with rheumatoid arthritis or osteoarthritis, tolmetin reduced disease progression as effectively as aspirin and indomethacin, while the incidence of minor gastrointestinal adverse reactions and tinnitus was lower than in patients treated with aspirin, and the occurrence of central nervous system adverse reactions was lower than in patients treated with indomethacin.

9.30.2 DOSAGE ADMINISTRATION

The recommended dosage is 200–600 mg three times per day. The recommended daily dosage is 1800 mg. Tolmetin should be taken with food and 8–12 ounces of water to avoid stomach upset.

9.30.3 DRUG INTERACTIONS

Because of the increased risk of bleeding, tolmetin is typically used with caution in people taking blood-thinning medications like warfarin. Toxic serum levels of either drug may concentrate in patients taking lithium (Eskalith, Lithobid) or methotrexate, as tolmetin may prevent the kidney from removing the drug from the body. Tolmetin can amplify the side effects of cyclosporin. Because it causes or exacerbates high blood pressure, tolmetin can reduce the effectiveness of antihypertensives. NSAIDs will reduce the antihypertensive effects of ACE inhibitors. In patients who are old, volume-depleted (including some on diuretic therapy), or have poor kidney function, combining NSAIDs with angiotensin receptor blockers (e.g., valsartan [Diovan], losartan [Cozaar], irbesartan [Avapro]) or angiotensin-transforming enzyme inhibitors (e.g., enalapril [Vasotec], captopril). These kinds of consequences are normally reversible.

9.30.4 PHARMACOLOGICAL USES

- Inflammatory Diseases: Symptomatic treatment of osteoarthritis and rheumatoid arthritis, developmental rheumatoid arthritis in children 2 years of age and older, effective in the treatment of ankylosing spondylitis (late and early stages), adhesive capsulitis treatment of the shoulder (frozen shoulder), radiohumeral bursitis (tennis elbow), and local fracture (e.g., recent sprains).
- Pediatric Patients: Juvenile Rheumatoid Arthritis With Inflammatory Diseases: Infants >2 years of age: Originally, 20 mg/kg daily in three to four separated doses. Based on reaction and resistance, change the dose. Normal effective dosage: daily, 15–30 mg/kg.
- Adults: Inflammatory conditions, osteoarthritis or rheumatoid arthritis, initially 400 mg 3 times daily, usually including the onset and bedtime doses. Change the reaction and tolerance dosage (after 1–2 weeks). Normal effective dosage: daily doses of 600 mg–1.8 g in three separated doses.
- Ankylosing Spondylitis: Shoulder adhesive capsulitis (frozen shoulder), radiohumeral bursitis (tennis elbow), local trauma (recent sprains, e.g.): split doses were used.
- Renal Impairment: Reduce dosage if necessary.

9.30.5 CONTRAINDICATIONS

- Hypersensitivity to tolmetin or any of the additives in the formulation is known.
- History of aspirin or other NSAIAs precipitated asthma, urticaria, or other sensitivity response.
- Management of perioperative pain in CABG surgery conditions.

9.31 VALDECOXIB[31]

Valdecoxib belongs to the NSAID class of drugs (NSAIDs). Valdecoxib works by lowering the body's inflammatory, painful, and fever-inducing chemicals. Valdecoxib is used to alleviate the pressure, fatigue, and discomfort associated with osteoarthritis and adult rheumatoid arthritis. Valdecoxib is also prescribed to avoid painful menstruation. Valdecoxib cannot be used to ease inflammation during CABG. Valdecoxib usage has been related to an increased incidence of cardiac attacks, severe surgical infections, and wound injuries in these patients.

9.31.1 IMPORTANT INFORMATION

Bextra was taken off the market in the United States in 2005. The manufacturer of Bextra has confirmed that the drug would be voluntarily removed from the market in the United States. Protection concerns about an increased risk of medical complications (including cardiac failure and stroke) in patients taking NSAIDs, as well as safety concerns about an increased risk of rare to severe skin reactions in patients taking Bextra, have led to its elimination. If you are experiencing an allergic reaction to sulfa-based drugs like sulfonamethoxazole (Bactrim, Septra, Gantanol, etc.) or sulfonamethoxazole (Gantrisin), aspirin, or other NSAIDs like celecoxib (Celebrex), ibuprofen (Motrin, Advil, Nuprin, etc.), naproxen (A (Tolectin), do not take Bextra, and first consult the doctor. It is in the FDA's pregnancy Category C. This means that it is uncertain if it will be dangerous to an unborn child. Valdecoxib may not be taken during the third trimester of pregnancy since it may influence the infant's heart formation. It is uncertain if Bextra gets into breast milk.

9.31.2 PHARMACOLOGICAL ACTIONS

Valdecoxib is a NSAID that comes in the form of an oral tablet (NSAIDs). Ibuprofen (Motrin), indomethacin (Indocin), naproxen (Naprosyn), Diclofenac (Voltaren), sulinac (Clinoril), ketoprofen (Orudis), aspirin and aspirin-related drugs, and some are examples of NSAIDs. Valdecoxib acts by influencing the synthesis of prostaglandins, which are body-produced chemicals that promote arthritis inflammation and cause discomfort, swelling, and tenderness in arthritic joints. Valdecoxib blocks one of the enzymes that generates prostaglandins

(COX-2), close to the newer celecoxib (Celebrex) and rofecoxib (Vioxx) NSAIDs, resulting in lower prostaglandin concentrations. As a consequence, arthritis causes less pain, stiffness, and tenderness of the joints. Valdecoxib (along with celecoxib and rofecoxib) varies from most other NSAIDs in that it causes less abdominal and intestinal discomfort and ulceration (at least in the short term) and does not affect blood clotting.

9.31.3 DOSAGE ADMINISTRATION

For osteoarthritis or rheumatoid arthritis, the recommended regular dose of valdecoxib is 10 mg once a day. For dysmenorrhea, the dose is 20 mg twice days. Valdecoxib may be taken either with or without food.

9.31.4 DRUG INTERACTIONS

NSAIDs decrease the action of diuretics including furosemide (Lasix) and hydrochlorothiazide (Hydrodiuril) and lead to water retention in certain patients. Fluconazole (Diflucan), ketoconazole (Nizoral), and lithium (Eskalith) can minimize valdecoxib accumulation in the body by inhibiting liver breakdown.

9.31.5 SIDE EFFECTS

The more prominent side effects of valdecoxib involve migraine, abdominal discomfort, diarrhea, fatigue, flatulence, and insomnia. Any potential side effects include fainting, organ failure, heart loss, blood bleeding, increased hypertension, chest pain, ringing in the ears, deafness, stomach and intestinal ulcers, coughing, distorted vision, anxiety, photosensitivity, weight gain, fluid retention, flu-like symptoms, drowsiness, and fatigue.

9.31.6 PHARMACOLOGICAL USES

The FDA has licensed valdecoxib for the care of osteoarthritis, adult rheumatoid arthritis, and primary dysmenorrhea in the United States (FDA). Valdecoxib was commonly used off-label for the prevention of severe pain and different cases of surgical pain.

9.32 SALSALATE[32]

Salsalate is a drug that belongs to the salicylate and NSAID classes. In comparison to other NSAIDs, salsalate has a minor inhibitory impact on the COX enzyme and prevents the production of certain proinflammatory chemical signals including interleukin-6, TNF-alpha, and C-reactive protein. Salsalate can be used for autoimmune conditions, such as rheumatoid arthritis or noninflammatory conditions such as osteoarthritis in certain instances.

9.32.1 PHARMACOLOGICAL ACTIONS

Salsalate is an anti-inflammatory NSAID that is used to treat fever, pain, and inflammation in the body. Salsalate is transformed to salicylic acid, which is its active source and directly linked to aspirin, in the liver. NSAIDs include Ibuprofen (Motrin), indomethacin (Indocin), nabumetone (Relafen), naproxen (Aleve), among others. They act by lowering the levels of prostaglandins, which are the chemicals that cause discomfort, fever, and inflammation in the body. The prostaglandin-making enzyme (COX) is inhibited by salsalate, resulting in lower prostaglandin concentrations. As a consequence, inflammation and fever are reduced. Salsalate is as powerful as aspirin at treating discomfort, but it has less impacts on blood clotting than aspirin.

9.32.2 DOSAGE ADMINISTRATION

Salsalate is normally taken in 3000 mg daily in 2–4 doses. To relieve abdominal pain induced by cooking, salsalate should be taken.

9.32.3 DRUG INTERACTION

Salsalate, like aspirin, converts to salicylic acid. Using aspirin to salsalate may also induce salicylic acid toxicity. NSAIDs can raise lithium levels in the blood by decreasing lithium excretion by the kidneys (Eskalith, Lithobid). Lithium toxicity may be worsened by elevated lithium levels. NSAIDs will minimize the effectiveness of blood pressure medicines that lower blood pressure. Since prostaglandins have a function in blood pressure management, this may happen (lowering). When NSAIDs are paired with methotrexate (Rheumatrex, Trexall) or aminoglycoside antibiotics (e.g., gentamicin), blood levels of methotrexate or aminoglycoside will grow, likely due to a decrease in methotrexate or aminoglycoside elimination. This could contribute to further methotrexate or aminoglycoside side effects. People taking oral blood thinners or anticoagulants, such as warfarin (Coumadin) should avoid taking NSAIDs when they thin the blood, and excessive blood thinning can cause bleeding. People who consume more than three alcoholic beverages a day while taking NSAIDs could be at an increased risk of developing stomach ulcers. There is no relevant study on salsalate in pregnant people. Salicylic acid levels in breast milk are close to maternal blood levels. This may have negative consequences in youth. When taking or utilizing replacement drugs with salsalate, nursing mothers should avoid nursing.

9.32.4 PHARMACOLOGICAL USES

Salsalate is used to alleviate discomfort and inflammation triggered by disorders like rheumatoid arthritis, osteoarthritis, and other rheumatic disorders. In patients with minor gastrointestinal bleeding or disturbed stomach, doctors can recommend salsalate as an anti-inflammatory alternative to naproxen and ibuprofen. It has also been used as an antidote to narcotic pain control in patients with spinal disk protrusion. As applied to aspirin, salsalate causes no further fecal gastrointestinal blood depletion than placebo.

Diabetes mellitus (DM) is a form of diabetes that affects the Salsalate has been proposed as a valuable medication in the prevention or management of type 2 diabetes mellitus because of its capacity to decrease insulin resistance and potential efficacy for prediabetes. Salsalate was first suggested as a possible diabetes treatment in 1876. The anti-inflammatory properties of salsalate may help to reverse the persistent inflammation that is believed to be the source of obesity-related diabetes. Only a few studies have looked at the efficacy of salsalate in stopping the progression from prediabetes to type 2 diabetes.

9.33 NAPROXEN[33]

Naproxen works by reversibly inhibiting both COX-1 and COX-2 enzymes. Naproxen may have antiviral effects against influenza. It blocks the virus's nucleoprotein's RNA-binding groove, preventing the ribonucleoprotein complex from forming and thereby excluding the necessary nucleoproteins from circulation. Sodium salt naproxen sodium is a NSAID that belongs to the propionic acid family (the same as ibuprofen). It is used to relieve a variety of symptoms, such as pain, headache, swelling, and stiffness. It is the preferred NSAID for long-term usage in people with a high risk of cardiovascular (heart accidents or strokes) risks owing to its relatively low risk of triggering such issues. When opposed to the low-risk ibuprofen and the high-risk indometacin, naproxen has a slight risk of developing gastric ulcers. To decrease the incidence of stomach ulcers, NSAIDs are often paired with proton pump inhibitors during long-term care of patients with stomach ulcers or a history of stomach ulcers who are also taking NSAIDs. Naproxen is widely used to alleviate nausea, headache, inflammation, and weakness of migraine, osteoarthritis, kidney stones, rheumatoid arthritis, psoriatic arthritis, gout, ankylosing spondylitis, menstrual cramps, tendinitis, and bursitis, among other conditions. It is also used to treat primary dysmenorrhea. Naproxen has been used to discriminate between infectious fevers and fevers induced by cancer or connective tissue disorder.

9.33.1 ADVERSE EFFECTS

The usage of COX-2 selective and nonselective NSAIDs has been related to a rise in the risk of serious and possibly fatal cardiovascular accidents such as myocardial infarctions and strokes. However, naproxen was linked with the lowest overall cardiovascular threats. Cardiovascular risk must be taken into account before using any nonsteroidal anti-inflammatory medication. When opposed to ibuprofen, the medication was correlated with almost half the incidence of stroke and a reduced number of myocardial infarctions. As with other non-COX-2 selective NSAIDs, naproxen may trigger gastrointestinal issues including heartburn, constipation, diarrhea, ulcers, and stomach bleeding. People who already suffered ulcers or inflammatory bowel disorder should obtain medical care before taking naproxen. Other rising NSAID regimens have seen only intermittent effects on COX-1 platelets and have been identified with "a small but substantial vascular hazard," while elevated naproxen has shown to cause near-complete repression of platelet thromboxane over the entire dosing period, with no apparent increase in the occurrence of cardiovascular disease (CVD), whereas other rising NSAID regimens have seen only intermittent effects on COX-1 platelets and have been identified with "a small but substantial vascular hazard."

KEYWORDS

- **non-steroidal anti-inflammatory drugs**
- **inflammation**
- **classifications**
- **pharmacological activity**
- **mechanism of action**
- **drug interactions**
- **adverse effects**

REFERENCES

1. Waller, D. G.; Sampson, A. *Medical Pharmacology and Therapeutics E-Book*; Elsevier Health Sciences, Jun 4, 2017.
2. Satoskar, R. S.; Bhandarkar, S. D. *Pharmacology and Pharmacotherapeutics*; Elsevier India, 2020.
3. Boothe, D. M. *Small Animal Clinical Pharmacology and Therapeutics-E-Book*; Elsevier Health Sciences, 2011.
4. Wecker, L.; Crespo, L.; Dunaway, G.; Faingold, C.; Watts, S. *Brody's Human Pharmacology-E-Book*; Elsevier Health Sciences, 2009.
5. Mycek, M. J.; Finkel, R.; Clark, M. A.; Cubeddu, L. X.; Howland, R. D. *Pharmacology*; Lippincott Williams & Wilkins: Philadelphia, PA, 2000.
6. Edmunds, M. W. *Introduction to Clinical Pharmacology-e-Book*; Elsevier Health Sciences, 2015.
7. Katzung, B. G. *Basic and Clinical Pharmacology*; Mc Graw Hill, 2012.
8. Aldred, E. M. *Pharmacology E-Book: A Handbook for Complementary Healthcare Professionals*; Elsevier Health Sciences, 2008.
9. Fulcher, E. M.; Fulcher, R. M.; Soto, C. D. *Pharmacology-E-Book: Principles and Applications*; Elsevier Health Sciences, 2014.
10. Golan, D. E.; Tashjian, A. H.; Armstrong, E. J., Eds. *Principles of Pharmacology: The Pathophysiologic Basis of Drug Therapy*; Lippincott Williams & Wilkins, 2011.
11. Offermanns, S.; Rosenthal, W., Eds. *Encyclopedia of Molecular Pharmacology*; Springer Science & Business Media, 2008.
12. Beaulieu, P.; Lussier, D.; Porreca, F.; Dickenson, A. *Pharmacology of Pain*; Lippincott Williams & Wilkins, 2015.
13. Wecker, L. *Brody's Human Pharmacology E-Book*; Elsevier Health Sciences, 2018.
14. Finkel, R.; Clark, M. A.; Cubeddu, L. X., Eds. *Pharmacology*; Lippincott Williams & Wilkins, 2009.
15. Dickenson, A.; Besson, J. M., Eds. *The Pharmacology of Pain*; Springer Science & Business Media, 2012.

16. Tallarida, R. J.; Raffa, R. B.; McGonigle, P. *Principles in General Pharmacology*; Springer Science & Business Media, 2012.
17. Bowman, W. C.; Rand, M. J. *Textbook of Pharmacology*; Blackwell Scientific Publications, 1980.
18. Srivastava, A. K. *Pharmacology: A Book of Achieving Knowledge for Drugs*; Educreation Publishing, 2017.
19. Clark, A. J. *General Pharmacology*; Springer Science & Business Media, 2012.
20. Seifert, R. *Basic Knowledge of Pharmacology*; Springer International Publishing: Cham, 2019.
21. Barber, P.; Robertson, D. *Essentials of Pharmacology for Nurses*; 4th ed.; McGraw-Hill Education, 2020.
22. Bacq, Z. M., Ed. *Fundamentals of Biochemical Pharmacology*; Elsevier, 2013.
23. Levine, R. R.; Walsh, C. T.; Schwartz-Bloom, R. D., Eds. *Pharmacology: Drug Actions and Reactions*; CRC Press, 2000.
24. Wermuth, C. G., Ed. The Practice of Medicinal Chemistry; Academic Press, 2011.
25. Kar, A. *Medicinal Chemistry*; New Age International, 2005.
26. Patrick, G. L. An Introduction to Medicinal Chemistry; Oxford University Press, 2013.
27. Kadam, S. S.; Bothara, K. G. *Principles of Medicinal Chemistry*, Vol. I; Pragati Books Pvt. Ltd., 2013.
28. Alagarsamy, V. *Textbook of Medicinal Chemistry Vol I-E-Book*; Elsevier Health Sciences, 2013.
29. Nadendla, R. R. *Principles of Organic Medicinal Chemistry*; New Age International, 2007.
30. Foye, W. O. *Foye's Principles of Medicinal Chemistry*; Lippincott Williams & Wilkins, 2008.
31. Beale, J. M.; Block, J. H.; Hill, R. A. *Organic Medicinal and Pharmaceutical Chemistry*; Lippincott Williams & Wilkins, Wolters Kluwer Business: China, 2011.
32. Davis, A.; Ward, S. E., Eds. *The Handbook of Medicinal Chemistry: Principles and Practice*; Royal Society of Chemistry, 2014.
33. Korolkovas, A. *Essentials of Medicinal Chemistry*; John Wiley & Sons, 2008.

PART III

Modern Approaches in Small Molecule Drug Discovery

CHAPTER 10

Biological Databases: Tool for Low Molecular Weight Ligand Discovery Process

SANTOSH S. CHHAJED[1*], VIRUPAKSHA A. BASTIKAR[2], PRAMODKUMAR P. GUPTA[3], and SHITAL CHASKAR[1]

[1]Department of Pharmaceutical Chemistry, MET Institute of Pharmacy, Bhujbal Knowledge City, Nashik 422003, Maharashtra, India

[2]Amity Institute of Biotechnology, Amity University, Mumbai, Maharashtra, India

[3]School of Biotechnology & Bioinformatics, D. Y. Patil University, CBD Belapur, Navi-Mumbai 400614, Maharashtra, India

[*]Corresponding author. E-mail: chhajedss@gmail.com

ABSTRACT

This chapter describes the concept of meta-databases and explores the wide categories of biological databases. This chapter enlists several well-known biological databases that serve as a repository for vital information as well as majorly contribute as a knowledge source too in biological and pharmaceutical science experiments. This content is helpful to beginners as well as an expert in life sciences, pharmaceutical, and allied fields.

10.1 INTRODUCTION

The recent trends and boom in biological analysis, specifically in the event of sequencing data have raised the challenges of data storage. The sequencing

Biologically Active Small Molecules: Modern Applications and Therapeutic Perspectives.
Debarshi Kar Mahapatra, & Sanjay Kumar Bharti (Eds.)

experiments have given rise to specialized concepts like genome, proteome, transcriptome, and metabolome leading to omics analysis via genomics, proteomics, transcriptomics, and metabolomics. As biology has more and more transformed into a data-rich scientific field, the need to stockpile and correspond to a large datasets has developed enormously. The apparent instances are the protein and nucleotide sequences, and the 3-D structural data are generated by macromolecular NMR and X-ray crystallography. An emerging field, that is, bioinformatics deals with challenges, issues, and new potentials generated by these databases.[1] Bioinformatics is the appliance of information technology (IT) to stockpile, systematize, and investigate the immense quantity of biological data which are available in the form of structures and sequences of proteins and nucleic acids. The biological information of the nucleic acids is obtainable as sequences while the data of proteins are available as sequences and structures. Sequences are represented in 1-D whereas the structure comprises 3-D data of sequences (Figure 10.1).

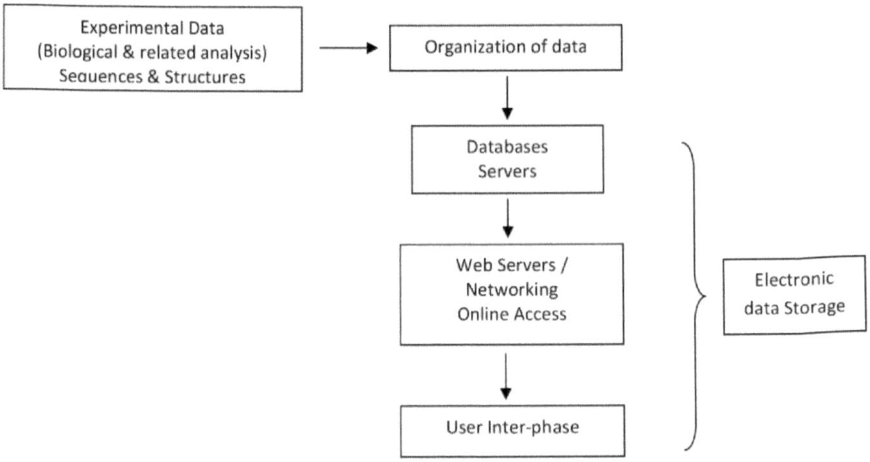

FIGURE 10.1 Schematics flow and layout of data flow and network for production, construction, and the accession of a database.

10.2 BIOLOGICAL DATABASES

Basically, let us technically attempt to understand databases. What is a database? As far as computational systems are concerned, a database is typically great compilation of information organized particularly for swift exploration and recovery. Now a biological database is a compilation of information that is systematized so that its information can effortlessly be accessed, supervised, and renewed.

The activity of preparing a database can be divided into:

- Collection of data in a form which can be easily accessed.
- Making it available to a multiuser system, that is making it always available for the users.

Biological databases are collection of information related to life sciences, gathered from various scientific experiments, computational analysis, high-throughput experiment technology, and reported papers.[2] They include information from critical areas including phylogenetics, proteomics, genomics, microarray gene expression, and metabolomics.[3] The biological databases provide information which includes structure, gene function, clinical effects of mutations, localization (both cellular and chromosomal), and similarities of biological sequences.

The biological databases can be categorized essentially as the primary database and secondary database. This categorization is based on the bases of its data type or the record configuration. Primary databases usually store first-hand information about the biological analysis done. They upload latest information explored in experiments and renew the information to make sure the data quality is retained. In short, it is a raw data or unprocessed data. Secondary databases are the databases which make the application of an another database for retaining the information. Therefore, they retrieve information by demanding the required data from other databases. They frequently process the data corresponding to the requested results.[4] Accordingly, the biological databases can be classified as:

- Meta-databases
- Nucleic acid databases
- Protein databases
- Application-based databases/specialized databases

10.3 META-DATABASES

Metadata is the data information that provides information about other data.

Three distinct types of metadata exist:

- descriptive metadata
- structural metadata
- administrative metadata[5]

These types of databases are integrative types of databases wherein they collect data from different sources and make them available in a new and more convenient form.

Some distinctive biological databases of this type are:

- Pathogen Portal
- MetaBase
- Metascape
- Entrez

10.3.1 ENTREZ

The Entrez Global Query Cross-Database Search System is a exploration engine (related to structures, references, sequences, chromosome maps, and genes), integrated search, a web portal, retrieval system with a single query string, user interface that permits consumers to explore numerous distinct health sciences databases simultaneously at the National Center for Biotechnology Information (NCBI) website, a part of the National Library of Medicine (NLM).[6]

10.3.2 METABASE

MetaBase is a user-made and user-contributed database, which provides information about all the existing biological databases on the Internet. Initially, it was developed using the data from Nucleic acid Research (NAR) database. Currently, MetaBase comprises about 1000 biological databases and over 800 "web services" derived from NAR, as well as more than 50 "user-contributed" databases.

10.4 NUCLEIC ACID DATABASES

They consist of the newly sequenced and already existing nucleic acid sequences compiled together with proper annotation. They consist of DNA databases, gene expression databases, genome databases, phenotype databases, and RNA databases.

10.4.1 DNA DATABASES

There are three primary DNA databases that contain original sequence data:

- GenBank (managed by the National Center for Biotechnology Information).[7]
- EMBL (managed by European Bioinformatics Institute)[8]
- DNA Data Bank of Japan (managed by National Institute of Genetics)[9]

These are storehouses of nucleotide sequence data from all living organisms. All three accept nucleotide sequence data submissions and exchange new and updated data on a daily basis to achieve optimum synchronization between them.

10.4.2 GENE EXPRESSION DATABASES

They mainly consist of expression data, especially the microarray data.

The most commonly used gene expression databases are:

- BioGPS (The Scripps Research Institute)
- ArrayExpress (European Bioinformatics Institute)[10]
- Gene Expression Omnibus (GEO)[11]

10.4.3 GENOME DATABASES

These databases have genome sequences, which they annotate and analyze, and make them publicly accessible. Some are curated with experimental literature data to improve computed annotations. These databases may contain many species genomes, or a single model organism genome.

Examples of these databases are:

- Ensembl
- FlyBase
- EcoSync

10.4.4 PHENOTYPE DATABASES

These databases contain phenotypic data of the organisms and its correlation with the genotypic data can be provided in some databases.

Common examples of these databases are:

- PhenCode
- Phi-Base[12]

10.4.5 RNA DATABASES

These databases contain information related to RNA sequences, that is, mRNA, rRNA, tRNA, and their related expression data. There are many commonly used databases like the following

- C-It-Loci—a database of RNA expression and conserved loci for studying lncRNAs across species.[13]
- microRNA database.
- LncRNAWiki—a wiki-based database for community curation of known human long noncoding RNAs, Rfam, a database of RNA families.[14]
- miRbase.[15]
- snoRNAdb—a database of snoRNAs.[16]

10.5 AMINO ACID/PROTEIN DATABASES

These databases contain the data of:

- amino acid sequences
- proteomics databases
- protein interaction data
- protein structures
- protein models

10.5.1 PROTEIN SEQUENCE DATABASES

The databases contain amino acid sequence data, its annotation and curation, and additional literature information. These are one of the widely used databases as the information they provide have a wide range of application in various fields of protein studies.

Common examples of these databases are:

- UniProt Universal Resource (managed by EBI, Swiss Institute of Bioinformatics, PIR).[17]

- Protein Information Resource (Georgetown University Medical Center (GUMC)).[18]
- Swiss-Prot Protein Knowledgebase (Swiss Institute of Bioinformatics).[19]
- PROSITE Database of Protein Families and Domains.[20]
- Pfam Protein families database of alignments and HMMs (Sanger Institute).[21]
- PRINTS—a compendium of protein fingerprints from (Manchester University).[22]

10.5.2 PROTEIN STRUCTURE DATABASES

Protein structural information constitutes the major information source for these databases.

The primary protein structure database is:

- PDB—3D structure of proteins obtained from X-ray crystallography and NMR experiments is stored.[23]

Secondary structural databases include:

- SCOP Structural Classification of Proteins
- CATH Protein Structure Classification

10.5.3 PROTEIN MODEL DATABASES

Examples of protein model databases include:

- Swiss-model Server and Repository for Protein Structure Models.
- ModBase Database of Comparative Protein Structure Models (Sali Lab, UCSF).
- Protein Model Portal (PMP).

Protein–Protein and Other Molecular Interactions:

- BioGRID—A General Repository for Interaction Datasets (Samuel Lunenfeld Research Institute).
- CCSB Interactome.
- IntAct molecular interaction database: a central, standards-compliant repository of molecular interactions, including protein–protein, protein–small molecule, and protein–nucleic acid interactions.
- BIND Biomolecular Interaction Network Database.
- DIP Database of Interacting Proteins.

10.6 ADDITIONAL DATABASES

In addition to the common nucleic acid and protein databases, there are other biological information databases such as the following:

- Signal Transduction Pathway databases
- Carbohydrate Structure databases
- Metabolic Pathway databases
- Taxonomic databases
- Protein Function databases
- PCR and Quantitative PCR Primer databases
- Mathematical Model databases

10.6.1 CARBOHYDRATE STRUCTURE DATABASES

- EuroCarbDB—A repository for both carbohydrate sequences/structures and experimental data.
- Signal Transduction Pathway databases include:
- NCI-Nature Pathway Interaction Database
- Cancer Cell Map
- Reactome—Navigable map of human biological pathways, ranging from metabolic processes to hormonal signaling
- Netpath—A curated resource of signal transduction pathways in humans
- SignaLink databases

10.6.2 METABOLIC PATHWAY AND PROTEIN FUNCTION DATABASES

These are important pathway databases useful in pathway modeling and mathematical simulation. These databases include:

- BioCyc Database Collection[24]
- EcoCyc and MetaCyc, BRENDA[25]
- The Comprehensive Enzyme Information System includes:
- KENDA, FRENDA, DRENDA, and AMENDA
- KEGG PATHWAY Database (Univ. of Kyoto), Reactome[26]

Navigable map of human biological pathways, ranging from metabolic processes to hormonal signaling (Cold Spring Harbor Laboratory, European Bioinformatics Institute, Gene Ontology Consortium).

10.6.3 METABOLOMIC DATABASES

- DrugBank
- Yeast Metabolome Database (YMDB)
- MetaboLights
- Human Metabolome Database (HMDB)
- NIH Common Funds Metabolomics Database
- *E. coli* Metabolome Database (ECMDB)

10.6.4 MATHEMATICAL MODEL DATABASES

Published mathematical models describing biological processes, build and simulate large-scale models in real-time and in a highly collaborative fashion. The databases include:

- CellML The Cell Collective
- Biomodels database

10.7 CONCLUSION

The various nucleic acid databases, gene databases, protein/amino acid databases, protein databases, RNA databases, metabolic databases, carbohydrate databases, metabolomic databases, mathematical model databases, signal transduction pathway databases, phenotype databases, and so on will provide the updated information and support to the emerging researchers across the globe for future development and research.

KEYWORDS

- **database**
- **meta-database**
- **nucleic acid database**
- **protein database**
- **metabolic database**
- **mathematical model database**

REFERENCES

1. Brief Bioinform. Published online November 19, 2016. DOI: 10.1093/bib/bbw114
2. Attwood, T. K.; Gisel, A.; Eriksson, N-E.; Bongcam-Rudloff, E. Concepts, Historical Milestones and the Central Place of Bioinformatics in Modern Biology: A European Perspective. *Bioinform.—Trends Methodol.*. InTech, 2011. Retrieved Jan 8, 2012.
3. Altman, R. B. Building Successful Biological Databases. *Brief. Bioinform.* Mar **2004,** *5* (1), 4–5. DOI: 10.1093/bib/5.1.4. PMID 15153301.
4. Nucleic Acids Research Database Issue 2016. Nucleic Acids Research. Oxford University Press. Retrieved Oct 26, 2016.
5. Zeng, M. *Metadata Types and Functions*. NISO, 2004. Retrieved Oct 5, 2016.
6. NCBI Resource Coordinators. Database Resources of the National Center for Biotechnology Information. *Nucl. Acids Res.* **2012,** *41* (Database issue), D8–D20. DOI: 10.1093/nar/gks1189. PMC 3531099. PMID 23193264
7. Benson, D; Karsch-Mizrachi, I.; Lipman, D. J.; Ostell, J.; Wheeler, D. L. et al. GenBank. *Nucl. Acids Res.* 2008, *36* (Database), D25–D30. DOI: 10.1093/nar/gkm929. PMC 2238942. PMID 18073190.
8. EMBL Member States. European Molecular Biology Laboratory, 2015.
9. Tateno, Y.; Imanishi, T.; Miyazaki, S.; Fukami-Kobayashi, K.; Saitou, N.; Sugawara, H.; et al. DNA Data Bank of Japan (DDBJ) for Genome Scale Research in Life Science.. *Nucl. Acids Res.* **2002,** *30* (1), 27–30. DOI: 10.1093/nar/30.1.27. PMC 99140. PMID 11752245.
10. http://www.ebi.ac.uk/arrayexpress/
11. Barrett, T.; Edgar, R. Gene Expression Omnibus (GEO): Microarray Data Storage, Submission, Retrieval, and Analysis. *Methods Enzymol.* **2006,** *411*, 352–369. DOI: 10.1016/S0076-6879(06)11019-8
12. Winnenburg, R.; Baldwin, T.K.; Urban, M.; Rawlings, C.; Köhler, J.; Hammond-Kosack, K.E. PHI-Base: A New Database For Pathogen Host Interactions (PDF). *Nucl. Acids Res.* **2014,** (Database Issue), D459–464. DOI: 10.1093/nar/gkj047. PMC 1347410. PMID 16381911
13. http://c-it-loci.uni-frankfurt.de/
14. Lina, Ma; Vladimir. B. B.; Zhang, Z. On the Classification of Long Non-Coding RNAs. (2013) *RNA Biol.* **2013,** *10* (6), 924–933.
15. Meyers, B. C.; Axtell, M. J.; Bartel, B.; Bartel, D. P.; Baulcombe, D.; Bowman, J. L.; Cao, X.; Carrington, J. C.; Chen, X.; Green, P. J.; Griffiths-Jones, S.; Jacobsen, S. E.; Mallory, A. C.; Martienssen, R. A.; Poethig, R. S.; Qi, Y.; Vaucheret, H.; Voinnet, O.; Watanabe, Y.; Weigel, D.; Zhu, J. K. *Plant Cell.* **2008,** *20* (12), 3186–3190.
16. http://lowelab.ucsc.edu/snoRNAdb/
17. Dayhoff, M. O. *Atlas of Protein Sequence and Structure*; Silver Spring, MD: National Biomedical Research Foundation, 1965.
18. Wu, C.; Nebert, D. W. Update on Genome Completion and Annotations: Protein Information Resource. *Human Genom.* 2004, *1* (3), 229–233. DOI: 10.1186/1479-7364-1-3-229. PMC 3525084. PMID 15588483.
19. Apweiler, R.; Bairoch, A.; Wu, C. H.; Barker, W. C.; Boeckmann, B.; Ferro, S.; Gasteiger, E.; Huang, H.; Lopez, R.; Magrane, M.; Martin, M. J.; Natale, D. A.; O'Donovan, C.; Redaschi, N.; Yeh, L. S. UniProt: The Universal Protein Knowledgebase. *Nucl. Acids Res.* **2004,** *32* (90001), 115D–1119. DOI: 10.1093/nar/gkh131. PMC 308865. PMID 14681372.

20. De Castro, E.; Sigrist, C. J. A.; Gattiker, A.; Bulliard, V.; Langendijk-Genevaux, P. S.; Gasteiger, E.; Bairoch, A.; Hulo, N. scanprosite: Detection Of PROSITE Signature Matches And Prorule-Associated Functional And Structural Residues In Proteins. *Nucl. Acids Res.* **2006,** *34* (Web Server issue), W362–W365. DOI: 10.1093/nar/gkl124. PMC 1538847. PMID 16845026.
21. Finn, R. D.; Tate, J.; Mistry, J.; Coggill, P. C.; Sammut, S. J.; Hotz, H. R.; Ceric, G.; Forslund, K.; Eddy, S. R.; Sonnhammer, E. L.; Bateman, A. The Pfam Protein Families Database. *Nucl. Acids Res.* **2008,** *36* (Database issue): D281–D288. DOI: 10.1093/nar/gkm960. PMC 2238907. PMID 18039703
22. Attwood, T. K.; Bradley, P.; Flower, D. R.; Gaulton, A.; Maudling, N.; Mitchell, A. L.; Moulton, G.; Nordle, A.; Paine, K.; Taylor, P.; Uddin, A.; Zygouri, C. PRINTS and Its Automatic Supplement, prePRINTS. *Nucl. Acids Res.* **2003,** *31* (1), 400–402. DOI: 10.1093/nar/gkg030. PMC 165477. PMID 12520033
23. Kleywegt, G. J.; Harris, M. R.; Zou, J. Y.; Taylor, T. C.; Wählby, A.; Jones, T. A. The Uppsala Electron-Density Server. *Acta Crystallogr D.* Dec **2004,** *60* (Pt 12 Pt 1), 2240–2249. DOI: 10.1107/S0907444904013253. PMID 15572777.
24. Caspi, R.; Altman, T.; Dreher, K.; Fulcher, C. A.; Subhraveti, P.; Keseler, I. M.; Kothari, A.; Krummenacker, M.; Latendresse, M.; Mueller, L. A.; Ong, Q.; Paley, S.; Pujar, A.; Shearer, A. G.; Travers, M.; Weerasinghe, D.; Zhang, P.; Karp, P. D. The Meta Cyc Database of Metabolic Pathways and Enzymes and the Bio Cyc Collection of Pathway/Genome Databases. *Nucl. Acids Res.* 2011, *40* (Database issue), D742–D753. DOI: 10.1093/nar/gkr1014. PMC 3245006. PMID 22102576.
25. Chang, A.; Scheer, M.; Grote, A.; Schomburg, I.; Schomburg, D. Brenda, Amenda and Frenda the Enzyme Information System: New Content and Tools in 2009. *Nucl. Acids Res.* **2008,** *37* (Database issue), D588–D592. DOI: 10.1093/nar/gkn820. PMC 2686525. PMID 18984617.
26. Kanehisa, M.; Goto, S. KEGG: Kyoto Encyclopedia of Genes and Genomes. *Nucl. Acids Res.* **2000,** *28* (1), 27–30. DOI: 10.1093/nar/28.1.27. PMC 102409. PMID 10592173.

CHAPTER 11

High-Throughput Screening Technique: Role in the Drug Discovery of Low Molecular Weight Ligands

SANTOSH S. CHHAJED* and PARAG A. PATHADE

Department of Pharmaceutical Chemistry, MET Institute of Pharmacy, Bhujbal Knowledge City, Nashik 422003, Maharashtra, India

**Corresponding author. E-mail: chhajedss@gmail.com*

ABSTRACT

Swift enhancement in molecular biology techniques has led to the growth of efficient, rapid, and drug testing systems. The techniques employed by these systems are together recognized as high-throughput screening (HTS). HTS is one of the latest method utilized in drug design and may be pertained in chemical sciences and biological sciences. This process, due to the exploitation of detectors, robots, and software that control the entire course, allows a sequence of chemical compounds analyses to be performed in a very diminutive time and the similarity of biological structures, which is frequently related to the toxicity. The extremely efficient environment of HTS screening for the recognition of decidedly targeted precise molecules is due to its accurate focus on a single mechanism. This logical development of receptor technology is closely connected with the changes in the strategy of chemical synthesis. The vast number of compounds produced by combinatorial chemistry and the possibility of testing many compounds, including natural products, in a short period of time by HTS attracted the attention of many workers. Various detection techniques like homogeneous time-resolved fluorescence, fluorescence resonance energy transfer, and

Biologically Active Small Molecules: Modern Applications and Therapeutic Perspectives.
Debarshi Kar Mahapatra, & Sanjay Kumar Bharti (Eds.)

so on, are accessible, and the screening more than 100,000 samples/day is possible. With the advent of robotics, automation, and miniaturization techniques, it is possible to monitor 50,000 molecules/day with complex workstations. HTS methods are also employed to typify pharmacokinetic and metabolic data about new drugs. With the application of cassette-dosing techniques, even the pharmacokinetic data can be evaluated for numerous drug candidates, though these are not free from drawbacks, yet an effectual technique to additional augment the drug discovery and development rate. The aim of this chapter is to provide a summary of the HTS methodology applications in industries as well as in academia.

11.1 INTRODUCTION

Biological "trial and error" testing of a huge compilation of low molecular weight molecules for a precise pharmacological effect is a typical way to find out original lead compounds which subsequently serve as templates for further optimization in medicinal chemistry programs. In the last two decades, the advent of recombinant DNA technologies together with improved assay techniques and high-performance laboratory automation noticeably altered pharmacological screening processes. Nowadays, high-throughput screening (HTS) has gained much attention.

The rapid increase in molecular biology techniques has resulted in the growth of efficient, rapid drug testing systems. The techniques used by these systems are collectively known as HTS. HTS methods give accurate results even when extremely small amounts of the test substance are available. However, if it is to be used in an economic fashion as well as efficiently, this technology requires the fast production of numerous molecules for testing that cannot be met by the conventional approach to organic synthesis, which is regularly geared to the production of one molecule at a time (Figure 11.1).

It is fundamentally a process of screening and assaying an enormous number of biological modulators and effectors against selected and specific targets. The methods and principles of HTS find their application for the screening of genomics, combinatorial chemistry, peptide, and protein libraries. The main reason of this method is to accelerate the process of drug discovery by screening massive molecule libraries with a speed which may go beyond a few thousand compounds/day or compounds/week. For any assay or screening by HTS to be successful, several steps like reagent preparation, target identification, assay development, compound management, and high-throughput library screening should be performed with paramount concern and exactitude.

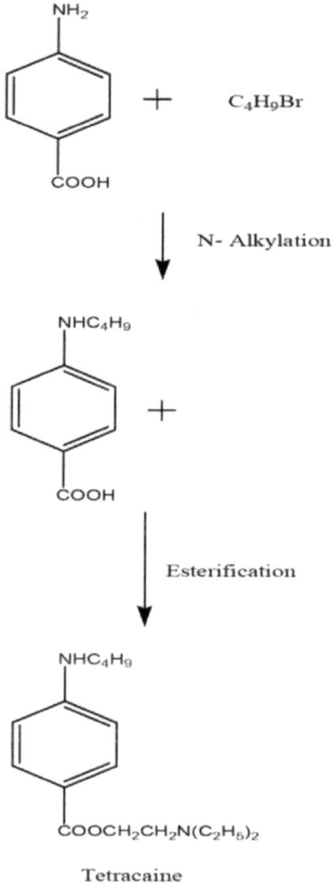

FIGURE 11.1 A traditional stepwise organic synthesis scheme illustrated by the synthesis of the local anesthetic tetracaine.

HTS is a very vast area of study and development that has deep scopes including whole organ testing, enzyme testing, and whole animal testing via cassette dosing. Cassette dosing is a process in HTS that enables to quickly determine the pharmacokinetics of numerous drug candidates. Contrasting other techniques to assess the pharmacokinetics in this procedure single animal is provided concurrently and blood samples were collected to estimate the same. The chief benefit is that the pharmacokinetics of a large number of compounds can be assessed accurately and rapidly. But the main shortcoming is that immediate administration can result in drug–drug interaction. HTS is a new technique for drug discovery, but it is not the only method. HTS not only helps in drug discovery but also in developing present drug moieties to optimize their activity. In the previous years, a lot of development in science and technology took place and economic pressures forced researchers to build up precise and speedy drug development and screening technologies to fight against the ever-increasing diseases and several resistant microbial pathogens to the existing accessible drugs. This also pertains to screening the ever-increasing compound libraries waiting to be screened due to the increase in the parallel and combinatorial chemistry synthesis. Research is also performed so as to reduce the drug development expenses so that industries keep side by side with ever-rising competition. It is predictable that with the introduction of human genomes as prospective candidates, the molecule library will be as huge as 100 million candidates that would necessitate nearly 1012 assays to set up their structure–activity relationship. Originally, the assays were performed in 96-well plates, but with progression, now there are also 1586-well plates available. Typical HTS programs have a great possibility to screen up to 10,000 compounds/day, while some laboratories with ultra-high-throughput screening can perform 100,000 assays/day.

11.2 ASSAY DESIGN

HTS is the name given to the quick semiautomated simultaneous prime screening of large numbers of molecules, mixtures, and plant extracts. The process is based on the use of biomicro assays that are swift to perform and necessitate a quite small amount of the reagents and test molecules. These assays are carried out in 96-well plates and bigger well plates using specialized handling equipment. They are based on the test compound interacting with a target, such as an enzyme, a cell membrane receptor, hormone, nuclear receptors, and DNA, that is related to the disease state under investigation. As a result, it may be essential to recognize, purify, and segregate the intention before a library can be monitored. Additionally, assays are particularly designed for analysis and so they have to be authenticated. These screenings would be costly both in context of time and money.

The participation of biological targets in the assays means that assays are usually performed in the water environment. As a result, an assay may be effectual if a considerable quantity of the molecule of interest dissolves completely in the water environment. Dimethylsulphoxide (DMSO) is repeatedly adjoined to the assay mixtures to improve the aqueous solubility of the testing molecule.

The price of the particular reagents employed for screening the huge libraries of particular molecules is extremely costly. Therefore, a lot of corporations and investigators decrease the expenses by investigating huge libraries as a mixture of molecules. Though, this can result in deceptive consequences; for example, false-positive results may happen when the mixture under test contains several individual molecules with a feeble activity. As a result, the mixture provides a superior largely reaction to the test. This effect would be wrongly understood by the researchers as the mixture having a sturdily active molecule. A false-negative result may arise if the dormant components of a mixture connect in inclination to the assay target. This averts the active molecule attachment with the target and giving a good assay response. The concentration of the test compound employed in the assay may also provide an increase in false positives and false negatives. Employing a very high concentration of the test molecule can also provide a false positive due to nonselective concentration-driven binding with the desired target.

On the other hand, employing too much amount can provide a false-negative result when an inadequate quantity of the active molecule for testing is there for binding with the target. Other sources of imprecision also happen in detailed forms of microassay.

The microassays used in HTS may be classified for expediency as either cell-based assay or biochemical assays. The biochemical assays are based on the interaction of the investigated molecule with distinct chemical units isolated from the cells, such as a hormone, enzyme, receptor, whereas entire cell assays are based on the application of whole cells. Though, it is emphasized that HTS is used as a chief screen for active molecules. Any active compounds (hits) that appear worthy of additional analysis must be subjected to a greater range of activity tests before they might be believed for clinical development.

11.2.1 BIOCHEMICAL ASSAYS

Biochemical assays (mechanism-based assays) are typically based on connecting a compound to a receptor or the inhibiting an enzyme-catalyzed reaction employing a target that has been recognized as being pertinent to a particular disease condition. This target has typically been isolated from a cell and is no longer fraction of the cell. The test compound binding with the desired target is determined by employing radioactive isotopes and conventional investigative techniques like spectroscopy by employing a mixture of protocols like by determining fluorescence in the scintillation proximity assays (SPA). This assay utilizes engineered resin beads so that it is proficient in binding with substances extensively. The bead comprises a scintillant that fluoresces in contact with low-energy radioactive source comes in contact within ~2 cm of the beads surface. The radioactive isotopes employed in SPA assays release low energy discharge that has a very little path in the water environment (Table 11.1).

TABLE 11.1 Examples of the Radioactive Isotopes Used in SPA Assays.

Element	Isotope	Half-life	Pathway in aqueous solution
Carbon	14C	5730 years	–
Hydrogen	3H	12.26 years	<1 μm
Iodine	125I	60 days	~1 μm

Several SPA-enzyme-based assays are performed by using radiolabeled substrates or ligands. Consider, for example, an enzyme inhibition assay, based on the use of a substrate A–B for the enzyme, where B is the part of the substrate that comprises the radioactive isotope and A contains a so-called capture group, which comprises structures that bind to the SPA beads. B does not restrain a capture group. The action of the substrate A–B with the enzyme and any crucial coenzymes in the absence of the inhibitor results in cleavage of all of the substrate A–B. When the SPA beads are added no fluorescence is observed as only the nonradioactive A binds to the beads. The radioactive B remains in the solution too far from the beads to cause fluorescence (Figure 11.2a). However, when a test molecule is present, the degree of the cleavage depends on the potency of the inhibition displayed by the test compound: the superior the inhibition, the less the cleavage of the substrate A–B. As a result, when the SPA beads are supplementary added, the unreacted radioactive substrate A–B and the nonradioactive A attaches to the beads. The emissions from the unreacted A–B bound to the bead are closely sufficient to promote bead fluorescence (Figure 11.2b). However, B, the radioactive product of the enzyme-catalyzed reaction, still continues in the solution too far from the beads to cause fluorescence. As a result, the intensity of the fluorescence is completely relative to the bead-bound A–B. It is noticed that the superior the fluorescence, the better the degree of enzyme inhibition of the substrate A–B by the test molecule.

Without any inhibition the substrate completely reacts

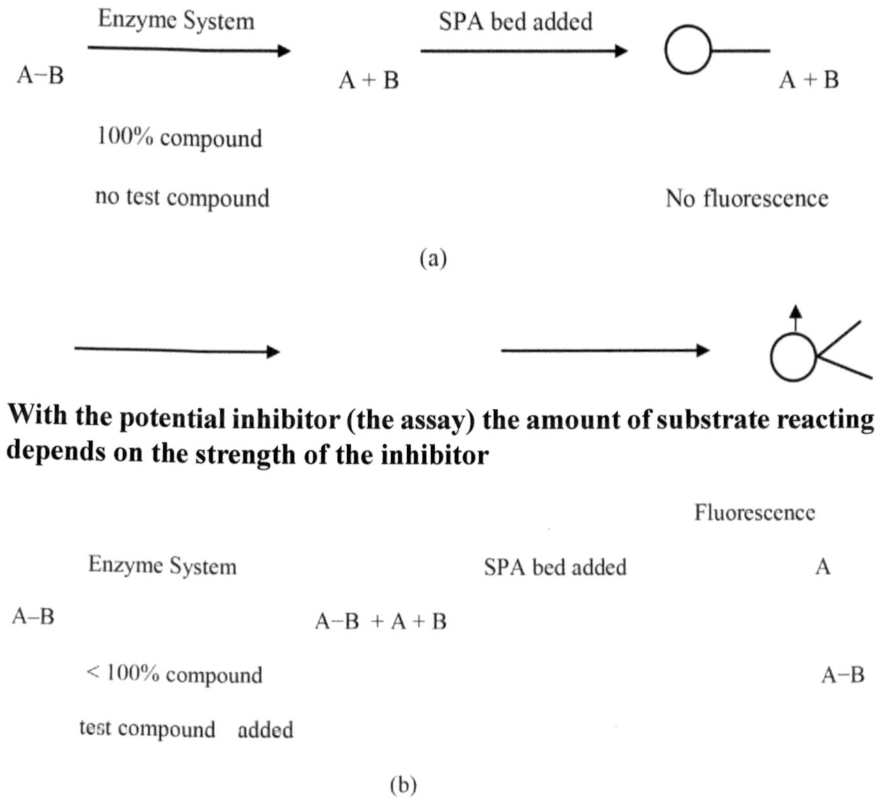

FIGURE 11.2 A schematic representation of an enzyme HTS microassay. Ideal situations are illustrated. (a) No fluorescence is observed as there are no A–B molecules to bind to the SPA beads. (b) The cleavage of only a fraction of the A–B substrate molecules results in the unreacted molecules binding to the SPA beads, resulting in the fluorescence. The fragment A of the substrate also binds to the beads.

11.2.2 WHOLE-CELL ASSAYS

Whole-cell assays are chosen when the mechanism of the disease state has not been well defined. They also suggest several other benefits over biochemical tests. For example, whole-cell tests may recognize a molecule that acts on sites other than the target site. They are typically performed under circumstances that are more like those encountered if the test molecules were applied in a patient. Consequently, test compounds that are either too hydrophobic, and as a result connect too sturdily to serum albumin or do not cross cell membranes without usually be active. Therefore, it is relatively easy to identify these molecules and eradicate them from the investigation. Moreover, test compounds that are toxic are regularly recognized because of their consequence on the cells employed in the test. These benefits reflect that many medicinal chemists prefer whole-cell assays to biochemical assays.

Cell-based assays for HTS can be classified as second messenger assays and reporter gene assays.

11.2.2.1 SECOND MESSENGER ASSAYS

It checks the signal transduction from the activated cell-surface receptors. The second messenger assays typically measure fast, transient fluorescent signals that occur in matter of seconds or milliseconds. Many fluorescent molecules are known to respond to the changes in intracellular calcium ion concentration, membrane potential, and various other parameters, hence they are used in the development of second messenger assays for receptor stimulation and ion-channel activation. The development of hydrophobic voltage-sensitive probes and fluorescence resonance energy transfer (FRET)-compatible microplate instrumentation has been helping the advancement of the screening technique for ion-channel drug discovery.

11.2.2.2 REPORTER GENE ASSAYS

It monitors the cellular reaction at the transcription/translation level. It point toward the existence or nonexistence of a gene product that in turn reflects an alteration in a signal transduction pathway. The quantification of the reporter is typically performed by biochemical methods, namely, by enzymatic activity determination. The plasmids are distinctive reporter genes utilized. An exclusive in vitro investigation was achieved by Rungpragayphan et al. for the screening and generation of combinatorial protein library in an array format. The exploration used in vitro coupled reporter gene assay and polymerase chain reaction (PCR).

11.3 REAGENTS IN HTS

In any chemical synthesis or testing and screening, reagents play the main function, and HTS is not the exemption in this case. Reagents must be described and optimized before utilization. In the investigations, it was established that nucleic acids aptamers that attach to other molecules with a greater affinity may be employed as a versatile reagents in competition binding HTS assays to recognize and optimize low molecular weight ligands to the desired targets. The chief benefits of employing aptamers in HTS assays are the high affinity of aptamers for protein targets, speed of aptamer identification, compatibility with various labeling/detection strategies, and relatively large aptamer–protein interaction surfaces. Aptamers may be chiefly functional in HTS assays with the protein targets that have no identified binding partners, such as orphan receptors.

The enzymes are regularly employed as reagents in HTS; for example, tyrosine kinase was utilized to determine its inhibitors. In this regard, care should be taken in the preparation of reagent where there should not be any contamination with other phosphatases, peptidases, and kinases which could contend with tyrosine kinase to produce false results. Other than kinase enzymes, generic reagents like streptavidine-allophycocyanine, streptavidine-europium, and biotinylated deoxyuridine triphosphate were utilized and extended for estimating the activity of HIV-reverse transcriptase. DMSO is an additional extensively employed reagent as it is the preferred vehicle for compound/sample delivery. The important point to remember during the use of DMSO is that its tolerance should be found out earlier at the assay development stage so as to perform additional optimization at the screening stage.

11.4 HITS AND HIT RATES

A hit occurs when the pharmacological potentials of a compound of screening have a value higher than a random lowest value fixed by the researchers while employing a particular assay. For example, in an enzyme inhibition assay, a hit may be scored when the enzyme activity gets inhibited by 50% preagreed value. It is significant to set the condition for hit prior to performing the assay as hit rates are frequently utilized as a determinant of the strength of an assay procedure. Hit rates are defined as the number of active samples discovered by an assay expressed as a percentage of the total samples used in that screen. The assays with values ~0.1% hits are usually considered as valid. The higher values may arise due to a number of reasons, for example, a series of compounds with similar structures and hence similar activities are being tested, the assay is not precisely sufficient or the hit criteria are not accurate adequate. However, high hit rates may be of utilize while assay developing.

Imprecise hit values may be acquired due to false positives and false negatives. In single compound libraries, false positives are detected as the inactive molecules show positive results due to unsuitable interaction with the chemical reagents and the biological material employed in the assay. Instead, they may be as a result of the molecule has passed an introductory screening but are not being dynamic sufficient to pass further scrupulous follow-up tests. Conversely, negative hits may also be due to compounds that are active but do not register as such in the assay. The screening of mixtures libraries may also demonstrate in false positives and false negatives A hit is only converted into a lead when it exhibits a superior activity profile in follow-up assays and its structure is confirmed.

11.5 AUTOMATIC METHODS OF LIBRARY GENERATION AND ROBOTICS IN HTS

The traditional organic synthesis and assay procedures are exhaustively laborious. The arrival of combinatorial chemistry and HTS has led to enhanced utilization of automation in the medicinal chemistry with several organizations producing HTS analyzers, automatic compound synthesizers, and custom-built machines. The synthesis is typically brought about in a series of computer-controlled steps using an appropriate array of reaction vessels at the so-called stand-alone workstations. The procedures in the synthesis-like reaction vessels loading with the solvents and reagents, washing, heating, pipetting, filtration, and so on, are generally conducted by means of a robotic arm bounded by the software controlling the synthesizer. Track systems and robotic arms are exploited to move microplates between the workstations. The autoanalysis is performed in the same universal way at the workstations to relocate the reagents and samples.

The combination of HTS and robotics has been significant to attain the preferred screening rates, as well as helping researchers from dull work. In the current years, researchers could argue that robotics for screening has been more of a research effort than a true execution of a stable technology. The problems associated with screening robotics have included long manual to automated method transfer time, long design and implementation time, limited error recovery abilities, and nonstable robotic operation. These problems can be characterized by poor software design, robot integration architectures, and robot–workstation compatibility issues. Conventionally, these integrated robot architectures have engaged different operating systems, multiple layered computers, the necessity of complex scheduling software, and a single central robot servicing all peripheral devices to coordinate all of the above. The currently exploited robot-centric HTS systems comprise a centralized robotic system that has a gripper placed so that it can selectively choose and put microplate (40–100 in number) around a platform in a single run. The screener loads the robotic platform with microplates and reagents at the beginning of the experiment and the assay is then processed unattended. Robotic HTS systems often possess humidified CO_2 incubators and are enclosed for tissue culture work. Similar to assembly-line manufacturing, microplates are passed down a line in serial fashion to consecutive processing modules. Each component has its individual plain pick and places the robotic arm to move plates to the next component and microplate processing device. Consequently, at every module, one step of the assay is finished which is attached with Windows NT™ (Microsoft, Redmond, WA) and an Ethernet TCP/IP link between the modules, which offers a quite simpler and additional stable platform than that of the robot-centric HTS systems.

11.6 PROFILING

HTS has its chief intention, the recognition of a few "VALIDATED HITS" with large compound libraries. The decision as to whether a particular "hit" is worth a practice as a chemical lead in a drug discovery project depends on numerous aspects, considerable ones being chemical attributes and its pharmacokinetic and pharmacodynamics properties. The technology entails automation, miniaturization, and assay readouts needed for HTS is ongoing to expand swiftly, and as it does so, the laboratory setups established in HTS facilities are progressively augmenting their potentials afar their key function of identifying hits. As this takes place, it becomes probable for HTS techniques to be concerned to more varied compound profiling assays relating not only to the target selectivity of the compound libraries but also to their pharmacokinetic attributes. Increasingly, therefore, early compound profiling tasks on "hit" compounds are being performed in the HTS laboratory where the necessary technical expertise is concentrated. Such assays are also very helpful in the "lead identification" stage of a project, where dedicated synthetic compound libraries based on the initial hits need to be evaluated, as this work usually necessitates testing small compound libraries, usually fewer than 1000 molecules simultaneously, in a number of diverse assays. Small bestowed robotic workstations are required, as compared to the fast, but the inflexible factory-style robotic assembles employed for large-scale HTS. It is clear that pharmacological profiling will be an increased activity of HTS units in the upcoming era, and will assist to add additional worth in the drug discovery process.

11.7 SCREENING EXPENSE AND OUTSOURCING SCREENING

Hardly ever any organization wishes to screen 100,000 molecules/day in house. The reason behind this includes equipment/robotic requirements, restrictions of any drug discovery process, a limited need to invest in changing technologies, and infrastructure investment. Some precise costs associated with the screening include microplate costs, assay reagent costs (cell-culture expenses, reagents, etc.), screening employee costs, pipette tip box costs, database costs, data handling/analysis time, laboratory space costs, and robot purchase costs. Due to the united complexity with the above mentioned, a rising amount of contract screening companies are budding such as *PanLabs, Tropix,* and *Evotec*. The services offered by these corporations typically include the data analysis, assay development and screening, and additional library support needs. A few contract screening organizations are also being used for their capability to offer assay data with very fast turnaround times. They attain this by conducting 24 h shifts and using HTS robotic technologies. Additionally, some companies choose to outsource primary screening, since they are discovering the requirement to shift a few of their screening personnel to the growing secondary screening programs. This allows for additional proprietary secondary screening to be done in-house, as well as continuous high-rate "hit" creation from outsourced initial screening. The price of screening a wholly huge molecule library may amount more than $300,000.

11.8 CONCLUSION

The emerging technique of drug design HTS finds immense applications in both biological and chemical sciences. With the introduction of automatic compound synthesizers, HTS robotics, analyzers, miniaturization techniques, and automation, it has become possible to screen 50,000 molecules/day with complex workstations. HTS is a very vast area of study and development having deep scopes including enzyme testing, whole organ testing, and even whole animal testing via cassette dosing. The economic technique will speed up the drug discovery process by rapid molecules screening.

KEYWORDS

- **high-throughput screening**
- **automation and robotics**
- **miniaturization**
- **robotics**
- **library generation**

REFERENCES

1. Carnero, A. High Throughput Screening in Drug Discovery. *Clin. Transl. Oncol.* **2006,** *8*, 482–490.
2. Broach, J. R.; Thorner, J. High-Throughput Screening for Drug Discovery. *Nature* **1996,** *384*, 14–16.
3. Du, G.; Fang, Q.; den Toonderb Jaap, M. J. Microfluidics for Cell-Based High Throughput Screening Platforms—A Review. *Anal. Chim. Acta*. **2016,** *903*, 36–50.
4. Alderman, E.; Elands, J. A. Novel Approach to Ultra-High Throughput Screening. *Genet. Eng. News*. **1998,** *18*.
5. Walter, G.; Bussow, K.; Lueking, A.; Glokler, J. High-Throughput Protein Arrays: Prospects for Molecular Diagnostics. *Trends. Mol. Med.* **2002,** *8*, 250–253.
6. Rose, D.; Lenmo, T. Challenges in Implementing High-Density Formats for High Throughput Screening. *Lab. Automat. News*, **1997,** 2.
7. Li Pira, G.; Ivaldi, F.; Moretti, P.; Manca, F. High Throughput T Epitope Mapping and Vaccine Development. *J. Biomed. Biotechnol.* **2010,** *2010*, 325720.
8. Hann, M. M.; Oprea, T. I. Pursuing the Lead Likeness Concept in Pharmaceutical Research. *Curr. Opin. Chem. Biol.* **2004,** *8*, 255–263.
9. Jo Wildey, M.; Haunso, A.; Tudor, M.; Webb, M.; Connick, J.H. Chapter Five—High Throughput Screening. *Annu. Rep. Med. Chem.* **2017,** *50*, 149–195.
10. Mayr, L. M.; Bojanic, D. Novel Trends in High-Throughput Screening. *Curr. Opin. Pharmacol.* **2009,** *9*, 580–588.
11. Liu, B.; Li, S.; Hu, J. Technological Advances in High-Throughput Screening. *Am. J. Pharmacogen.* **2004,** *4*, 263–276.

CHAPTER 12

Homology Modeling: Applications in the Low Molecular Weight Ligand Designing

SANTOSH S. CHHAJED[1*], PRAMODKUMAR P. GUPTA[2], VIRUPAKSHA A. BASTIKAR[3], and VAIBHAV PATIL[1]

[1]*Department of Pharmaceutical Chemistry, MET Institute of Pharmacy, Bhujbal Knowledge City, Nashik 422003, Maharashtra, India*

[2]*School of Biotechnology & Bioinformatics, D. Y. Patil University, Plot no 50, Sector 15, CBD Belapur, Navi-Mumbai 400614, Maharashtra, India*

[3]*Amity Institute of Biotechnology, Amity University, Mumbai, Maharashtra, India*

**Corresponding author. E-mail: chhajedss@gmail.com*

ABSTRACT

Owing to a large and significant genome data outcome from sequencing projects in the last three decades from many organisms with 100 million of nucleotide sequence has been deposited in major DNA sequence databases. Numerous millions of nucleotide bases have been translated into amino acid sequences and deposited in UniprotKB and UniprotTrEMBL databases. However, the biological functions of protein are inherently dependent on its three-dimensional structure, the deposited and novel protein sequences are insufficient to determine the most appropriate functions of them. In contrast, the X-ray crystallography and NMR techniques thrive to be the most accurate structural characterization of biomolecules. Due to the technical complications and labor intensiveness of these methods, various protein structures unraveled by experimental methods (NMR and X-ray) lag far behind the

Biologically Active Small Molecules: Modern Applications and Therapeutic Perspectives.
Debarshi Kar Mahapatra, & Sanjay Kumar Bharti (Eds.)

accumulation of protein sequences. Technical advancement in the last few decades, with major efforts in protein structure determination, the concept of protein structure determination and homology modeling has speeded up in protein three-dimensional structural modeling. The significant goal of homology modeling technique is to predict a structure from its sequence by comparing to its most optimum aligned protein structural template. This technique allows users to speedily generate in silico protein models in all the contexts, where experimental methods like X-ray and NMR have failed to date.

12.1 INTRODUCTION

Since 1958, when Kendrew et al. reported the first atomic-level resolution of a protein structure of myoglobin, and the field of structural biology dramatically expanded its horizon with the progress of new techniques and tools to gain access into the atomic details of the biological molecule such as nucleic acids and protein.[1,2] The gateway of solving biological data opened an entirely emerging area of understanding and knowledge to the scientific community, as the study and analysis of protein structures provide fundamental insight of biochemical behavior and functions into the cause of disease and their treatment. The core development in the last few years in the field of structural biology is now identified as a fundamental step to understanding life at the molecular level.[1] Solving the structures of every existing proteins is presently a tailback for the genomics studies. To overcome such situation, protein structure initiative (PSI) aims to determine ~100,000 3D structures in a decade.[1]

However, protein sequence databank (UniProt/TrEMBL) is emerging swiftly, with more than 10 million sequences available from November 2016 are Swiss-Prot: 552,884 and TrEMBL: 70,656,157.[3]. Moreover, the Protein Data Bank (PDB) includes 124,029 3D structures till November 2016 but owing to the high degree of similarity in the sequences, the decrease in uniqueness of each protein is observed (once we remove these "redundant" proteins that have >95% sequence identity with the protein in the PDB).[4]

The human genomes have ~21,000 protein-encoding genes for each proteome of ~1,000,000 proteins.[5] Due to the experimental limitations, the bulk solved structures are <50 kDa excluding the majority if they are larger proteins. However, larger proteins represent significant fractions of the proteins situated in any living organism.[6] The structures of larger proteins can be solved with the help of sophisticated electron microscopy system at a low resolution.[7,8] Majority of these structures have been resolved by either the nuclear magnetic resonance (NMR) or X-ray crystallography techniques. But still, most of the proteins are out of reach due to various technical difficulties. This remark a vast gap between the world of a collection of recognized structures and universe of identified sequences of protein.[1]

To overcome these problems one can adopt computational methods to predict protein 3D structures. Computational/in silico techniques for predicting the protein 3D structure can be broadly classified into two major classes: ab initio folding methods and homology modeling. The homology modeling provides the base for the protein structure prediction by dealing with the inference from a structurally familiar homologous protein. The accomplishment of the 3D protein structure produced from the homology modeling techniques exclusively depends on following factors: (1) the subsistence of a homolog with already recognized protein structure, (2) our capability to identify/detect this homolog from biological database, and (3) the model building quality when we carry the process, once the homolog is detected.[1]

In recent, CASP experiments targets with 6% sequence identity to their templates was included in the homology modeling category.[10–12] With an increase in technological advancements in the near prospect homology modeling will achieve its definitive objective in the generation of model protein structures as correct as can compete with those resolve by high-resolution experiments.[1]

12.2 STEPS INVOLVED IN HOMOLOGY MODELING

- Template identification
- Template and sequence alignment
- Alignment correction
- Backbone generation
- Generation of loops
- Model optimization
- Estimating model quality.

12.2.1 TEMPLATE IDENTIFICATION

Template identification is the most important and crucial part of homology modeling. In this, we compare the sequence of unknown structure with all known structures stored in Research Collaboratory for Structural Bioinformatics (RCSB) PDB or any other structural database. A search with most popular homology search server is BLAST (Basic Local Alignment Search Tool)[13] (http://www.ncbi.nlm.nih.gov/blast/) against the structural database will give a directory of recognized protein structures that match with the given sequence known as the query. In case if the sequence identity is below to 30% BLAST fails to identify the optimum template and results are not reliable. BLAST cannot find a template when the sequence identity is well below 30%; homology hits from BLAST are not reliable. Various algorithms and methods can be used in the recognition of the template such as gapped BLAST, iterative profile BLAST, and HH search. Gapped Blast query is used for the identification of closely related homologous, Iterative profile BLAST is used for the detection of distantly associated sequences, whereas HH search uses hidden Markov model-based template library, for the identification of distantly related sequences.[14–16]

12.2.2 TEMPLATE AND SEQUENCE ALIGNMENT

Once the template is recognized the immediate next step is to set up the configuration between the template and the target amino acid sequence. When the alignment aligns a substring of the query sequence to a substring of the target sequence it is known as local alignment. In case, if all substrings of both the sequences are included in the alignment is called global alignment. Local alignment is widely used for pattern recognition and global alignment is widely employed for model construction. Sequential identities >25% suggest that the target and template have similar 3D structure and is appropriate for modeling whereas sequential identity >60% indicates the correctness of the resultant model is analogous to the experimental structure, as the structural conformations of protein is highly conserved than the amino acid sequence.[17] Few listed online server in Table 12.1.

12.2.3 ALIGNMENT CORRECTION

It is possible that the alignment between the sequence and template structure residues has to be corrected. Even a single amino acid change, that is, Ala to Glu cannot align though they found in hydrophobic core or in-del process may cause a shift in the amino acid sequence and is prone to serious error to the model. Employing the multiple sequence alignment method, utilizing the tools such as Clustal W, Clustal X, Clustal Omega, and so on, the regions with conserved residues can be simply found and are accepted as per modeling requirement. Insertion and deletion sites of the gap in the arrangement can be easily detected by employing the multiple sequence alignment analysis. Gaps have to be shifted around until they are as small as possible. The correction of alignment is generally performed by the manual process.[15]

TABLE 12.1 Alignment Server.

Sr. No.	Name	Address
1	BLAST	https://blast.ncbi.nlm.nih.gov/Blast.cgi
2	Clustal Omega	http://www.ebi.ac.uk/Tools/msa/clustalo/
3	Expresso	http://www.tcoffee.org/
4	MUSCLE	http://www.drive5.com/muscle/
5	PRALINE	http://www.ibi.vu.nl/programs/pralinewww/
6	PROBCONS	http://probcons.stanford.edu/
7	PROMALS3D	http://prodata.swmed.edu/promals3d/
8	SAM	https://compbio.soe.ucsc.edu/sam.html
9	VAST, Cn3D	https://www.ncbi.nlm.nih.gov/Structure/

12.2.4 BACKBONE GENERATION

The backbone generation is the base of the model initiation step which follows the path of correct alignment in the 3D protein model development. When the configuration is accurate, the coordinates of the template structure backbone are copied to the target sequence (C, N, O, Cα) and if the residues are identical, the side chain coordinates are also copied to the target sequence. If two aligned residues are different, only the backbone coordinates can be copied. The majority of the protein structure determined experimentally is not ideal in PDB. There are countless sources of errors, ranging from poor electron density in the X-ray diffraction map to common human-made errors in experimental design as well as in PDB file submission. In case, if two dissimilar templates are obtainable and each has a partial or poorly determined section of the target sequence, one should clearly follow multiple template homology modeling concept; whereas the good parts of both the templates are considered in one model. In principle, the multiple templates modeling concept is simple, but practically it is hard to attain outcomes that are really close to the true structure as compared to single-template modeling.[15,18]

12.2.5 GENERATION OF LOOPS

Many times the alignment consists of gaps as a result of insertions or deletions in the target sequence. When there is a gap due to a deletion in the target sequence, one can simply delete the corresponding residue from the template structure which leaves a hole in the model. But in case of insertion in the target sequence, the template contains a gap and there are no backbone coordinates to model this stretch of residue from template structure. Such large changes cannot be modeled in secondary structure elements and hence, can be placed in loops.[18,19]

The basic goal is to envisage the conformation of a loop that is set aside at both the ends by the protein backbone. Database methods and ab initio (de novo) are the two methods extensively applied in the loop prediction. A few are listed in Table 12.2.

TABLE 12.2 Loop Prediction Server.

Sr. No.	Name	Address
1	RCD+ server	http://rcd.chaconlab.org/
2	DISGRO	http://tanto.bioe.uic.edu/DiSGro/
3	ModLoop	http://modbase.compbio.ucsf.edu/modloop/
4	ArchPRED Server	http://manaslu.fiserlab.org/loopred/
5	SuperLooper	http://bioinf-applied.charite.de/superlooper/

12.2.5.1 DATABASE SEARCH

The database technique attempts to discover a section of a protein with an identified 3D structure that fits the main regions of a loop. In such case, the most efficient manner is to take some residue before and after the insertion as "anchor" residues and search the PDB for loops with same anchor residues.[19]

12.2.5.2 AB INITIO (DE NOVO)

Ab initio methods of application for the prediction of loop involve the production of a huge amount of arbitrarily selected candidate conformations. In the ab initio method, data from the available database is regularly included, for example, psi-maps and phi-maps of the identified loops. Once the loops are produced in this approach, the energetic criteria (or sequence resemblance for database technique) are usually in applications to opt for the most probable compounds.[19–22]

12.2.6 MODEL OPTIMIZATION

Model optimization or refinement is a very important task that involves efficient sampling for conformational space and a means to accurately identify near-native structures.[14,23] As the model building procedure involves a series of amino acid residue substitution, insertion, and deletions which possibly affects the conformations of the model. Even the side chains can affect the backbone and simultaneously have an effect on the rotamers too. Optimization can be done by performing energy refinements using molecular dynamics simulations. The modeled protein is placed in a force-field and the dynamic movement of the protein molecule is followed in time, this mimics the folding of the protein. Types of molecular dynamics algorithms such as Monte Carlo and genetic algorithm are widely used. Force field such as AMBER, GROMACS, GROMOS, CHARMM, OPLS are widely used.[23–27]

12.2.7 ESTIMATING MODEL QUALITY

The 3D protein model quality can be classified into two types:

- The stereochemical quality of the structural model
- The accuracy of the homology-based structural model with respect to its experimental structure.[28]

Due to the lack of experimentally solved structure for the query sequence, the real accuracy of a homology-based structural model cannot be assessed. In case where experimental structures are known, several measures are considered to estimate the model's quality. Root mean square deviation (RMSD) is a widely used measure to estimate the structural similarity between any two 3D structures. However, one can find a large difference in the RMSD value in a short region of protein structure due to mismatch of residual information, thus not reflecting the most of the region where the structure is highly similar. Hence, CASP uses another measure called Global Distance Test (GDT) or Global Distance Test "total Score" (GDTTS). GDT measures the similarity of two structures with similar amino acid residual information but different tertiary structures. GDT compares the experimentally resolved structures with that of the protein structure determined by prediction techniques such as X-ray crystallography or protein NMR data. GDT score is calculated by considering the largest set of corresponding amino acid residue alpha carbon atom of the compared structure falling within the defined distance cutoff value. Usually, an average 1, 2, 3, 4, and 8 Å is used to measure the structure quality.[28–35]

However, quality assessment programs assess a predicted 3D structural model by itself (single model)[36–38] or a large set of the reference set of structure models generated for a given set of the sequence by different methods (consensus).[39–41]

In single-model method assessment, it compares various structural parameters such as solvent accessibility (buried or exposed residue) and secondary structure that are predicted for a given sequence with the corresponding 3D predicted model. The scoring function such as sequence-dependent torsion term (three residues), distance, and cutoff-based residue–residue interaction potentials and all-atom interaction potentials are considered.[28,42]

The consensus quality assessment program mostly relies upon that if a diverse set of the method were implemented to generate many structural models for a given query sequence, then the model that fits the closest to an experimental data is the best among all. How one can achieve its goal by the consensus method using these criteria? Using a large reference dataset of models for a given sequence from a various structure prediction server, they determined the average distance of a given model to all other models. In most of the cases, the GDT_TS score is calculated for all the models; however, specific servers apply modifications to the distance to obtain better predictive power. Interestingly, determining the average distance with single model scores a better prediction of the GDT_TS of a model in contrast to the experimental structure.[28,42] With several rounds of CASP, the model with least average distance to the rest of the reference model set features is the best GDT_TS when compared with the experimental structure.[28]

Despite model accuracy, another important criterion for 3D model quality is the stereochemical quality of the predicted model. The quality of the stereochemical model vastly defines the acceptable quality of the covalent geometry and the packing of the 3D structural model.[43]

The covalent geometry is assessed by comparing all its bond-angle, bond-length, and torsions to its standard values, which are either obtained from studies on the model of the small molecules[43] or through review of high-resolution crystal structures.[44] The backbone torsion comprises phi-si and omega angles. The rotamers libraries are considered for side chain torsions.[45] Considering all the above factors ensures that the covalent geometry of the 3D model is physically acceptable. Significantly, the quality of the stereochemical of the structural model is necessary for further studies including molecular simulation studies.[43–47] A few are listed below in Table 12.3.

TABLE 12.3 Model Validation Server.

Sr. No.	Name	Address
1	PROCHECK	https://services.mbi.ucla.edu/SAVES/ (SAVE server)
2	WHAT_CHECK	
3	ERRAT	
4	VERIFY_3D	
5	PROVE	
6	CRYST1 record matches	
7	Ramachandran plot	
8	ProSA-Web	https://prosa.services.came.sbg.ac.at/prosa.php

12.3 CONCLUSION

Protein three-dimensional structure plays a major function in the study and considering the mechanism of protein functions. Due to the exponential expansion of new protein sequences everyday, experimentally solving a protein structure is an exhaustive process and lack to meet the demand on time. Protein structure prediction methods have stood a promising tool not only to structural biologists but also to chemist, physicist, and others who are working in protein structure related phenomenon. In the next few years the outcomes from structural genomics and proteomics projects could probably map out all the closely and distinctly related folds of protein nature, which may serve as optimum templates and training dataset to solve the problem of protein structure prediction in the consistent use of homology method.

KEYWORDS

- **homology modeling technique**
- ***Ab initio***
- **database search**
- **model validation server**
- **loop prediction server**

REFERENCES

1. Luccio, E. D.; Koeh, P. A Quality Metric for Homology Modeling: The H-factor. *BMC Bioinform.* **2011,** *12*, 48.
2. Kendrew, J. C.; Bodo, G.; Dintzis, H. M.; Parrish, R. G.; Wyckoff, H.; Phillips, D. C. A Three-Dimensional Model of the Myoglobin Molecule Obtained by X-ray Analysis. *Nature* **1958,** *181*, 662-666.
3. http://www.uniprot.org/ (accessed June 30, 2018).
4. http://www.rcsb.org/pdb/home/home.do (accessed June 30, 2018).
5. Jensen, O. N. Modification-Specific Proteomics: Characterization of Posttranslational Modifications by Mass Spectrometry. *Curr. Opin. Chem. Biol.* **2004,** *8*, 33–41.
6. Warringer, J.; Blomberg, A. Evolutionary Constraints on Yeast Protein Size. *BMC Evol. Biol.* **2006,** *6*, 61.
7. Zhang, X.; Settembre, E.; Xu, C.; Dormitzer, P. R.; Bellamy, R.; Harrison, S. C.; Grigorieff, N. Near-Atomic Resolution Using Electron Cryomicroscopy and Single-Particle Reconstruction. *Proc. Natl. Acad. Sci.* **2008,** *105*, 1867–1872.
8. Yu, X.; Jin, L.; Zhou, Z. H. 3.88 A Structure of Cytoplasmic Polyhedrosis Virus by Cryo-Electron Microscopy. *Nature* **2008,** *453*, 415–419.
9. Moult, J.; Pedersen, J. T.; Judson, R.; Fidelis, K. A Large-Scale Experiment to Assess Protein Structure Prediction Methods. *Proteins* **1995,** *23*, 2–5.
10. Tramontano, A.; Morea, V. Assessment of Homology-Based Predictions in CASP5. *Proteins* **2003,** *53*, 352–368.
11. Tress, M.; Tai, C. H.; Wang, G.; Ezkurdia, I.; Lopez, G.; Valencia, A.; Lee, B.; Dunbrack, R. L. Jr. Domain Definition and Target Classification for CASP6. *Proteins* **2005,** *61*, 8–18.
12. Tress, M.; Ezkurdia, I.; Grana, O.; Lopez. G.; Valencia, A. Assessment of Predictions Submitted for the CASP6 Comparative Modeling Category. *Proteins* **2005,** *61*, 27–45.
13. Stephen, A.; Warren, G.; Webb, M.; Eugene, M.; David, L. Basic Local Alignment Search Tool. *J. Mol. Biol.* **1990,** *215*, 403–410.
14. Vyas, V. K.; Ukawala, R. D.; Ghate, M.; Chintha, C. Homology Modeling a Fast Tool for Drug Discovery: Current Perspectives. *Indian J. Pharm. Sci.* **2012,** *74*, 1–17.
15. Introduction to Homology Modeling. http://www.cmbi.ru.nl/~hvensela/EGFR-verslag/homology.html (accessed July 25, 2018).
16. Structural Bioinformatics: Practical Guide. http://www.proteinstructures.com/Modeling/Modeling/modeling-tutorial-2.html (accessed July 26, 2018).
17. França, T. C. Homology Modeling: An Important Tool for the Drug Discovery. *J. Biomol. Struct. Dyn.* **2015,** *33*, 1780–1793.
18. Krieger, E.; Nabuurs, S. B.; Vriend, G. Homology Modeling. http://www.cmbi.ru.nl/edu/bioinf4/articles/homologymodeling.pdf (accessed July 10, 2018).
19. Xiang, Z. Advances in Homology Protein Structure Modeling. *Curr. Protein Pept. Sci.* **2006,** *7*, 217–227.
20. Fiser, A.; Gian, Do. R.; Sali, A. Modeling of Loops in Protein Structures. *Protein Sci.* **2000,** *9*, 1753–1773.
21. Deane, C. M.; Blundell, T. L. CODA: A Combined Algorithm for Predicting the Structurally Variable Regions of Protein Models. *Protein Sci.* **2001,** *10*, 599–612.
22. Tosatto, S. C.; Bindewald, E.; Hesser, J.; Manner, R. A Divide and Conquer Approach to Fast Loop Modeling. *Protein Eng.* **2002,** *15*, 79–86.

23. Van-Gelder, C. W.; Leusen, F. J.; Leunissen, J. A.; Noordik, J. H. A Molecular Dynamics Approach for the Generation of Complete Protein Structures from Limited Coordinate Data. *Proteins* **1994,** *18*, 174–185.
24. Zhu, J.; Fan, H.; Periole, X.; Honig, B.; Mark, A. E. Refining Homology Models by Combining Replica-Exchange Molecular Dynamics and Statistical Potentials. *Proteins* **2008,** *72*, 1171–1188.
25. Cornell, W. D.; Cieplak, P.; Bayly, C. I.; Gould, I. R.; Merz, K. M.; Ferguson, D. M. et al. A Second Generation Force Field for the Simulation of Proteins, Nucleic Acids, and Organic Molecules. *J. Am. Chem. Soc.* **1995,** *117*, 5179–5197.
26. Das, R.; Qian, B.; Raman, S.; Vernon, R.; Thompson, J.; Bradley, P. et al. Structure Prediction for CASP7 Targets Using Extensive All-Atom Refinement with Rosetta. *Proteins* **2007,** *69*, 118–128.
27. Han, R.; Leo-Macias, A.; Zerbino, D.; Bastolla, U.; Contreras-Moreira, B.; Ortiz, A. R. An Efficient Conformational Sampling Method for Homology Modeling. *Proteins* **2008,** *71*, 175–188.
28. Homology Modeling: Generating Structural Models to Understand Protein Function and Mechanism. In *Computational Modeling of Biological Systems, Biological and Medical Physics, Biomedical Engineering*; Srinivas, R.; Nikolay, V. D., Eds.; Springer, 2012.
29. Zhang, Y.; Skolnick, J. Scoring Function for Automated Assessment of Protein Structure Template Quality. *Proteins* **2004,** *57*, 702–710.
30. Sadreyev, R. I.; Shi, S.; Baker, D.; Grishin, N. V. Structure Similarity Measure with Penalty for Close Non-Equivalent Residues. *Bioinformatics* **2009,** *25*, 1259–1263.
31. Zemla, A.; Venclovas, C.; Moult, J.; Fidelis, K. Processing and Analysis of CASP3 Protein Structure Predictions. *Proteins* **1999,** *S3*, 22–29.
32. Zemla, A.; Venclovas, C.; Moult, J.; Fidelis, K. Processing and Evaluation of Predictions in CASP4. *Proteins* **2001,** *45*, 13–21.
33. Zemla, A. LGA: A Method for Finding 3D Similarities in Protein Structures. *Nucl. Acids Res.* **2003,** *31*, 3370–3374.
34. Read, R. J.; Chavali, G. Assessment of CASP7 Predictions in the High Accuracy Template-Based Modeling Category. *Proteins* **2007,** *69*, 27–37.
35. Sadreyev, R. I.; Shi, S.; Baker, D.; Grishin, N. V. Structure Similarity Measure with Penalty for Close Non-Equivalent Residues. *Bioinformatics* **2009,** *25*, 1259–1263.
36. Eramian, D.; Eswar, N.; Shen, M. Y.; Sali, A. How Well Can the Accuracy of Comparative Protein Structure Models Be Predicted? *Protein Sci.* **2008,** *17*, 1881–1893.
37. Benkert, P.; Tosatto, S. C.; Schomburg, D. QMEAN: A Comprehensive Scoring Function for Model Quality Assessment. *Proteins* **2008,** *71*, 261–277.
38. Wallner, B.; Elofsson, A. Can Correct Protein Models Be Identified? *Protein Sci.* **2003,** *12*, 1073–1086.
39. Wallner, B.; Elofsson, A. Pcons5: Combining Consensus, Structural Evaluation and Fold Recognition Scores. *Bioinformatics* **2005,** *21*, 4248–4254.
40. McGuffin, L. J.; Roche, D. B. Rapid Model Quality Assessment for Protein Structure Predictions Using the Comparison of Multiple Models without Structural Alignments. *Bioinformatics* **2010,** *26*, 182–188.
41. Cheng, J.; Wang, Z.; Tegge, A. N.; Eickholt, J. Prediction of Global and Local Quality of CASP8 Models by MULTICOM Series. *Proteins* **2009,** *77*, 181–184.

42. Benkert, P.; Schwede, T.; Tosatto, S. C. QMEANclust: Estimation of Protein Model Quality by Combining a Composite Scoring Function with Structural Density Information. *BMC Struct. Biol.* **2009,** *9*, 35.
43. Hooft, R. W. W.; Vriend, G.; Sander, C.; Abola, E. E. Errors in Protein Structures. *Nature* **1996,** *381*, 272–272.
44. Kota, P.; Ding, F.; Ramachandran, S.; Dokholyan, N. V. Gaia: Automated Quality Assessment of Protein Structure Models. *Bioinformatics* **2011,** *27*, 2209–2215.
45. Dunbrack, R. L. Jr.; Cohen, F. E. Bayesian Statistical Analysis of Protein Side-Chain Rotamer Preferences. *Protein Sci.* **1997,** *6*, 1661–1681.
46. Davis, I. W.; Leaver-Fay, A.; Chen, V. B.; Block, J. N.; Kapral, G. J.; Wang, X.; Murray, L. W.; Arendall, W. B.; Snoeyink, J.; Richardson, J. S. et al. MolProbity: All-Atom Contacts and Structure Validation for Proteins and Nucleic Acids. *Nucl. Acids Res.* **2007,** *35*, W375–W383.
47. Laskowski, R. A.; MacArthur, M. W.; Moss, D. S.; Thornton, J. M. PROCHECK: A Program to Check the Stereochemical Quality of Protein Structures. *J. Appl. Crystallogr.* **1993,** *26*, 283–291.62.

PART IV

Biopharmaceutics Approaches for Low Molecular Weight Ligands

CHAPTER 13

Drug Metabolism: Detoxification and Xenobiotic Biotransformation

JAYAKUMAR SURENDRADOSS[1], ALICE VARGHESE[2], and SUBRATA DEB[3*]

[1]*The Centre for Drug Research and Development, Vancouver, Canada*

[2]*Apotex Inc., Toronto, Canada*

[3]*Department of Pharmaceutical Sciences, College of Pharmacy, Larkin University, Miami 33169, FL, USA*

[*]*Corresponding author. E-mail: sdeb@alumni.ubc.ca*

ABSTRACT

This chapter focuses on the drug metabolism pathways [phase-I reactions; oxidation (aliphatic and aromatic hydroxylation, epoxidation, heteroatom dealkylation, heteroatom oxidation, oxidative deamination), reduction (azo reduction, nitro reduction, carbonyl reduction), hydrolysis (ester hydrolysis, amide hydrolysis, epoxide hydrolysis), and phase-II reactions (glucuronidation, sulfation, acetylation, methylation, glucuronide conjugation, amino acid conjugation)]. An insightful focus on nomenclature, classification of CYP enzymes, tissue-specific metabolism, pharmacological aspects, and toxicological effects of metabolism is the heart of this chapter. Various factors affecting metabolism such as diet, age, stereochemistry, gender, stereochemistry, pregnancy, temporal cycle, species, and genetics are described.

Biologically Active Small Molecules: Modern Applications and Therapeutic Perspectives.
Debarshi Kar Mahapatra, & Sanjay Kumar Bharti (Eds.)

13.1 INTRODUCTION

Physiologically, humans are not designed to allow facile excretion of lipophilic compounds. Hence, lipophilic endogenous and xenobiotic compounds need to be biotransformed to more hydrophilic substances before they are eliminated from the body. Drug metabolism or biotransformation (both the terms are interchangeably used) converts lipophilic xenobiotics and endogenous compounds into more water-soluble metabolites by exposing or introducing a functional group into the lipophilic substrate molecule.[68] Pharmacologically active organic molecules are usually lipophilic and relatively unionized at physiological pH.[213] The basic premise of drug metabolism is the conversion of lipophilic drugs (which are not easily excreted) to more water-soluble hydrophilic metabolites (amenable to excretion). The polar and ionized metabolites are less able to cross cellular membranes, leading to decreased tubular reabsorption, increased renal elimination, and facile excretion from the body.[112] Drugs are administered to exert their pharmacological effects and after which they are expected to get clear from the body. In the deficiency of metabolism process, the drug gathers in the body, prolonging the duration of drug action and potential toxicity.[112] Thus, metabolism plays a key role in the designing and pharmacological use of drugs.

13.2 PHASE-I AND PHASE-II METABOLISM PATHWAYS

Drug metabolism phenomenon is divided into phase-I and phase-II reactions. In phase-I reactions, the polar functional groups are added by various biochemical conversions, such as oxidation, hydrolysis, and reduction.[68] The phase-I metabolites either get excreted directly or undergo subsequently phase-II conjugation reactions, forming more polar water-soluble products. This facilitates the excretion of drug molecules.[68] Phase-I metabolites are mostly inactive (but activation does occur) and are sufficiently polar to be excreted readily.[68,69] Phase-II reactions most often abolish biological activity and through conjugation reactions insert polarity in the compound. In mammals, cytochrome P450 (CYP) proteins are the predominant phase-I enzymes involved in the biochemical conversion that facilitates the addition of various functional groups typically rendering the molecule inactive and more water-soluble.[149] However, multiplicity in physicochemical and pharmacological outcomes is also possible and examples are abundant. CYP enzymes are a superfamily of hemeproteins that are present in a wide variety of organisms including bacteria, plants, fishes, birds, and mammals. CYP enzymes are localized predominantly in the smooth endoplasmic reticulum of cells. A subset of CYP enzymes is found in the mitochondria of cells.[147] CYP enzymes demonstrate complex sex-, tissue-, and development-dependent expression patterns that are regulated by hormones or growth factors.[4,112,157,207]

In phase-II reactions, the transfer of a moiety from an endogenous substance occurs to the phase-I metabolites and thus phase-II enzymes are also known as transferases.[89,106–108] However, unlike phase-I enzymatic reactions, phase-II conjugation reactions require energy input. There are six major phase-II enzymes, namely, uridine diphosphate-glucuronosyltransferase (UGTs), sulfotransferase (SULTs), *N*-acetyltransferase (NATs), glutathione transferases (GSTs), methyl transfereases, and amino acid transferases. Among these, qualitatively and quantitatively UGTs are the most important phase-II enzymes.[89,106–108] UGTs conjugate with drugs having –OH, $-NH_2$, –COOH, –SH functional groups. The most common phase-II conjugation reaction is catalyzed by UGTs that actuate the shift of glucuronic acid from uridine diphosphate to the substrate.[40,106–108,] Similarly, SULTS, NATs, and other phase-II enzymes have specific functional group targets and can conjugate a diverse array of substrates. Interestingly, GSTs have the exclusive capability to conjugate with reactive metabolites such as epoxides or organic halides, thus neutralizing and eliminating them from the body.[89,192,208]

13.3 NOMENCLATURE AND CLASSIFICATION OF CYP AND UGT ENZYMES

The CYP nomenclature system was first proposed by Nebert and Gonzalez.[136] The CYP nomenclature system is based on amino acid sequence identity, rather than activities, and organizes CYP enzymes into families and subfamilies. The nomenclature of the CYP enzymes starts with the CYP abbreviation followed by an Arabic number (e.g., 1, 2, 3) denoting the family, a letter defining the subfamily (e.g., A, B, C), and a numeral (e.g., 1, 2, 3) denoting the individual protein.[139] For example, CYP1A1 is a member of family 1, a member of subfamily A, and is identified within this subfamily as enzyme 1. According to the classification system, CYP enzymes with amino acid sequences that exhibit greater than or equal to 40% identity are assigned to the same family.[138,139] The same subfamily is assigned if more than 55% identity is observed in the amino acid sequences of CYP enzymes.[138,139] At present, 57 CYP enzymes have been identified in humans and classified into

18 families and 43 subfamilies.[74,79,87,137] CYP enzymes belonging to families 1–4 are involved in the xenobiotics biotransformation, including drugs, environmental toxicants, and endogenous compounds, including hormones, vitamins, and fatty acids.[21,37,39,46,79] CYP enzymes belonging to families 5, 7, 8, 11, 17, 19, 21, 27, and 51 play important roles in the biosynthesis and breakdown of endogenous substances (e.g., hormones, bile acids, eicosanoids).[132,159]

Similarly, there are 19 functional UGT enzymes classified into two families (UGT1 and UGT2) and three subfamilies (UGT1A, UGT2A, UGT2B).[81] Enzymes are placed in the same family if they have more than 40% DNA sequence similarity and assigned the same subfamily if the enzyme has more than 60% DNA sequence similarity. The nomenclature of the UGT enzymes is very similar to the CYP enzymes. However, a significant difference lies in the fact that CYP classification is based on amino acids, whereas UGT classification is based on gene sequences (Mackenzie et al., 2005). The Arabic number that follows the abbreviation UGT indicates the family, the letter designates the subfamily and the second Arabic number denotes the individual UGT isoform.[121]

13.4 CYP-MEDIATED CATALYTIC BIOTRANSFORMATION REACTIONS

CYP enzymes contain a hemeprosthetic group, present as protoporphyrin IX, that is attached to a polypeptide chain (apoprotein) consisting of approximately 500 amino acids. The iron atom in the heme prosthetic group is located in the center of the protoporphyrin ring and is part of the catalytic site of CYP enzymes.[74] The iron atom in the resting state is in the oxidized or ferric form (Fe^{3+}). The reduced or ferrous form (Fe^{2+}) readily binds to oxygen or carbon monoxide. The N-terminal domain of CYP enzymes consists of hydrophobic amino acids that anchor the protein in the membrane.[111] CYP enzymes contain a preserved cysteine residue in the C-terminal region of the polypeptide chain to which the iron atom of the heme prosthetic group is coordinated. The heme (Fe^{2+})–thiolate complex is responsible for the signature absorbance band at 450 nm of the reduced CYP–carbon monoxide complex.[111] Hence, CYP enzymes are also known as hemethiolate proteins.

CYP enzymes use molecular oxygen and reduced nicotinamide adenine dinucleotidephosphate (NADPH), or in some cases, reduced nicotinamide adenine dinucleotide (NADH), as reducing agents in the oxidative biotransformation of drugs (RH). The products of the reaction are the respective monooxygenated products of drug molecule (ROH) and water,[74,111] as shown by the following general reaction equation:

$$RH + O_2 + 2H^+ + 2\ NADPH \xrightarrow[\text{CYP reductase}]{\text{CYP}} ROH + H_2O + 2\ NADP^+$$

CYP enzymes play important roles in the reductive, oxidative, and peroxidative metabolism of xenobiotic compounds, such as environmental toxicants, drugs, and endogenous compounds, such as sex steroids (testosterone, estrogen), bile acids, vitamins, and fatty acids.[16] Most of the CYP-catalyzed reactions are oxidative in nature and thus, oxidation is the main focus of this discussion. CYP enzymes catalyze a wide variety of oxidative biotransformation reactions including carbon hydroxylation, heteroatom oxidation, heteroatom dealkylation, and epoxidation of aliphatic and aromatic substances.[68] In CYP-catalyzed oxidative reactions, an oxygen atom is incorporated in the molecule. Aliphatic and aromatic hydroxylation reactions catalyzed by CYP enzymes involve the formation of primary, secondary, and tertiary alcohols. In heteroatom oxidation reactions, the substrate is converted to a mono-oxygenated product where an atom of oxygen is attached to nitrogen or sulfur. In epoxidation reactions, the substrate is converted to a highly reactive electrophile, which can potentially bind to endogenous nucleophilic moieties such as DNA and proteins to activate toxic pathways.[68,69,149] The substrates of CYP-catalyzed epoxidation reactions include alkenes and arenes. For epoxidation reactions, one proposed mechanism is oxidation of arenes, leading to epoxide formation, followed by generation of a ketone intermediate, which subsequently undergoes a rearrangement reaction to give an aromatic phenol.[68,149]

CYP enzymes are also capable of catalyzing reductive reactions in mammals. Reduction reactions are unusual considering the fact that CYP (Fe^{2+}) exhibits high binding affinity for molecular oxygen.[68] The CYP-catalyzed reductive reactions take place when the oxygen tension in the tissue is very low.[68,149] It has been suggested that the binding of certain substrates to the CYP enzyme may block the interaction of the oxygen atom with the CYP enzyme.[68] Substrates that undergo CYP-catalyzed reduction include halocarbons, azo, and nitro compounds. In these reactions, electrons are shifted to the substrate from the CYP (Fe^{2+}), leading to reductive dehalogenation, aromatic nitro reduction, or azo reduction of the substrates.[68] Occasionally, the epoxides of alkenes are reduced back to the parent compounds.[187]

13.5 TISSUE-SPECIFIC DRUG METABOLISM

The mainstream CYP enzymes are expressed primarily in the liver, which is the chief site for in vivo drug metabolism. The protein levels of enzymes in the CYP1A, CYP2C, and CYP3A subfamilies represent approximately 60% of the total CYP content of liver in humans and rodents.[112] In the human liver, CYP1A2, CYP2C8, CYP2C9, and CYP3A4 are expressed at much higher levels as compared to the other CYP enzymes. These four classes of enzymes metabolize >55% of the present marketed drugs.[112] Hepatic CYP3A4 (and CYP3A5) protein levels alone constitute >28% of the total CYP content of the liver and catalyze the process of xenobiotic biotransformation of several agents. Interindividual difference in the hepatic CYP2C, CYP2D, and CYP3A enzymes expression in humans is large and can result in a difference in the range of 20- to 1000-fold in the levels of a single CYP enzyme. For example, hepatic CYP2D6 protein levels differ by 1000-fold in humans.[165]

After a drug is administered orally, it is absorbed in the intestinal tract and biotransformation takes place partly in the gut wall during absorption.[96] Following absorption, the parent drug goes into the liver via hepatic portal system before the rest of the body gets exposed to it. Majority of the clinically used drugs are metabolized by the liver, occasionally only a small fraction of administered drug reaches to the site of action.[96] Typically, enzyme systems in gastrointestinal lumen, gut wall, and liver primarily influence the first pass clearance of the drugs.[158] This phenomenon of drug metabolism in the gut (partial) and liver (majority) before reaching the systemic circulation is known as first-pass effect and significantly decreases the oral bioavailability of susceptible drugs. Luminal enzymes responsible for presystemic metabolism are the peptidases and esterases.[62] Inactivation of alcohol occurs primarily in the stomach by the enzyme alcohol dehydrogenase. Both small intestine and colon express phase-I and phase-II enzymes, of which enzymes of the proximal small intestine are most active.[62] CYP expression levels and catalytic activities in the small intestine are much lower than the liver.[114] About 82% of the intestinal CYP enzymes are CYP3A isoforms; however, intestinal CYP3A content is only 1% of hepatic CYP3A levels.[151] Thus, the overall contribution of first-pass oxidation is less important than the oxidative clearance by the liver. The variable and low oral bioavailability of some of the CYP3A4 substrates, such as cyclosporine and verapamil, is due to the intra- and interindividual variation of intestinal enzyme activity.[114] In addition to oxidative metabolism, conjugation reactions (glucuronidation and sulfation) play an imperative role in the detoxification of drugs in the liver and small intestine.[188] The activity of gut UGT enzymes has been reported for estradiol, 17β-estradiol, ethinyl estradiol, acetaminophen, *p*-nitrophenol, bilirubin, morphine, and propofol.[158] Interestingly, olfactory epithelium has about 20% of hepatic UGT content.[53] Adipose tissue is another site where UGT is expressed at a higher level.[53] There are significant ontogenic differences between adult and fetal tissues in UGT expression and activity. UGT2B4 has the highest level of mRNA expression, which is about 40% of total hepatic UGT expression. In comparison, fetal UGT2B4 is about 90% of the total fetal hepatic UGT.[53]

Although CYP enzymes are predominantly expressed in the liver and intestine, they are also active in the lung, kidney, adrenal glands, and skin.[41] In the lungs, several phase-I and phase-II enzymes are expressed that catalyze the local metabolism of inhaled drugs.[41] Pulmonary expression of drug-metabolizing enzymes is lower than the intestine and liver.[41] UGT enzyme levels are also low in the lungs; however, SULT activity appears to be the same as that in the liver.[182] Thus, biotransformation of drugs in the lungs is substantially less and has minimal contribution to systemic clearance of the drugs.[182] Similarly, renal biotransformation contributes minimally to the metabolic clearance of drugs. Knowledge of renal CYP and UGTs expression and their role in biotransformation are limited as compared to hepatic and intestinal drug metabolism.[41,53] In the normal human skin, most CYPs are localized within the epidermis and sebaceous glands. GST and SULT are reported to be present on the human foreskin, UGT is present on the stratum corneum.[143] Due to the cutaneous expression of vital drug-metabolizing enzymes, it has been hypothesized that cutaneous biotransformation may affect the efficacy and toxicity of drugs that are topically used.[189]

Drugs in the bloodstream are eliminated mainly by the liver and kidneys. However, drugs that are eliminated via the bile can be reabsorbed from the gastrointestinal tract into the bloodstream and again excreted into the bile, known as enterohepatic recirculation, resulting in multiple peaks in the plasma-concentration time profile and an increase in apparent elimination half-life of the drugs.[60] Thus, enterohepatic recycling can increase the duration of drug action and potentiate therapeutic effects.[168] Also, glucuronide conjugates of drugs can be deconjugated by bacterial enzymes present in the gut leading to higher concentrations of the free-parent drug available for reabsorption. The molecular weight threshold for biliary excretion of drugs is generally estimated to be in the range of 500–600 Da as compounds with lower molecular weights than this are primarily excreted in the urine.[110] Examples of some of the drugs undergoing enterohepatic recirculation include mifepristone, metapristone, meloxicam, lorazepam, colchicine, piroxicam, ezetimibe, bacailin, chloramphenicol, digitoxin, azithromycin, rifampicin, and methotrexate.[168] The impact of enterohepatic circulation on the pharmacokinetics and pharmacodynamics of drugs depends solely on the level of biliary excretion as compared to renal and metabolism and also on the efficiency with which the compound is reabsorbed from the gastrointestinal tract.

13.6 PHARMACOLOGICAL AND TOXICOLOGICAL EFFECTS OF METABOLISM/OUTCOMES OF DRUG METABOLISM

Following absorption of drugs in the system circulation, the drug can have three different fates: either it can reach to the site of action (e.g., sedative drugs reach to the brain), distributed in various tissues, or metabolized and excreted. The distribution and transportation to the site of action steps are reversible, whereas metabolism and excretion are unidirectional.[112] Thus, biotransformation causes loss of drug from the systemic circulation and eventually affects the drug concentration at the site of action and its pharmacological and clinical effects including toxicity. Interestingly, the pharmacological and toxicological outcomes of drug metabolism of drugs can be multifaceted. Depending on the nature of the drug metabolites, the outcomes can potentiate therapeutic benefit, terminate drug action, or cause a toxic reaction to the humans[68,69,149] (Figure 13.1).

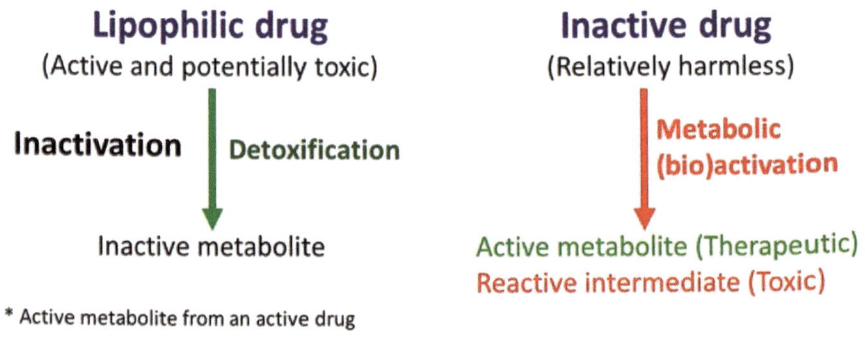

FIGURE 13.1 Pharmacological outcomes of drug metabolism.

Drug metabolism potentially exerts two types of effects, inactivation or detoxification and bioactivation. In the majority of the cases, the administered drugs are lipophilic, active, and potentially toxic at higher concentrations. Metabolism of such drugs by CYP enzymes leads to the termination of activity of the parent drug and excretion of the more water-soluble inactive metabolite, reducing/removing any toxic effect, thus also known as detoxification. Most drugs are rendered inactive following a biotransformation, for example, ibuprofen, lidocaine, and so on. In certain cases, one or more pharmacologically active metabolite can also form from an active-parent drug. An augment in the extent of action as well as the potentiation of pharmacotherapeutic activity is general consequence when a drug has active metabolite(s). Interestingly, in some cases, both the parent drug and active metabolite are marketed, for example, amitriptyline and nortriptyline, codeine and morphine.[112]

In contrast, bioactivation of inactive and relatively harmless drugs can have two opposite effects: active metabolite formation exhibiting therapeutic effects or formation of reactive intermediate eliciting toxic effects. Pharmacologically inert prodrugs are metabolically bioactivated in the body to one or more active metabolites that exert therapeutic effects. Quite a few anticancer and antiviral drugs are prodrugs that are more stable, more soluble, better absorbed, and possess fewer side effects.[149] In fact designing of a prodrug is an effective drug development strategy.

Bioactivation of harmless and inert drugs leads to the formation of metabolites that are more reactive than the parent compound. These are known as reactive toxic intermediates or reactive metabolites. Ironically, some of the most significant/serious toxicological effects of drugs are from reactive metabolites that have the ability to cause numerous toxic effects including carcinogenesis and hepatotoxicity.[68,69,149] Reactive metabolites can react with nucleophilic sites such as sulfhydryl groups in cysteine amino acid and glutathione, as well as amino/carboxyl groups on DNA, RNA, and protein molecules. Toxic metabolites or their intermediates have specific structural features, such as highly reactive epoxides and free-radical species that lead to adverse drug effects.[68] Epoxides and free-radical species (unpaired electrons) are highly reactive and hence are short lived. Toxicological studies have determined that phase-I enzymes are principally responsible for catalyzing the formation of reactive metabolites with few phase-II enzymes (e.g., NAT and UGT) having minor roles. Subsequently, reactive metabolites covalently bind to nucleic acids, enzymes, and exert a variety of toxic effects including teratogenicity (birth deformity), carcinogenicity, cell or tissue death, and hypersensitivity reactions.[68,69,149]

The precise mechanism of reactive metabolite-related hepatotoxic effects is not known; however, the majority of them are the consequence of immunogenic conjugates formed from the reactive metabolites of drugs with cellular proteins.[185] Some of the functional groups that have been responsible for hepatotoxicity include thiophene, p-hydroxy aniline, nitrobenzene, furan, arylacetic acid, aniline, catechol, olefin.[68,69,149,185] Some of the structural features of the drugs (e.g., amiodarone, isoniazid, ketoconazole, tienilic acid, procainamide, sulfamethoxazole, and troglitazone) that are associated with metabolism-based black box warnings or withdrawal from the market are *p*-alkylaniline, *p*-alkoxyaniline, nitrobenzene, aniline, and anilide.[185]

13.7 SUBSTRATES OF CYP ENZYMES

About 90% of the drugs in the commercial market are substrates of human CYP enzymes.[102] CYP1A2, CYP2C9, CYP2C19, CYP2D6, CYP2E1, and CYP3A4 are the major CYP isoforms involved in drug biotransformation.[70] Despite the observed overlap in substrate specificity between the major CYP isoforms, there are certain broad structural characteristics that define the substrates of each of these isoforms based on the nature of their binding cavity and the access channel.[98] The substrates of CYP1A2, in general, are neutral or basic lipophilic planar molecules with one or more putative hydrogen bonding sites, for example, phenacetin, caffeine, and theophylline,[180] which is in line with the broad and flat binding pocket of CYP1A2.[98] CYP2B6 substrates tend to be nonplanar, lipophilic, and neutral to weakly basic molecules, for example, bupropion, efavirenz, and ketamine.[133] Lipophilic and basic arylalkylamines, such as chlorpheniramine, bufuralol, and propranolol, with the site of oxidation away from the protonated nitrogen, are the preferred substrates for CYP2D6.[180] CYP2C9, which contains a positively charged region at the access channel to its binding cavity,[98] naturally prefers neutral or acidic molecules such as the NSAIDs and sulfonyl urea-based hypoglycemic drugs.[180] CYP3A4 has a large binding pocket and can, therefore, accept large bulky substrates.[98] Accordingly, CYP3A4 shows a broad range of substrate specificity covering drugs from the majority of the therapeutic areas and its substrates include lipophilic acidic, neutral, or basic molecules with sites of oxidation on the nitrogen atom or allylic sites.[180] On the contrary, CYP2E1 has a small binding cavity[98] and, therefore, accepts small and cyclic substrates with low molecular weight (< 200 Da) such as the volatile anesthetics.[180] Eicosanoids, saturated and unsaturated fatty acids are the preferred substrates for CYP4 enzymes.[70,153]

13.8 SUBSTRATES OF PHASE-II ENZYMES

Major phase-II drug-metabolizing enzymes include uridine-5'-diphospho-glucuronosyl transferases (UGTs), *N*-acetyltransferases (NATs), sulfotransferases (SULTs), and glutathione-S-transferases (GSTs) and they catalyze glucuronidation, sulfonation, acetylation, and glutathione conjugation, respectively.[89,157,177] The substrates of phase-II enzymes include parent drugs and/or their phase-I metabolites; however, it must be emphasized that not all phase-I metabolites undergo phase-II metabolism. Likewise, phase-I metabolism does not always precede phase- II metabolism.[98] UGTs have a broad substrate specificity and are engaged in the metabolism of drugs from almost all therapeutic classes.[157] The substrates of UGT enzymes include various structurally unrelated drugs that contain nucleophilic functional groups such as alcohols (e.g., acetaminophen and chloramphenicol), phenols (e.g., morphine and mycophenolic acid), aliphatic amines (e.g., amitriptyline, imipramine, and tamoxifen), aromatic amines (e.g., retigabine), and carboxylic acids (e.g., diclofenac, ibuprofen, and valproic acid).[89,95,198] Human SULTs also exhibit broad substrate specificity owing to the multiple isoforms of this enzyme and the plasticity of the binding sites of some of the isoforms of SULT.[40] The substrates of human SULTs include phenols, alcohols, aryl amines, heterocyclic amines, and alicyclic hydroxysteroids.[59] NATs catalyze the acetylation of substrates such as aromatic amines, arylhydrazines, and sulfonamides.[63,157] The substrates of GST include various electrophilic xenobiotics such as acetaminophen (NSAID), cyclophosphamide (cancer chemotherapeutic), simvastatin (antihyperlipidemic), and phase-I metabolites of polycyclic aromatic hydrocarbons (e.g., phenanthrene epoxide).[12]

13.9 PHASE-I REACTIONS

Phase-I metabolic reactions involve the introduction of new hydrophilic (polar) functional groups, such as –OH, $-NH_2$, –SH, or –COOH, and are therefore called "functionalization reactions." Alternatively, these reactions could also result in exposing or unmasking of the polar functional groups already present in compounds.[102] Phase-I reactions can be broadly classified as reduction, oxidation, and hydrolysis reactions.[203] Common subtypes of phase-I reactions are described here.

13.9.1 OXIDATION

13.9.1.1 ALIPHATIC AND AROMATIC HYDROXYLATION

Hydrocarbon hydroxylation is one of the most commonly observed phase-I oxidation reactions catalyzed by CYP, with aromatic hydroxylation being more common than aliphatic hydroxylation.[107] Generic reactions for aliphatic and aromatic hydroxylation are shown in Figure 13.2. Aliphatic hydroxylation involves the abstraction of the hydrogen atom from the carbon leading to the development of a carbon radical intermediate, which is followed by oxygen rebound resulting in incorporation of the hydroxyl group.[66,68] Although carbon hydroxylation can occur at different positions, it is site-dependent with benzylic carbon being the most favored, followed by allylic carbon and then the aliphatic carbon atoms.[98] Hydroxylation of aliphatic carbon atoms is relatively rare and the usual sites are the terminal (ω) and the penultimate (ω-1) carbon atoms.[98] Different mechanisms have been proposed to explain aromatic hydroxylation, one of which involves the formation of an arene oxide intermediate, which then undergoes rearrangement via "NIH shift" to form a phenol derivative.[129] Hydroxylation reactions mostly result in polar and inactive metabolites.[141]

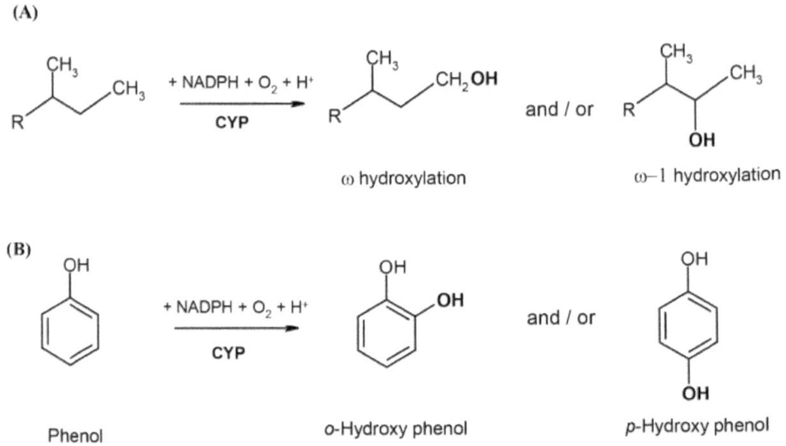

FIGURE 13.2 Generic reactions for aliphatic and aromatic hydroxylation.

13.9.1.2 EPOXIDATION

Oxidation of unsaturated hydrocarbons such as alkenes (olefins) or arenes by CYP enzymes results in the formation of three-member cyclic ethers called epoxides (Figure 13.3),[45,69] which are unstable electrophilic metabolites.[97] Epoxides are of toxicological interest because of their reactive nature and potential covalent interaction with the nucleophilic groups of cellular macromolecules and have been associated with mutagenic and carcinogenic effects of xenobiotics.[45,97]

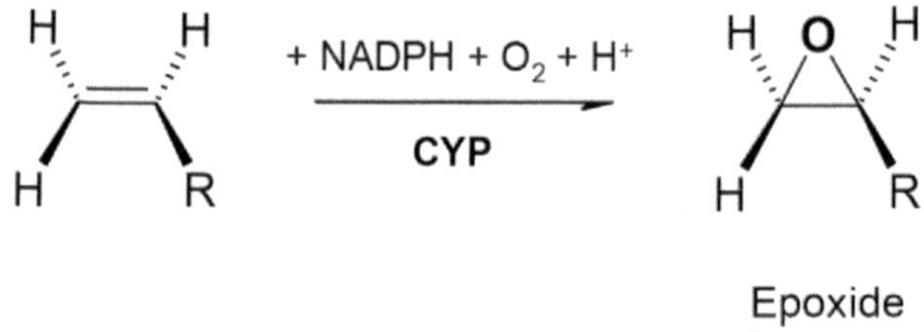

FIGURE 13.3 Generic reaction for epoxidation.

13.9.1.3 HETEROATOM DEALKYLATION

Dealkylation is another common reaction catalyzed by CYP enzymes, wherein the alkyl group situated adjacent to a heteroatom such as nitrogen, oxygen, and sulfur is removed and eliminated as an aldehyde (Figure 13.4). According to the heteroatom involved (i.e., nitrogen, oxygen, or sulfur), the reaction is termed as *N*-, *O*-, or *S*-dealkylation.[63,98] There are two mechanisms proposed to explain dealkylation reactions: (1) formal hydroxylation of the carbon atom adjacent to a heteroatom forming an unstable intermediate that decomposes and cleaves the carbon–heteroatom bond, and (2) stepwise single-electron oxidation of the heteroatom itself.[68,129] As a result of dealkylation, the heteroatom gets exposed as amino, hydroxyl, or thiol groups, thereby becoming substrates of the phase-II enzymes.[25]

13.9.1.4 HETEROATOM OXIDATION

Heteroatoms such as nitrogen and sulfur can also undergo direct oxygenation catalyzed by CYP enzymes and flavin-containing mono-oxygenases (FMOs) to form *N*- and *S*-oxides, respectively.[129] The mechanism of heteroatom oxidation involves a stepwise electron transfer followed by an oxygen rebound.[68] Whereas tertiary amines are oxidized to *N*-oxides,[171] oxidation of primary and secondary amines yields respective *N*-hydroxylamines (Figure 13.5), which are unstable and undergo further oxidation to nitroso/nitro and nitrone forms, respectively.[10,98] In the case of sulfur-containing compounds, *S*-oxidation of thioethers results in sulfoxide or sulfones (Figure 13.6), and that of thiols produces disulfide-derivatives.[107] Heteroatom oxygenation mostly results in detoxification of the parent compounds.[68]

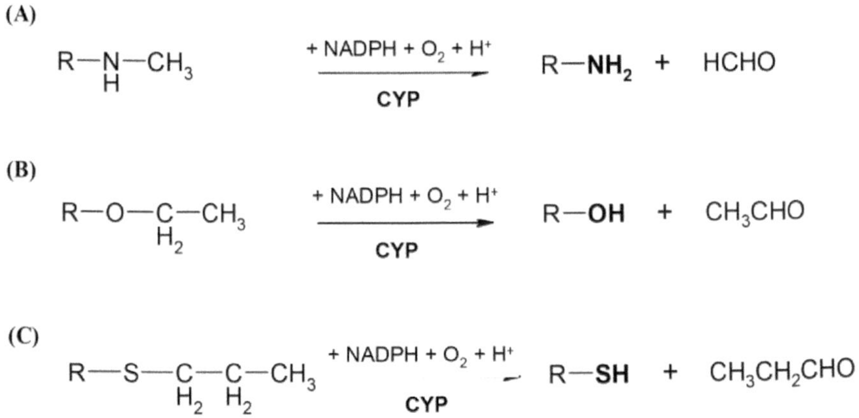

FIGURE 13.4 Generic reactions for heteroatom dealkylation.

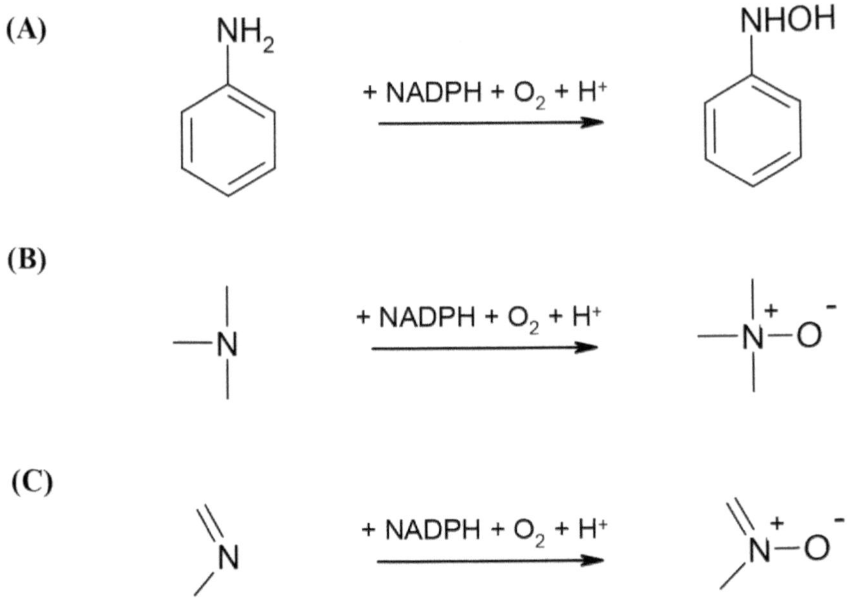

FIGURE 13.5 Generic reactions for *N*-oxidation.

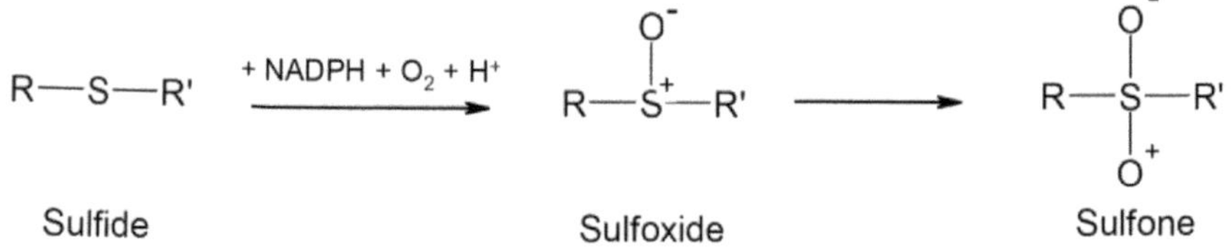

FIGURE 13.6 Generic reactions for *S*-oxidation.

13.9.1.5 OXIDATIVE DEAMINATION

Primary amines get deaminated by CYP enzymes resulting in the creation of a ketone and the loss of amino group as NH_3. The mechanism of oxidative deamination reactions catalyzed by CYP enzymes is identical to *N*-dealkylation reactions and involves the hydroxylation of the α-carbon attached to the amino group, forming an unstable intermediate which spontaneously decomposes causing the breakage of the C–N bond (Figure 13.7).[63] In addition to CYP enzymes, monoamine oxidases A and B also catalyze the oxidative deamination of various drugs containing primary, secondary, or tertiary amine groups.[193] MAO-mediated deamination proceeds differently with the formation of a protonated imine complex, which undergoes nonenzymatic hydrolysis to form the corresponding aldehyde and NH_4^+.[44] Unlike CYP enzymes, the oxygen incorporated in the deaminated metabolite by MAOs is derived from water and not molecular oxygen.[6]

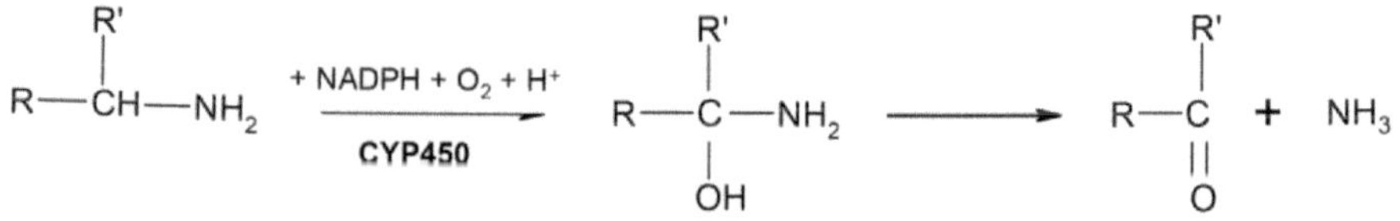

FIGURE 13.7 Generic reaction for oxidative deamination.

13.9.2 REDUCTION

The reductive reactions constitute an important set of biotransformation reactions and are catalyzed by both CYP and non-CYP enzymes. Most of these reactions require NADPH as a cofactor. Some of the major reductive reactions are listed below.

13.9.2.1 NITRO- AND AZO-REDUCTION

CYP enzymes, NADPH-cytochrome P450(c) reductase, and also the reductases present in gut microflora catalyze the reduction of nitro- and azo-compounds to primary amines (Figure 13.8). Nitroreduction is reported to occur in a stepwise fashion and involves the formation of nitrosoamine and hydroxylamine intermediates.[63,88,157]

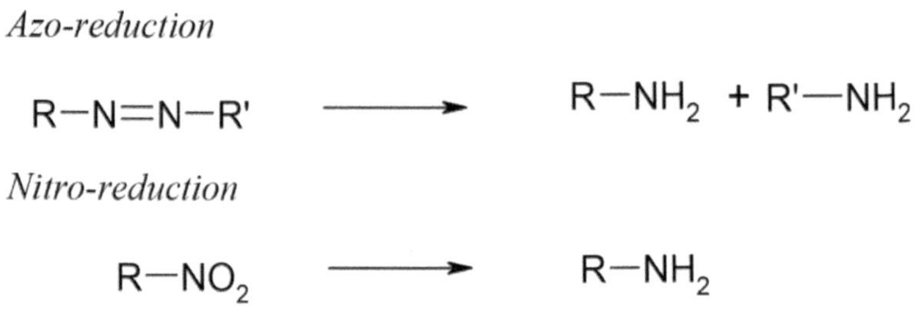

FIGURE 13.8 Generic reactions for reduction reactions.

13.9.2.2 CARBONYL REDUCTION

The reduction of the carbonyl group in the ketones and aldehydes to primary alcohols (Figure 13.9) is catalyzed by carbonyl-reducing enzymes (CREs), which are NADPH-dependent reductases and include enzymes such as alcohol dehydrogenase and aldo-keto reductase. In addition to CREs, NADPH-dependent quinone reductases (NQO) are also involved in the reduction of aromatic ketones, such as quinines.[123] Some of the other reductive reactions catalyzed by CYP enzymes include *N*-oxide reduction, epoxide reduction, and reductive dehalogenation.[208]

13.9.3 HYDROLYSIS

Drugs containing ester, amide, carbamate, or hydrazide moieties undergo metabolism via hydrolysis. Various enzymes that catalyze the hydrolytic reactions include carboxylesterases, cholinesterases, aryl esterases, serine endopeptidases, alkaline phosphatases, and epoxide hydrolases.[102,157] The next section describes some of the major hydrolytic biotransformation reactions and the readers are referred to the review by Testa and Kramer[191] for various hydrolytic reactions and the enzymes involved.

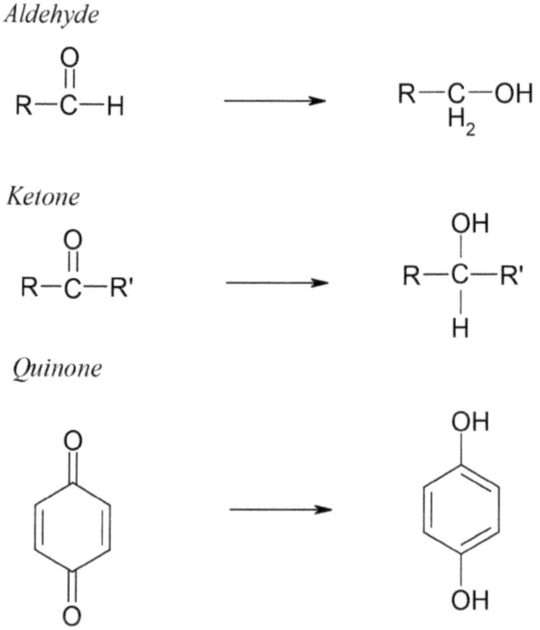

FIGURE 13.9 Generic reactions for carbonyl reduction.

13.9.3.1 ESTER HYDROLYSIS

Esters undergo rapid hydrolysis to corresponding carboxylic acid and alcohol (Figure 13.10), which could then become candidates for phase-II metabolic reactions.[88] The sites of ester hydrolysis include plasma, which contains various esterases, such as nonspecific acetylcholinesterases, butyrylcholinesterases, pseudocholinesterases, and paraoxonase among others, and organs, such as liver, small intestine, lung, and brain, which are rich in carboxylesterases and various other esterases mentioned above.[58,63] Ester hydrolysis plays an important role in the bioactivation of various prodrugs, such as enalapril, clofibrate, and oseltamivir.[103]

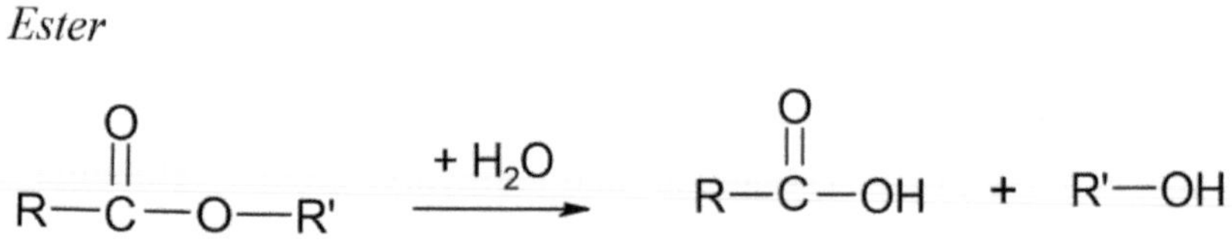

FIGURE 13.10 Generic reactions for ester hydrolysis.

13.9.3.2 AMIDE HYDROLYSIS

Amidases in the liver and various nonspecific esterases in the plasma can hydrolyze amides to corresponding carboxylic acid and amines (Figure 13.11).[88]

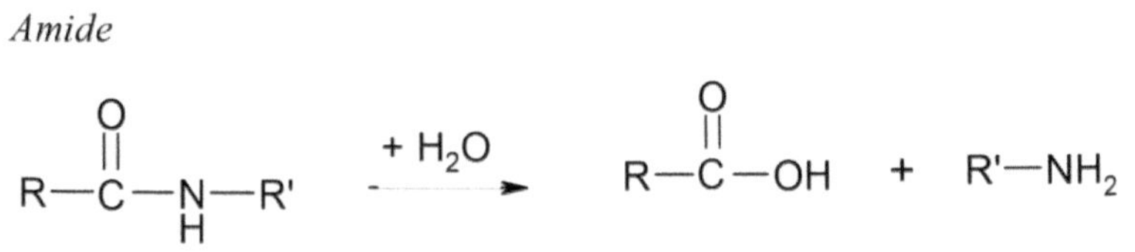

FIGURE 13.11 Generic reaction for amide hydrolysis.

13.9.3.3 EPOXIDE HYDROLYSIS

Epoxide hydrolases catalyze the hydrolysis of epoxides to dihydrodiols and thereby play an important role in the detoxification of reactive and electrophilic epoxide intermediates produced by CYP-mediated oxidation reactions of arenes and alkenes (Figure 13.12).[45,208]

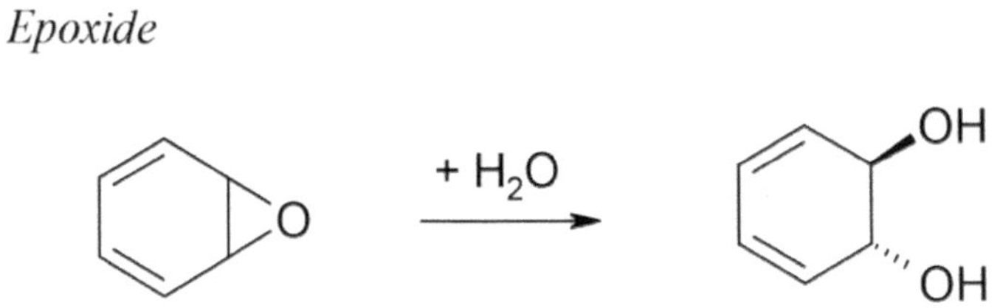

FIGURE 13.12 Generic reaction for epoxide hydrolysis.

13.10 PHASE-II REACTIONS

Phase-II biotransformation reactions involve the addition of an endogenous moiety to xenobiotics or their phase-I metabolites and are therefore called *conjugation* reactions. Glucuronidation, acetylation, sulfation, glutathione conjugation, methylation, and amino acid conjugation are the common phase-II reactions, and the functional groups that undergo these reactions include carboxyl, hydroxyl, amino, and sulfhydryl groups. Although there are exceptions, in most cases, the conjugated metabolites are pharmacologically inactive and therefore, these are detoxification reactions.[89,107]

13.10.1 GLUCURONIDATION

Glucuronidation involves the transfer of D-glucuronic acid moiety from the cofactor uridine-5'-diphospho-α-D-glucuronic acid (UDPGA) to drugs or their phase-I metabolites (aglycones) that contain electron-rich nucleophiles (N, O, or S), such as alcohols, phenols, thiols, amines, and carboxylic acids (Figure 13.13).[40,107] Glucuronidation reactions are catalyzed by a large family of enzymes called uridine-5'-diphospho-glucuronosyl transferases (UGTs 1 and 2), which are microsomal enzymes present in the small intestine, liver, kidney, and other extrahepatic tissues.[89,107] Glucuronidation, which is quantitatively the most important conjugation reaction in humans, is regarded as a high-capacity, low-affinity system.[208] Glucuronides are polar and hydrophilic metabolites, which get excreted via urine or bile, depending on their molecular weight and transporter affinity.[40] Although glucuronidation is generally a detoxification reaction, acyl glucuronides are considered as an exception. Acyl glucuronides, which are formed from carboxylic acid-containing drugs, are capable of undergoing intramolecular acyl migration from 1-*O*-β form to 2-, 3-, or 4-*O*-β glucuronides, which are positional isomers of 1-*O*-β-acyl glucuronide. Some of the acyl glucuronides or their positional isomers are unstable, potentially reactive, and electrophilic, and have been shown to undergo covalent binding with proteins and other macromolecules via transacylation or glycation mechanisms.[163] Despite a lack of evidence for the causal relationship, acyl glucuronides are implicated in the toxicities associated with some of the carboxylic acid drugs.[9,184]

$$\text{R}-\text{OH} \xrightarrow[\text{UGTs}]{+\ \text{UDPGA}} \text{R}-\text{O}-\text{glucuronide} + \text{H}^+$$

$$\text{R}-\text{COOH} \xrightarrow[\text{UGTs}]{+\ \text{UDPGA}} \text{R}-\text{CO}-\text{O}-\text{glucuronide} + \text{H}^+$$

$$\text{R}-\text{NH}_2 \xrightarrow[\text{UGTs}]{+\ \text{UDPGA}} \text{R}-\text{NH}-\text{glucuronide} + \text{H}^+$$

FIGURE 13.13 General reactions for glucuronidation of various functional groups.

13.10.2 SULFATION

Sulfation constitutes a second major conjugation pathway in humans for the biotransformation of drugs.[208] Sulfation reactions are catalyzed by enzymes called sulfotransferases (SULTs) and involve the transfer of a sulfonyl group from the cofactor 3'-phosphoadenosine 5'-phosphosulfate (PAPS) to drugs or their phase-I metabolites containing alcohol, phenol, amino, or thiol functional groups (Figure 13.14).[40] SULTs belong to two major classes—cytosolic SULTs and membrane-bound SULTs, with only the former being relevant in the conjugation reactions involving drugs and also, endogenous substances such as bile acids, thyroxine, steroids, and neurotransmitters.[40,89] Due to a great degree of overlap in substrate specificities of UGTs and SULTs, glucuronidation and sulfation reactions compete with each other for common substrates. Unlike glucuronidation, sulfation in humans is a high affinity, low capacity system, and therefore, sulfation predominates when substrate concentrations are low, while glucuronidation becomes the major pathway at higher substrate concentrations.[40]

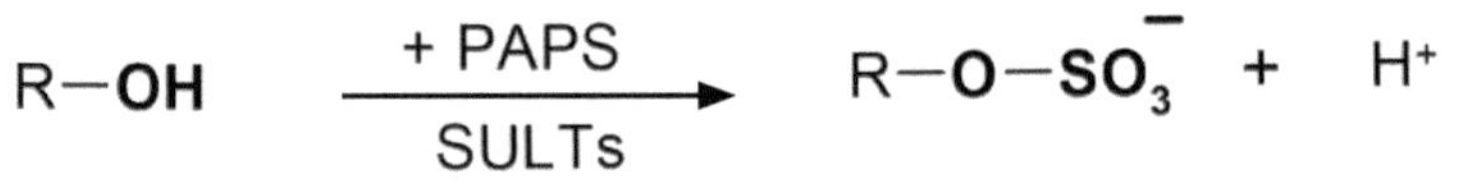

FIGURE 13.14 General reaction for sulfate conjugation.

13.10.3 ACETYLATION

N-acetyl transferases (NATs) catalyze the transfer of "acetyl" moiety from the cofactor acetyl CoA to substrates such as sulfonamides, aromatic amines, alcohols, carboxylic acids, and thiols (Figure 13.15). Human NATs—NAT1 and NAT2—are polymorphic enzymes and are located in the liver and reticuloendothelial cells of spleen, lung, and gut. Unlike glucuronidation and sulfation, N-acetyl conjugates are less water soluble than the parent and also may retain pharmacological activity. Accordingly, *N*-acetyl sulfonamides have been implicated in the renal toxicity of sulfonamides owing to the precipitation of these conjugates in the kidney.[63,89]

13.10.4 METHYLATION

Methylation reactions involve the transfer of methyl group from the cofactor *S*-adenosyl L-methionine (SAM) to substrates containing alcohols, phenols, amines, and thiols and are catalyzed by a family of O–, N–, and S–methyl transferases (Figure 13.16).[107,208] Although mostly involved in the metabolism of endogenous substrates, certain nonspecific methyltransferases and physiological methyltransferases can catalyze the methylation of some drugs with similar structures. Like acetylation, methylation yields metabolites that are less polar and less water-soluble than the parent.[63,107]

$$R{-}NH_2 \xrightarrow[\text{NATs}]{\text{+ Acetyl-CoA}} R{-}NH{-}COCH_3 \quad + \quad H^+$$

FIGURE 13.15 General reaction for *N*-acetylation.

$$R{-}OH \xrightarrow[\text{O-Methyl transferase}]{\text{+ SAM}} R{-}O{-}CH_3 \quad + \quad H^+$$

$$R{-}SH \xrightarrow[\text{S-Methyl transferase}]{\text{+ SAM}} R{-}S{-}CH_3 \quad + \quad H^+$$

FIGURE 13.16 General reaction for methylation.

13.10.5 GLUTATHIONE CONJUGATION

Glutathione (GSH), which is an endogenous tripeptide made up of γ-glutamine, cysteine, and glycine, protects cellular macromolecules from reactive and electrophilic metabolites.[89] GSH conjugation can either be spontaneous or enzymatic, mediated by glutathione *S*-transferases (GSTs).[192] GSTs are present in mitochondria, cytosol, and smooth endoplasmic reticulum of liver, kidney, intestines, and various other tissues.[192] Some of the substrates detoxified by GSH conjugation include epoxides, conjugated enones/enals, haloalkanes, and nitroalkanes (Figure 13.17).[63,208] GSH conjugates may undergo direct excretion in urine or bile, or be further metabolized *via* the mercapturic acid pathway earlier to excretion.[89]

13.10.6 AMINO ACID CONJUGATION

The amino acid conjugation is a detoxification pathway that defends against the reactive metabolites of carboxylic acid-containing drugs,[89] with glycine being the major acceptor amino acid in humans.[83] As shown in Figure 13.18, the mechanism for amino acid conjugation involves (1) an initial activation of carboxylic acid moiety with adenosine triphosphate to form acyl adenylate, (2) reaction of acyl adenylate with coenzyme A (CoASH) to form xenobiotic acyl-CoA thioester, (3) transfer of the acyl group from the activated xenobiotic acyl-CoA thioester to glycine forming *N*-acyl glycine conjugate and liberating CoASH.[101] The *N*-acyl glycine conjugates are nontoxic and excreted from the body in urine. Unlike other conjugation pathways, amino acid conjugation is unique in that the xenobiotic carboxylic acid, and not the cofactor, is activated to high-energy form.[101] Amino acid conjugation is reported to be a high affinity-low capacity process, being significant at low concentrations of xenobiotic carboxylic acids, whereas other pathways such as glutathione conjugation and acyl glucuronidation predominate at higher concentrations of carboxylic acid drugs.[179]

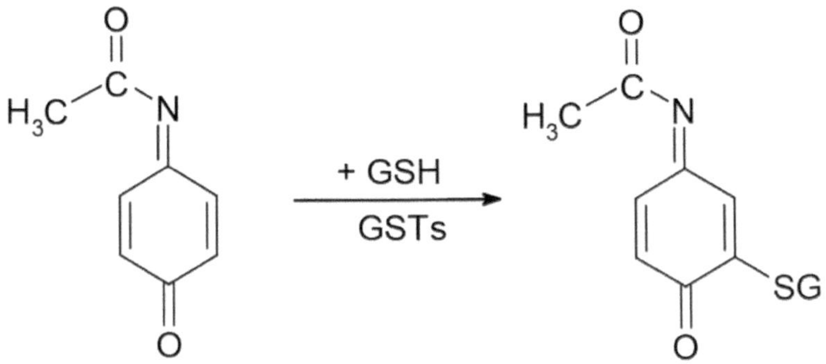

FIGURE 13.17 General reaction for GSH conjugation.

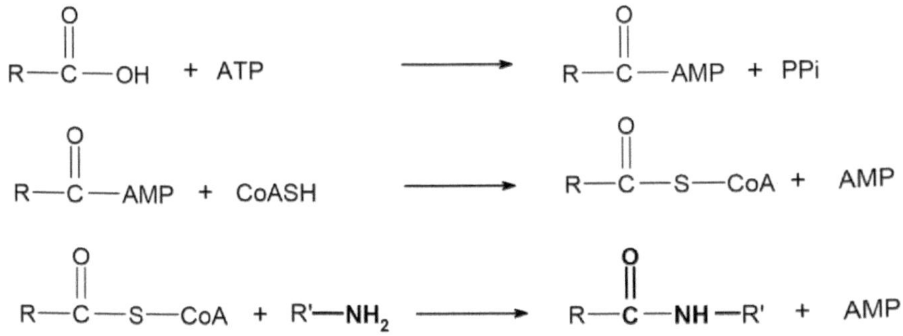

FIGURE 13.18 General reaction for amino acid conjugation of xenobiotic carboxylic acids.

13.11 REGULATION OF CYP EXPRESSION

The expression of human CYP enzymes is regulated by various factors including age, hormones (e.g., pituitary hormones, sex hormones), pregnancy, dietary components, nutritional status, and pathophysiological conditions. Altered CYP expression has been observed in humans after exposure to therapeutic agents (e.g., rifampicin, tamoxifen) and environmental toxicants (e.g., cigarette smoke).[17,31,32,86,201] Endogenous compounds (e.g., bile acids) and hormones are also involved in the modulation of CYP expression.[19,33,36,38] Differential exposure to xenobiotics or endogenous compounds may lead to variability in CYP-mediated metabolism, leading to altered pharmacokinetic and toxicological profiles of therapeutic agents.

Induction or increased protein expression is accompanied by increased enzymatic activity. CYP induction by concomitant drugs and dietary substances can cause a reduction in clinical response by augmenting the clearance of therapeutic agents, which explains some cases of drug–drug or drug–food interactions. CYP enzyme expression can also be suppressed or downregulated after exposure to inflammatory mediators, endogenous compounds, and xenobiotics including therapeutic agents and polycyclic aromatic hydrocarbons (PAHs). The suppression of CYP enzyme expression by xenobiotics has been linked to the inhibition of hormonal pathways involved in the constitutive expression of these enzymes.[86,167] Receptors play a key role in constitutive and inducible CYP expression in humans and rodents. The receptors involved in the regulation of CYP enzymes are localized in cytosol and act primarily in the nucleus, where the receptor–ligand complex binds to the CYP regulatory regions to enhance the gene transcription leading to altered protein levels and activities.[80,167,172] Inducers of CYP enzymes are structurally assorted and bind to receptors with different specificity and potency to initiate the transcription process. Regulation of CYP expression is controlled by at least five different receptors, specifically, pregnane X receptor (PXR), aryl hydrocarbon receptor (AhR), peroxisome proliferator-activated receptors (PPARs), constitutive androstane receptor (CAR), and glucocorticoid receptor.[112] The specific involvement of these receptors in CYP regulation has been elucidated using receptor–ligand binding assay and advanced molecular biology techniques including knockout mice and gene reporter assays. The literature review suggests that AhR is chiefly engaged in the regulation of the CYP1 enzymes, whereas PXR and CAR are engaged in the regulation of CYP2 and CYP3 family enzymes.[112] Expression of the CYP4 family enzyme(s) is controlled by PPARα.[112]

13.12 REGULATION OF UGT EXPRESSION

Human UGT enzymes are classified into three subfamilies namely, UGT1A, UGT2A, UGT2B with each of them consisting of nine (UGT1A1, UGT1A3, UGT1A4, UGT1A5, UGT1A6, UGT1A7, UGT1A8, UGT1A9, and 1A10), three (UGT2A1, UGT2A2, and UGT2A3), and seven (UGT2B4, UGT2B7, UGT2B10, UGT2B11, UGT2B15, UGT2B17, and UGT2B28) functional isoforms, respectively.[81,142] The liver has the highest expression levels of all UGTs, except UGT1A7, UGT1A8, and UGT1A10, which are present only in the gastrointestinal tract.[81] The expression of UGT enzymes is regulated at the transcriptional level by a combination of ligand-activated transcription factors and tissue-specific transcription factors. The constitutive (or basal) expression of UGTs is regulated mainly by tissue-specific transcription factors, such as HNF-1α and β in the liver, hepatocyte nuclear factors, and caudal-related homeodomain protein 2 (Cdx2) in the intestines.[61,81,121] Alternatively, various ligand-activated transcription factors such as nuclear factor erythroid 2-related factor 2 (Nrf2), aryl hydrocarbon receptor (AhR), pregnane X-receptor (PXR), constitutive androstane receptor (CAR), liver X receptor (LXR), farnesoid X receptor (FXR), and peroxisome proliferator-activated receptors α and γ (PPARα and PPARγ) regulate the inducible expression of different UGT enzymes.[121] Each UGT gene has distinct proximal "promoter" and distal "enhancer" regions, which contain "response elements" (binding sites) for these transcription factors.[81] Upon activation by corresponding ligands (endogenous or xenobiotic), these transcription factors heterodimerize with their partners, that is, Arnt (for AhR) and RXR (for CAR, PXR, FXR, LXR, PPARα, and PPARγ), translocate to the nucleus, and bind to their respective response elements on UGT genes to promote their transcription.[8] In addition to transcriptional regulation, UGT expression is also regulated at pretranscriptional level by epigenetic mechanisms such as histone modifications and DNA methylation, and post-transcriptionally by alternative splicing and microRNAs.[81] DNA hypermethylation, which represses gene transcription, is considered to be in charge for the tissue-specific expression of UGTs like UGT1A1 (liver) and UGTs 1A8 and 1A10 (intestines).[142] Genetic polymorphisms and the various factors engaged in the controlling of UGT expression are associated with the large interindividual variability reported for these enzymes.[8]

13.13 INHIBITION OF CYP AND UGT ACTIVITIES

The CYP isoforms metabolize a surprisingly a large number of drugs, which augments the latent unwanted drug interactions following concurrent administration of two or more drugs.[115] Inhibition of CYP enzymes is one of the chief factors leading to toxic drug–drug interactions, leading to a withdrawal of numerous drugs from the present market. The metabolic inhibition of the coadministered drugs may lead to enhanced drug concentrations, higher incidence of adverse effects, and toxicity due to accumulation.[156] In the case of prodrugs, inhibition of enzymes can lead to a decrease in efficacy because of decreased levels of active form of the drug.[156] The types of CYP inhibition can be classified into reversible inhibition, quasi-irreversible inhibition, and irreversible inhibition.[172] In the reversible mode of inhibition, there is competition for the active site of the enzyme between the inhibitor and substrate and the process occurs in the first catalytic step in most cases.[172] In reversible inhibitors, there are competitive, noncompetitive, uncompetitive, and mixed-type inhibitors. The drugs which are known as competitive inhibitors of CYP isoforms typically contend with the substrate for the active site of the enzyme.[115] The noncompetitive inhibitors bind to the diverse set of amino acids than that situated in the active site of the enzyme.[115] Uncompetitive inhibitors bind to the enzyme–substrate complex, but not to the free enzyme.[115] The clinically used drugs that are recognized as potent CYP inhibitors typically harbor one or more nitrogen-containing heterocyclic structures including imidazoles, pyridines, and quinolones. Inhibitors that bind to the prosthetic heme iron and the lipophilic amino acid core of the enzyme are very potent inhibitors.[169] Ketoconazole is an imidazole-containing potent CYP3A4 inhibitor, whereas quinoline derivatives, quinine, and quinidine are potent CYP2D6 inhibitors.[169] The reactive metabolites formation is principally accountable for quasi-irreversible and irreversible inhibition of CYP enzymes. Quasi-irreversible inhibitors form a metabolite intermediate (MI) complex with the prosthetic heme of the CYP protein.[115] Though the MI bond is noncovalent and does not denature the enzyme, the extended stability of the MI complex makes the enzyme unavailable for drug metabolism. Drugs containing functional groups including methylenedioxybenzenes, alkyamines, macrolide antibiotics, and hydrazines form MI complexes and are classified as quasi-irreversible inhibitors.[115] Troleandomycin, erythromycin (macrolide antibiotics), orphenadrine (alkyl amine), and isoniazid (hydrazide) are known as quasi-irreversible inhibitors.[116] Drugs can be oxidized to reactive metabolites by CYP enzymes which lead to irreversible inactivation of the enzyme. Since metabolic activation is a prerequisite for the inhibition of the enzymes, these inhibitors are categorized as mechanism-based inactivators or suicide substrates.[116] Drugs that have terminal double or triple bonds can be metabolized by CYP enzymes to reactive chemical species that have the prospective to alkylate the heme group of the CYP enzyme leading to irreversible inactivation of the enzyme. Allylisopropylacetamide and ethinyl estradiol are examples of a suicide substrate of CYP enzyme.[116] Metabolites of chloramphenicol and tienilic acid covalently bound to the apoprotein of CYP enzyme, thus irreversibly inhibiting the enzyme.[115] In metabolism-based drug interactions, the drug which is getting affected as a result of enzyme modulation is called "victim," whereas the drug which is inducing or inhibiting the enzyme is called "perpetrator." Clinical consequence of a drug–drug interaction depends on the therapeutic index of the "victim" drug and whether there is a considerable alteration in the active drug concentration at the site of action. Narrower the therapeutic window, higher the potential to cause clinically considerable drug–drug interactions.[115]

Since glucuronidation is the most general conjugation pathway and is entailed in the clearance process of all drugs classes,[117] inhibition of UGT activity leads to both drug–drug and drug–endobiotic interactions in vivo. Immunosuppressants (tacrolimus), nonsteroidal anti-inflammatory drugs (diclofenac), benzodiazepines (diazepam), and tricyclic antidepressants (nortryptyline) are reported to be potent in vitro inhibitors of UGT.[117] It has been observed that in vitro UGT inhibitory effects are not translated well into in vivo studies. For example, diclofenac shows potent inhibition of glucuronidation of codeine in vitro, but no significant alteration in pharmacokinetic properties in healthy human volunteers.[211] The location of UGT enzymes in the lumen of the endoplasmic reticulum prevents movement of substrates and conjugates to and from the active site, which can potentially be accountable for poor in vitro–in vivo correlation.[211] Though a number of drugs are reported to be in vitro UGT inhibitors, clinical studies are reported less frequently.

13.14 FACTORS AFFECTING DRUG METABOLISM

Among numerous factors that influence drug metabolism, few are important in controlling the extent of metabolism in humans and preclinical species.

13.14.1 PHYSICOCHEMICAL

For neutral drugs with log D values > 0, increase in lipophilicity increases the role of hepatic clearance to overall clearance. In these instances, the hepatic clearance can become the main route of elimination for highly lipophilic drugs. This phenomenon is prominent in β-adrenergic antagonists and calcium channel antagonists.[181] However, even highly lipophilic drugs need to have access to the active site of the drug-metabolizing enzyme. Steric hindrance plays a vital role in the accessibility to the active site of the drug-metabolizing enzyme.[82]

13.14.2 DIET

Several drugs display variation in pharmacokinetics depending on timing and content of the diet. Dietary components are capable of modulating the enzyme responsible for the drug metabolizing and disposition of the drugs. Since the CYP3A4 enzyme is promiscuous to substrates and highly expressed in the liver and gut wall, this crafts the enzyme more susceptible to modulation by dietary components.[73] Dietary components are known as important contributors to inter- and intra-individual variability in human drug metabolism.

13.14.3 STEREOCHEMISTRY

A vast number of the currently marketed drugs are chiral in nature and are developed as racemates. However, the enantiomers generally differ in pharmacodynamics and pharmacokinetic properties due to stereoselective interaction with the biological molecules. Metabolism of *S*-propafenone is retarded in the presence of *R*-propafenone.[113] Since the beta-blocking activity is due to *S*-propafenone, inhibition of its metabolism leads to side effects.[219] Metabolic interactions between enantiomer-to-enantiomer in racemates and enantiomers with other drugs is also influenced by polymorphism, age, species, and gender.[219]

13.14.4 AGE

The expression of drug-metabolizing enzymes is age-dependent and thus the degree of metabolism varies between ages. Individuals with advanced age experience a noteworthy reduction in hepatic blood flow, liver volume, and activity of certain drug-metabolizing enzymes.[209] Frailty (person above 65 years of age) has been known to reduce the plasma esterase and glucuronidase activity.[128] In comparison, pediatric population exhibit more contrasting differences from adults in CYPs, reductive enzymes, hydrolytic enzymes, *N*-methyl transferases, and glucuronosyl transferases.[99] Drugs that demonstrate age-specific differences in metabolism in children include paracetamol, chloramphenicol, cimetidine, salicylamide, theophylline, and valproic acid.[186] CYP3A4 is the prime enzyme expressed in the adult liver; however, CYP3A7 is highly expressed in the fetal liver.[144] Cisapride is primarily cleared by CYP3A4 and minimally by CYP3A7, which explains the cisapride-mediated cardiac toxicity in neonates.[197] Glucuronidation is impaired in neonates which explains the development of neonatal jaundice, since the bilirubin conjugation is catalyzed by UGT1A1.[186] Toxicity due to morphine and chloramphenicol in neonates is also attributed to impaired glucuronidation.[186]

13.14.5 GENDER

The gender disparities in the pharmacokinetics of drugs exist, chiefly, due to dissimilarity in body water, muscle mass, organ blood flow, organ function, and endocrine systems.[15] In general, females achieve higher plasma levels than males due to differences in the volume of distribution, body water, renal blood flow, and metabolic activity.[15] CYP2C19, a genetically polymorphic enzyme, metabolizes mephobarbital, and piroxicam rapidly in females than males. CYP3A4 substrates such as alprazolam, cyclosporine, and midazolam, have shown faster clearance in females than in males.[190] In contrast, CYP1A2 activity is more pronounced in males compared to that in females.[164] Plasma levels of CYP1A2 substrates such as olanzapine and clozapine are higher in females, leading to the potential for higher adverse effects.[47]

13.14.6 PREGNANCY

Due to altered physiology, the pharmacokinetics of drugs varies in pregnant women. Pregnancy influences hepatic blood flow, protein binding of drugs, and hepatic enzyme activity which can appreciably influence the hepatic clearance of a number of drugs.[91] Clinical pharmacokinetic studies have reported increased clearance of drugs metabolized by UGT1A4, UGT2B7, CYP2A6, CYP2C9, CYP2D6, and CYP3A4, whereas a reduction in CYP1A2 and CYP2C19 activities were examined during pregnancy. During pregnancy, circulating and tissue levels of hormones such as progesterone, estrogen, placental growth hormones, and prolactin are elevated, leading to differential expression of hormone-regulated drug-metabolizing enzymes and thereby, changes in pharmacokinetics and drug metabolism in pregnancy.[90]

13.14.7 SPECIES

Body size and weight are important to describe pharmacokinetics parameters of xenobiotics across species. The relative amount of hepatic enzymes (CYP/gram body weight) is higher in small animals than in humans, leading to faster drug elimination in animals. Interestingly, slight variation in the amino acid sequences of the CYP enzymes across species leads to profound differences in substrate specificity and catalytic activity.[125] CYP1A subfamily enzymes demonstrate more than 80% similarity between isoforms in rats, humans, dogs, mice, and monkeys. CYP2C is the largest subfamily in humans, rats, and mice; however, CYP2C is not expressed in dogs.[67] CYP2D inhibition profile is different among species, for example, quinidine potently inhibits CYP2D in humans, dog, and monkey, but not in rats and mice.[125] There are significant species differences in substrate specificities, inhibitory, and induction properties among human and rodent CYP enzymes.[125] Extrapolation of animal results to human without incorporating any scaling factor is not desirable.[125] Species differences also exist with respect to CYP induction. Omeprazole is a CYP1A2 inducer in humans, but has little effect on rats, mouse, and rabbit.[75] Rifampicin is a potent CYP3A inducer in humans but has very little or no effect in rats, whereas dexamethasone is a potent CYP3A inducer in rodents and little effect in humans.[65]

13.14.8 GENETICS

Genetic polymorphism occurs if a single gene responsible for producing the metabolizing enzyme is having a variant allele with a frequency of about 1%.[130] Genetic differences categorize the population into extensive metabolizers and poor metabolizers. Phase-I drug-metabolizing enzymes, such as CYP2D6, CYP2C8, CYP2C9, and CYP2C19, are genetically polymorphic and accounts for the majority of FDA drug label warnings related to polymorphism.[173] For example, ultrarapid metabolizers experience life-threatening respiratory depression following treatment with codeine which is a prodrug bioactivated by CYP2D6.[92] Higher metabolism of antidepressants by CYP2D6 ultrarapid metabolizers leads to nonresponsiveness of the treatment and suicidal tendencies in psychiatric patient population.[92]

13.14.9 TEMPORAL

Temporal cycles influence different physiological functions which in turn can persuade the drugs pharmacokinetics. For drugs with a high extraction ratio, metabolism depends on the blood flow and may be persuade by the disparities in blood flow during the day. Hepatic blood flow is elevated and low in the morning and night, respectively, and correlating with the hepatic flood flow the drug metabolism seems to get reduced at night.[109] Since hepatic blood flow and other physiological considerations are regulated by circadian rhythm and differ among humans, circadian rhythmicity has been identified as one of the factors for intraindividual dissimilarity in drug metabolism.[5] Flourouracil, ketoprofen, ranitidine, terbutaline, and vindesine are few examples of drugs that exhibit pharmacokinetic circadian modifications following intravenous infusion.[5]

13.15 DRUG–DRUG INTERACTIONS/HERB–DRUG INTERACTIONS FROM ALTERED DRUG METABOLISM

Several pharmacokinetic drug–drug interactions (DDIs) occur due to the variations in metabolic enzymes in the liver and extrahepatic tissues. Major pharmacokinetic interactions are the outcome of induction or inhibition of hepatic CYPs by coadministered drugs. Enzyme inhibition can lead to unanticipated adverse reactions, whereas enzyme induction can lead to therapeutic failure. CYP3A4 substrates such as cisapride, terfenadine, and astemizole were removed from the clinic following the development of life-threatening arrhythmias due to the inhibition of their oxidative metabolism by other CYP3A4 inhibitors.[120] Example of an in vivo competitive inhibition involves two- to fivefold increases in steady-state concentration of metoprolol after adding propafenone, a potent CYP2D6 inhibitor in the therapy.[43] The level to which an inhibitor affects the drug metabolism depends on the drug dose and inhibitor binding affinities. For example, sertraline is a moderate inhibitor at 50 mg dose, however, at 200 mg dose, it is a potent CYP2D6 inhibitor.[43] CYP enzymes also exhibit varied responses in drug metabolism due to genetic polymorphism, disease, age, and gender. CYP2C19 and CYP2C9 are genetically polymorphic enzymes allowing the classification of individuals as poor metabolizers and extensive metabolizers.[145] Some of the clinically important CYP3A4 inhibitors include ketoconazole, itraconazole, erythromycin, and grape fruit juice. Torsade de Pointes and venticular arrhythmia occur when these inhibitors are coadministered with terfenadine.[134] Coadministration of statins like lovastatin, simvastatin, atorvastatin with CYP3A4 inhibitors had been reported to cause rhabdomyolysis.[76] Symptomatic hypotension occurs on administering dihydropyridine calcium antagonists, for example, felodipine and nifedipine, with CYP3A4 inhibitors. In rare cases, DDIs can be helpful and clinically desired to achieve the target concentrations of the victim drug. For example, administration of a CYP3A4 inhibitor with cyclosporine leads to a reduction in dose and cost of treatment, classifying it as a therapeutically beneficial drug interaction.[42] Similarly, abirtaterone, a recently approved medication for advanced prostate cancer, can block CYP3A4-mediated calcitriol metabolism in the human liver and intestine,[34] potentially leading to additional benefits from calcitriol.

The initiation of drug-metabolizing enzymes is also a concern as CYP inducers augment drug elimination, decreased exposure, altered dug action, and therapeutic failure. Alternatively, induction can also result in an enhanced toxicity following an augmentation in reactive metabolite creation in biotransformation. Examples of clinically significant CYP inducers include phenobarbital, rifampin, phenytoin, ritonavir, carbamazepine, and St. John's wort.[85] In humans, often it is difficult to predict the time-course of enzyme induction as it depends on factors like half-life and enzyme turnover.[2] The short half-life of rifampicin leads to induction of CYP3A and CYP2C within 24 h; however, phenobarbital with a half-life to 3–5 days induces CYP3A4, CYP1A2, and CYP2C in 1 week.[2]

To reduce the risks connected with clinical DDIs, the pharmaceutical organizations are required to exemplified the drug inhibition, metabolite, and induction potentials in the early discovery stages. Guidelines for the industry designed by USFDA outline the drug interaction data analysis, study design, implications for dosing, and labeling recommendations.[52] Time-dependent inhibition of CYP enzymes is an active area of research, where preincubation of the test compound with enzyme leads to an amplified in the extent of inhibition. For example, tienilic acid forms a MI complex that inactivates human CYP2C9 by covalently binding to the apoprotein, leading to an immunogenic response-mediated hepatotoxicity.[55] The primary cultures of human hepatocytes are the gold standard to evaluate CYP induction in vitro. Limited availability of human tissue, stimulates the global use of human hepatocytes for early discovery and proof-of-principle clinical studies.[105]

Concomitant consumption of herbs and allopathic drugs is an important clinical concern. The need for polypharmacy in the management of many conditions further increases the chances of herb–drug interactions (HDIs). Structurally unrelated compounds induce or inhibit hepatic and intestinal CYP is accountable for the large number of reported drug–drug and drug–food interactions.[162] Active constituents of herbal products can be substrate, inhibitor, or inducer of multiple CYP isoforms. St. John's wort is a potent CYP3A4 inducer that significantly lowers plasma concentrations of common CYP3A4 substrates such as cyclosporine, simvastatin, indinavir, warfarin, and amitriptyline.[50] *Gingko biloba* has been reported to induce CYP2C19 and garlic has been reported to induce CYP3A4.[215] Grapefruit juice has been reported to contain flavonoids that have potent CYP inhibitory activity.[50] *Terminalia arjuna* has been reported to inhibit CYP1A2, CYP2C9, CYP2D6, and CYP3A4 in human liver microsomes.[202] Ginsenosides, the active components of ginseng herbal formulations, inhibit calcitriol (the active vitamin D) metabolism in human liver and intestine.[35] Thus, numerous natural products can potentially influence the pharmacokinetics of clinically important pharmaceutical drugs by interacting with the drug-metabolizing enzymes, and thus safety profiling of herbal botanicals is also essential.

13.16 CONCLUSION

The detoxification and xenobiotic modifications presented in the book chapter provided directions toward the probable mechanisms by which a drug can be modified into readily excretable from the human body. The factors affecting the metabolism and the various routes of detoxification may also inspire the researchers involved in metabolism-based drug discovery where active analogs can be developed from the metabolized products or through several other approaches. In addition to it, therapeutic regimen, drug interactions, and other pharmacy practice aspects can be associated with the metabolic consequences and can be studied for pharmaceutical interest.

KEYWORDS

- **metabolism**
- **phase-I reactions**
- **phase-II reactions**
- **CYP_{450} enzymes**
- **regulations**
- **drug-herb interactions**

REFERENCES

1. Abelo, A.; Andersson, T. B.; Antonsson, M.; Naudot, A. K.; Skanberg, I.; Weidolf, L. Stereoselective Metabolism of Omeprazole by Human Cytochrome P450 Enzymes. *Drug Metab. Dispos.* **2000,** *28*, 966–972.
2. Almond, L. M.; Yang, J.; Jamei, M.; Tucker, G. T.; Rostami-Hodjegan, A. Towards a Quantitative Framework for the Prediction of DDIs Arising from Cytochrome P450 Induction. *Curr. Drug Metab.* **2009,** *10*, 420–432.
3. Andersson, T.; Hassan-Alin, M.; Hasselgren, G.; Rohss, K.; Weidolf, L. Pharmacokinetic Studies with Esomeprazole, the (S)-isomer of Omeprazole. *Clin. Pharmacokinet* **2001,** *40*, 411–426.
4. Bandiera, S. Expression and Catalysis of Sex-Specific Cytochrome P450 Isozymes in Rat Liver. *Can. J. Physiol. Pharmacol.* **1990,** *68*, 762–768.
5. Baraldo, M. The Influence of Circadian Rhythms on the Kinetics of Drugs in Humans. *Expert Opin. Drug Metab. Toxicol.* **2008,** *4*, 175–192.
6. Benedetti, M. S. Biotransformation of Xenobiotics by Amine Oxidases. *Fundam. Clin. Pharmacol.* **2001,** *15*, 75–84.
7. Bertelsen, K. M.; Venkatakrishnan, K.; Von Moltke, L. L.; Obach, R. S.; Greenblatt, D. J. Apparent Mechanism-Based Inhibition of Human CYP2D6 in Vitro by Paroxetine: Comparison with Fluoxetine and Quinidine. *Drug Metab. Dispos.* **2003,** *31*, 289–293.
8. Bock, K. W. Functions and Transcriptional Regulation of Adult Human Hepatic UDP-glucuronosyl-transferases (UGTs): Mechanisms Responsible for Interindividual Variation of UGT Levels. *Biochem. Pharmacol.* **2010,** *80*, 771–777.
9. Boelsterli, U. A. Acyl Glucuronides: Mechanistic Role in Drug Toxicity? *Curr. Drug Metab.* **2011,** *12*, 213–214.
10. Bolleddula, J.; DeMent, K.; Driscoll, J. P.; Worboys, P.; Brassil, P. J.; Bourdet, D. L. Biotransformation and Bioactivation Reactions of Alicyclic Amines in Drug Molecules. *Drug Metab. Rev.* **2014,** *46*, 379–419.
11. Bolt, H. M. Rifampicin, a Keystone Inducer of Drug Metabolism: From Herbert Remmer's Pioneering Ideas to Modern Concepts. *Drug Metab. Rev.* **2004,** *36*, 497–509.

12. Bousova, I.; Skalova, L. Inhibition and Induction of Glutathione S-Transferases by Flavonoids: Possible Pharmacological and Toxicological Consequences. *Drug Metab. Rev.* **2012,** *44*, 267–286.
13. Buchanan, M. L.; Easterling, T. R.; Carr, D. B.; Shen, D. D.; Risler, L. J.; Nelson, W. L.; Mattison, D. R.; Hebert, M. F. Clonidine Pharmacokinetics in Pregnancy. *Drug Metab. Dispos.* **2009,** *37*, 702–705.
14. Caraco, Y.; Zylber-Katz, E.; Granit, L.; Levy, M. Does Restriction of Caffeine Intake Affect Mixed Function Oxidase Activity and Caffeine Metabolism? *Biopharm. Drug Dispos.* **1990,** *11*, 639–643.
15. Carrasco-Portugal, M. F-M. F. Gender Differences in the Pharmacokinetics of Oral Drugs. *Pharmacol. Pharm.* **2011,** *2*, 31–41.
16. Chang, G. W.; Kam, P. C. The Physiological and Pharmacological Roles of Cytochrome P450 Isoenzymes. *Anaesthesia* **1999,** *54*, 42–50.
17. Chang, T. K.; Chen, J.; Pillay, V.; Ho, J. Y.; Bandiera, S. M. Real-time Polymerase Chain Reaction Analysis of CYP1B1 Gene Expression in Human Liver. *Toxicol. Sci.* **2003,** *71*, 11–19.
18. Chen, Y.; Ferguson, S. S.; Negishi, M.; Goldstein, J. A. Induction of Human CYP2C9 by Rifampicin, Hyperforin, and Phenobarbital Is Mediated by the Pregnane X Receptor. *J Pharmacol. Exp. Ther.* **2004,** *308*, 495–501.
19. Chiang, J. Y. Bile Acid Regulation of Gene Expression: Roles of Nuclear Hormone Receptors. *Endocr. Rev.* **2002,** *23*, 443–463.
20. Chiba, M.; Hensleigh, M.; Lin, J. H. Hepatic and Intestinal Metabolism of Indinavir, an HIV Protease Inhibitor, in Rat and Human Microsomes. Major Role of CYP3A. *Biochem. Pharmacol.* **1997,** *53*, 1187–1195.
21. Chiu, N. T.; Tomlinson Guns, E. S.; Adomat, H.; Jia, W.; Deb, S Identification of Human Cytochrome P450 Enzymes Involved in the Hepatic and Intestinal Biotransformation of 20(S)-protopanaxadiol. *Biopharm. Drug Dispos.* **2014,** *35*, 104–118.
22. Cho, U. S.; Park, E. Y.; Dong, M. S.; Park, B. S.; Kim, K.; Kim, K. H. Tight-Binding Inhibition by Alpha-Naphthoflavone of Human Cytochrome P450 1A2. *Biochim. Biophys. Acta* **2003,** *1648*, 195–202.
23. Christians, U.; Jacobsen, W.; Floren, L. C Metabolism and Drug Interactions of 3–hydroxy-3-methylglutaryl Coenzyme A Reductase Inhibitors in Transplant Patients: Are the Statins Mechanistically Similar? *Pharmacol. Ther.* **1998,** *80*, 1–34.
24. Clifford, C. P.; Adams, D. A.; Murray, S.; Taylor, G. W.; Wilkins, M. R.; Boobis, A. R.; Davies, D. S. The Cardiac Effects of Terfenadine after Inhibition of Its Metabolism by Grapefruit Juice. *Eur. J. Clin. Pharmacol.* **1997,** *52*, 311–315.
25. Coleman, M. D. How Oxidative Systems Metabolize Substrates. In *Human Drug Metabolism: An Introduction*; Wiley-Blackwell: West Sussex, UK, 2010; pp 23–64.
26. Conney, A. H.; Pantuck, E. J.; Hsiao, K. C.; Garland, W. A.; Anderson, K. E.; Alvares, A. P.; Kappas, A. Enhanced Phenacetin Metabolism in Human Subjects Fed Charcoal-Broiled Beef. *Clin. Pharmacol. Ther.* **1976,** *20*, 633–642.
27. Cotreau, M. M.; von Moltke, L. L.; Beinfeld, M. C.; Greenblatt, D. J. Methodologies to Study the Induction of Rat Hepatic and Intestinal Cytochrome P450 3A at the mRNA, Protein, and Catalytic Activity Level. *J. Pharmacol. Toxicol. Methods* **2000,** *43*, 41–54.
28. Daly, A. K. Significance of the Minor Cytochrome P450 3A Isoforms. *Clin. Pharmacokinet.* **2006,** *45*, 13–31.

29. Dansette, P. M.; Rosi, J.; Bertho, G.; Mansuy, D. Cytochromes P450 Catalyze Both Steps of the Major Pathway of Clopidogrel Bioactivation, Whereas Paraoxonase Catalyzes the Formation of a Minor Thiol Metabolite Isomer. *Chem. Res. Toxicol.* **2012,** *25*, 348–356.
30. Davies, N. M. Clinical Pharmacokinetics of Ibuprofen. The First 30 Years. *Clin. Pharmacokinet.* **1998,** *34*, 101–154.
31. Deb, S.; Bandiera, S. M. Characterization and Expression of Extrahepatic CYP2S1. *Expert Opin. Drug Metab. Toxicol.* **2009,** *5*, 367–380.
32. Deb, S.; Bandiera, S. M. Characterization of a New Cytochrome P450 Enzyme, CYP2S1, in Rats: Its Regulation by Aryl Hydrocarbon Receptor Agonists. *Toxicology* **2010,** *267*, 91–98.
33. Deb, S.; Bandiera, S. M. Regulation of Cytochrome P450 1B1 Expression by Luteinizing Hormone in Mouse MA-10 and Rat R2C Leydig Cells: Role of Protein Kinase A. *Biol. Reprod.* **2011,** *85*, 89–96.
34. Deb, S.; Chin, M. Y.; Adomat, H.; Guns, E. S. Abiraterone Inhibits 1alpha,25-dihydroxyvitamin D3 Metabolism by CYP3A4 in Human Liver and Intestine In Vitro. *J. Steroid Biochem. Mol. Biol.* **2014,** *144* (Pt A), 50–58.
35. Deb, S.; Chin, M. Y.; Adomat, H.; Guns, E. S. Ginsenoside-Mediated Blockade of 1alpha,25-Dihydroxyvitamin D3 Inactivation in Human Liver and Intestine in Vitro. *J. Steroid Biochem. Mol. Biol.* **2014b,** *141*, 94–103.
36. Deb, S.; Kawai, M.; Chang, T. K.; Bandiera, S. M. CYP1B1 Expression in Rat Testis and Leydig Cells Is Not Inducible by Aryl Hydrocarbon Receptor Agonists. *Xenobiotica* **2010,** *40*, 447–457.
37. Deb, S.; Pandey, M.; Adomat, H.; Guns, E. S. Cytochrome P450 3A-Mediated Microsomal Biotransformation of 1alpha,25-dihydroxyvitamin D3 in Mouse and Human Liver: Drug-Related Induction and Inhibition of Catabolism. *Drug Metab. Dispos.* **2012,** *40*, 907–918.
38. Deb, S.; Tai, J. K.; Leung, G. S.; Chang, T. K.; Bandiera, S. M. Estradiol-Mediated Suppression of CYP1B1 Expression in Mouse MA-10 Leydig Cells Is Independent of Protein Kinase A and Estrogen Receptor. *Mol. Cell Biochem.* **2011,** *358*, 387–395.
39. Denisov, I. G.; Makris, T. M.; Sligar, S. G. Schlichting I Structure and Chemistry of Cytochrome P450. *Chem. Rev.* **2005,** *105*, 2253–2277.
40. Di, L. The Role of Drug Metabolizing Enzymes in Clearance. *Expert Opin. Drug Metab. Toxicol.* **2014,** *10*, 379–393.
41. Ding, X.; Kaminsky, L. S. Human Extrahepatic Cytochromes P450: Function in Xenobiotic Metabolism and Tissue-Selective Chemical Toxicity in the Respiratory and Gastrointestinal Tracts. *Annu. Rev. Pharmacol. Toxicol.* **2003,** *43*, 149–173.
42. Dresser, G. K.; Spence, J. D.; Bailey, D. G. Pharmacokinetic-Pharmacodynamic Consequences and Clinical Relevance of Cytochrome P450 3A4 Inhibition. *Clin. Pharmacokinet.* **2000,** *38*, 41–57.
43. Duricova, J.; Perinova, I.; Jurckova, N.; Kacirova, I.; Grundmann, M. Clinically Important Interaction Between Metoprolol and Propafenone. *Canadian Fam. Phys. Med. de famille canadien* **2013,** *59*, 373–375.
44. Edmondson, D, E. Hydrogen Peroxide Produced by Mitochondrial Monoamine Oxidase Catalysis: Biological Implications. *Curr. Pharm. Des.* **2014,** *20*, 155–160.
45. El Sherbeni, A. A.; El Kadi, A. O. The Role of Epoxide Hydrolases in Health and Disease. *Arch. Toxicol.* **2014,** *88*, 2013–2032.
46. Elbekai, R. H.; El-Kadi, A. O. Cytochrome P450 Enzymes: Central Players in Cardiovascular Health and Disease. *Pharmacol. Ther.* **2006,** *112*, 564–587.

47. Fabrazzo, M.; Esposito, G.; Fusco, R.; Maj, M. Effect of Treatment Duration on Plasma Levels of Clozapine and N-Desmethylclozapine in Men and Women. *Psychopharmacology* **1996,** *124*, 197–200.
48. Fareed, J.; Jeske, W.; Thethi, I. Metabolic Differences of Current Thienopyridine Antiplatelet Agents. *Expert Opin. Drug Metab. Toxicol.* **2013,** *9*, 307–317.
49. Farid, N. A.; Kurihara, A.; Wrighton, S. A. Metabolism and Disposition of the Thienopyridine Antiplatelet Drugs Ticlopidine, Clopidogrel, and Prasugrel in Humans. *J. Clin. Pharmacol.* **2010,** *50*, 126–142.
50. Fasinu, P. S.; Bouic, P. J.; Rosenkranz, B. An Overview of the Evidence and Mechanisms of Herb-Drug Interactions. *Front. Pharmacol.* **2012,** *3*, 69.
51. Faucette, S. R.; Wang, H.; Hamilton, G. A.; Jolley, S. L.; Gilbert, D.; Lindley, C.; Yan, B.; Negishi, M.; LeCluyse, E. L. Regulation of CYP2B6 in Primary Human Hepatocytes by Prototypical Inducers. *Drug Metab. Dispos.* **2004,** *32*, 348–358.
52. FDA. US Drug Interaction Studies-Study Design, Data Analysis, Impli-cations for Dosing, and Labeling Recommendations, Draft Guidance, 2012.
53. Fisher, M. B.; Paine, M. F.; Strelevitz, T. J.; Wrighton, S. A. The Role of Hepatic and Extrahepatic UDP-Glucuronosyltransferases in Human Drug Metabolism. *Drug Metab. Rev.* **2001,** *33*, 273–297.
54. Foti, R. S.; Wahlstrom, J. L. CYP2C19 Inhibition: The Impact of Substrate Probe Selection on in Vitro Inhibition Profiles. *Drug Metab. Dispos.* **2008,** *36*, 523–528.
55. Fowler, S.; Zhang, H. In Vitro Evaluation of Reversible and Irreversible Cytochrome P450 Inhibition: Current Status on Methodologies and Their Utility for Predicting Drug-Drug Interactions. *AAPS J.* **2008,** *10*, 410–424.
56. Franco, V.; Mazzucchelli, I.; Gatti, G.; Specchio, L. M.; La Neve, A.; Papantonio, A.; Ozkaynakci, A. E.; Perucca, E. Changes in Lamotrigine Pharmacokinetics during Pregnancy and the Puerperium. *Ther. Drug Monit.* **2008,** *30*, 544–547.
57. Franco, V.; Perucca, E. CYP2C9 Polymorphisms and Phenytoin Metabolism: Implications for Adverse Effects. *Expert Opin. Drug Metab. Toxicol.* **2015,** *11*, 1269–1279.
58. Fukami, T.; Yokoi, T. The Emerging Role of Human Esterases. *Drug Metab. Pharmacokinet.* **2012,** *27*, 466–477.
59. Gamage, N.; Barnett, A.; Hempel, N.; Duggleby, R. G.; Windmill, K. F.; Martin, J. L.; McManus, M. E. Human Sulfotransferases and Their Role in Chemical Metabolism. *Toxicol. Sci.* **2006,** *90*, 5–22.
60. Gao, Y.; Shao, J.; Jiang, Z.; Chen, J.; Gu, S.; Yu, S.; Zheng, K.; Jia, L. Drug Enterohepatic Circulation and Disposition: Constituents of Systems Pharmacokinetics. *Drug Discov. Today* **2014,** *19*, 326–340.
61. Gardner-Stephen, D. A.; Mackenzie, P. I. Liver-Enriched Transcription Factors and Their Role in Regulating UDP Glucuronosyltransferase Gene Expression. *Curr. Drug Metab.* **2008,** *9*, 439–452.
62. Gavhane, Y. N.; Yadav, A. V. Loss of Orally Administered Drugs in GI Tract. *Saudi Pharm. J.* **2012,** *20*, 331–344.
63. Gibson, G. G.; Skett, P. Pathways of Drug Metabolism. In *Introduction to Drug Metabolism*; Nelson Thornes: Cheltenham, 2001; pp 1–36.
64. Girre, C.; Lucas, D.; Hispard, E.; Menez, C.; Dally, S.; Menez, J. F. Assessment of Cytochrome P4502E1 Induction in Alcoholic Patients by Chlorzoxazone Pharmacokinetics. *Biochem. Pharmacol.* **1994,** *47*, 1503–1508.
65. Graham, M. J.; Lake, B. G. Induction of Drug Metabolism: Species Differences and Toxicological Relevance. *Toxicology* **2008,** *254*, 184–191.

66. Groves, J. T.; McClusky, G. A. Aliphatic Hydroxylation by Highly Purified Liver Microsomal Cytochrome P-450. Evidence for a Carbon Radical Intermediate. *Biochem. Biophys. Res. Commun.* **1978,** *81*, 154–160.
67. Guengerich, F. P. Comparisons of Catalytic Selectivity of Cytochrome P450 Subfamily Enzymes from Different Species. *Chem. Biol. Interact.* **1997,** *106*, 161–182.
68. Guengerich, F. P. Common and Uncommon Cytochrome P450 Reactions Related to Metabolism and Chemical Toxicity. *Chem. Res. Toxicol.* **2001,** *14*, 611–650.
69. Guengerich, F. P. Cytochrome P450 Oxidations in the Generation of Reactive Electrophiles: Epoxidation and Related Reactions. *Arch. Biochem. Biophys.* **2003,** *409*, 59–71.
70. Guengerich, F. P. Human Cytochrome P450 Enzymes. In *Cytochrome P450: Structure, Mechanism, and Biochemistry*; Ortiz de Montellano, P. R., Ed.; Kluwer Academic: New York, 2005; pp 377–530.
71. Gupta, S.; Nihalani, N.; Masand, P. Duloxetine: Review of Its Pharmacology, and Therapeutic Use in Depression and Other Psychiatric Disorders. *Ann. Clin. Psych.* **2007,** *19*, 125–132.
72. Halpert, J. R. Structural Basis of Selective Cytochrome P450 Inhibition. *Annu. Rev. Pharmacol. Toxicol.* **1995,** *35*, 29–53.
73. Harris, R. Z.; Jang, G. R.; Tsunoda, S. Dietary Effects on Drug Metabolism and Transport. *Clin. Pharmacokinet.* **2003,** *42*, 1071–1088.
74. Hasler, J. A. Pharmacogenetics of Cytochromes P450. *Mol. Aspects Med.* **1999,** *20*, 12–24, 25–137.
75. Hewitt, N. J.; Lecluyse, E. L.; Ferguson, S. S. Induction of Hepatic Cytochrome P450 Enzymes: Methods, Mechanisms, Recommendations, and in Vitro-in Vivo Correlations. *Xenobiotica* **2007,** *37*, 1196–1224.
76. Hino, I.; Akama, H.; Furuya, T.; Ueda, H.; Taniguchi, A.; Hara, M.; Kashiwazaki, S. Pravastatin-induced Rhabdomyolysis in a Patient with Mixed Connective Tissue Disease. *Arthritis Rheumatism* **1996,** *39*, 1259–1260.
77. Hirota, N.; Ito, K.; Iwatsubo, T.; Green, C. E.; Tyson, C. A.; Shimada, N.; Suzuki, H.; Sugiyama, Y. In Vitro/in Vivo Scaling of Alprazolam Metabolism by CYP3A4 and CYP3A5 in Humans. *Biopharm. Drug Dispos.* **2001,** *22*, 53–71.
78. Hoffart, E.; Ghebreghiorghis, L.; Nussler, A. K.; Thasler, W. E.; Weiss, T. S.; Schwab. M.; Burk, O. Effects of Atorvastatin Metabolites on Induction of Drug-Metabolizing Enzymes and Membrane Transporters through Human Pregnane X Receptor. *Br. J. Pharmacol.* **2012,** *165*, 1595–1608.
79. Hoffman, S. M., Nelson, D. R.; Keeney, D. S. Organization, Structure and Evolution of the CYP2 Gene Cluster on Human Chromosome 19. *Pharmacogenetics* **2001,** *11*, 687–698.
80. Honkakoski, P.; Negishi, M. Regulation of Cytochrome P450 (CYP) Genes by Nuclear Receptors. *Biochem. J.* **2000,** *347*, 321–337.
81. Hu, D. G.; Meech, R.; McKinnon, R. A.; Mackenzie, P. I. Transcriptional Regulation of Human UDP-Glucuronosyltransferase Genes. *Drug Metab. Rev.* **2014,** *46*, 421–458.
82. Hurst, S.; Loi, C. M.; Brodfuehrer, J.; El-Kattan, A. Impact of Physiological, Physicochemical and Biopharmaceutical Factors in Absorption and Metabolism Mechanisms on the Drug Oral Bioavailability of Rats and Humans. *Expert Opin. Drug Metab. Toxicol.* **2007,** *3*, 469–489.
83. Hutt, A. J.; Caldwell, J. Amino Acid Conjugation. In *Conjugation Reactions in Drug Metabolism: An Integrated Approach*; Mulder, G. J., Ed.; Taylor & Francis Ltd.: London, 1990; pp 273–305.

84. Hutzler, J. M.; Balogh, L. M.; Zientek, M.; Kumar, V.; Tracy, T. S. Mechanism-Based Inactivation of Cytochrome P450 2C9 by Tienilic Acid and (+/-)-Suprofen: A Comparison of Kinetics and Probe Substrate Selection. *Drug Metab. Dispos.* **2009,** *37*, 59–65.
85. Hutzler, J. M. C. J.; Fleishaker, J. C. Drug–Drug Interactions: Designing Development Programs and Appropriate Product Labeling. In *Pharmacokinetics in Drug Development: Advances and Applications*; Bonate, P. L. H. D., Ed.; Springer, 2011; pp 21–56.
86. Ickenstein, L. M.; Bandiera, S. M. Persistent Suppression of Hepatic CYP2A1 Expression and Serum Triiodothyronine Levels by Tamoxifen in Intact Female Rats: Dose-Response Analysis and Comparison with 4-hydroxytamoxifen, Fulvestrant (ICI 182,780), and 17beta-estradiol-3-benzoate. *J. Pharmacol. Exp. Ther.* **2002,** *302*, 584–593.
87. Ingelman-Sundberg, M. Human Drug Metabolising Cytochrome P450 Enzymes: Properties and Polymorphisms. *Naunyn Schmiedebergs Arch. Pharmacol.* **2004,** *369*, 89–104.
88. Ionescu, C.; Caira, M. R. Pathways of Biotransformation—Phase I Reactions. In *Drug Metabolism: Current Concepts*; Springer: Dordrecht, The Netherlands, 2005; pp 41–128.
89. Ionescu, C.; Caira, M. R. Pathways of Biotransformation—Phase II Reactions. In *Drug Metabolism: Current Concepts*; Springer: Dordrecht, The Netherlands, 2005; pp 129–170.
90. Isoherranen, N.; Thummel, K. E. Drug Metabolism and Transport during Pregnancy: How Does Drug Disposition Change during Pregnancy and What Are the Mechanisms That Cause Such Changes? *Drug Metab. Dispos.* **2013,** *41*, 256–262.
91. Jeong, H. Altered Drug Metabolism during Pregnancy: Hormonal Regulation of Drug-Metabolizing Enzymes. *Expert Opin. Drug Metab. Toxicol.* **2010,** *6*, 689–699.
92. Johansson, I.; Ingelman-Sundberg, M. Genetic Polymorphism and Toxicology—with Emphasis on Cytochrome p450. *Toxicol. Sci.* **2011,** *120*, 1–13.
93. Jones, D. R.; Miller, G. P. Assays and Applications in Warfarin Metabolism: What We Know, How We Know It and What We Need to Know. *Expert Opin. Drug Metab. Toxicol.* **2011,** *7*, 857–874.
94. Jones, D. R.; Moran, J. H.; Miller, G. P. Warfarin and UDP-glucuronosyltransferases: Writing a New Chapter of Metabolism. *Drug Metab. Rev.* **2010,** *42*, 55–61.
95. Kaivosaari, S.; Finel, M.; Koskinen, M. N-glucuronidation of Drugs and Other Xenobiotics by Human and Animal UDP-glucuronosyltransferases. *Xenobiotica* **2011,** *41*, 652–669.
96. Kato, M. Intestinal First-Pass Metabolism of CYP3A4 Substrates. *Drug Metab. Pharmacokinet.* **2008,** *23*, 87–94.
97. Kemper, R. A.; Hayes, J. R.; Bogdanffy, M. S. Metabolism: A Determinant of Toxicity. In *Principles and Methods of Toxicology*; Hayes, A. W., Ed.; Informa Healthcare USA, Inc.: New York, 2008; pp 103–178.
98. King, R. S. Biotransformations in Drug Metabolism. In *Drug Metabolism Handbook: Concepts and Applications*; Nassar, A. F., Hollenberg, P. F., Scatina, J., Eds; John Wiley & Sons, Inc.: New Jersey, 2009; pp 17–40.
99. Kinirons, M. T.; O'Mahony, M. S. Drug Metabolism and Ageing. *Br. J. Clin. Pharmacol.* **2004,** *57*, 540–544.
100. Knadler, M. P.; Lobo, E.; Chappell, J.; Bergstrom, R. Duloxetine: Clinical Pharmacokinetics and Drug Interactions. *Clin. Pharmacokinet.* **2011,** *50*, 281–294.
101. Knights, K. M.; Sykes, M. J.; Miners, J. O. Amino Acid Conjugation: Contribution to the Metabolism and Toxicity of Xenobiotic Carboxylic Acids. *Expert Opin. Drug Metab. Toxicol.* **2007,** *3*, 159–168.

102. Kwon, Y. Metabolism. In *Handbook of Essential Pharmacokinetics, Pharmacodynamics, and Drug Metabolism for Industrial Scientists*; Kluwer Academic Publishers: New York, 2002; pp 121–168.
103. Laizure, S. C.; Herring, V.; Hu, Z.; Witbrodt, K.; Parker, R. B. The Role of Human Carboxylesterases in Drug Metabolism: Have We Overlooked Their Importance? *Pharmacotherapy* **2013,** *33*, 210–222.
104. Laizure, S. C.; Parker, R. B. A Comparison of the Metabolism of Clopidogrel and Prasugrel. *Expert Opin. Drug Metab. Toxicol.* **2010,** *6*, 1417–1424.
105. LeCluyse, E. L. Human Hepatocyte Culture Systems for the in Vitro Evaluation of Cytochrome P450 Expression and Regulation. *Eur. J. Pharm. Sci.* **2001,** *13*, 343–368.
106. Lemke, T. L. Aldehydes and Ketones. In *Review of Organic Functional Groups: Introduction to Medicinal Organic Chemistry*; Lippincott Williams & Wilkins: Baltimore, MD, 2012; pp 42–48.
107. Lemke, T. L. Drug Metabolism. In *Review of Organic Functional Groups: Introduction to Medicinal Organic Chemistry*; Lippincott Williams & Wilkins, Baltimore, MD, 2012; pp 176–183.
108. Lemke, T. L. Phenols. In *Review of Organic Functional Groups: Introduction to Medicinal Organic Chemistry*; Lippincott Williams & Wilkins: Baltimore, MD, 2012; pp 29–35.
109. Lemmer, B.; Nold, G. Circadian Changes in Estimated Hepatic Blood Flow in Healthy Subjects. *Br. J. Clin. Pharmacol.* **1991,** *32*, 627–629.
110. Levine, W. G. Biliary Excretion of Drugs and Other Xenobiotics. *Annu. Rev. Pharmacol. Toxicol.* **1978,** *18*, 81–96.
111. Lewis, F. *Cytochrome P450: Structure, Function and Mechanism*; Taylor & Franscis: New York, 1996.
112. Lewis, F. *Guide to Cytochrome P450: Structure and Function*; Taylor & Francis Inc.: New York, 2001.
113. Li, G.; Gong, P. L.; Qiu, J.; Zeng, F. D.; Klotz, U. Stereoselective Steady State Disposition and Action of Propafenone in Chinese Subjects. *Br. J. Clin. Pharmacol.* **1998,** *46*, 441–445.
114. Lin, J. H.; Chiba, M.; Baillie, T. A. Is the Role of the Small Intestine in First-Pass Metabolism Overemphasized? *Pharmacol. Rev.* **1999,** *51*, 135–158.
115. Lin, J. H.; Lu, A. Y. Inhibition and Induction of Cytochrome P450 and the Clinical Implications. *Clin. Pharmacokinet.* **1998,** *35*, 361–390.
116. Lin, J. H.; Lu, A. Y. Interindividual Variability in Inhibition and Induction of Cytochrome P450 Enzymes. *Annu. Rev. Pharmacol. Toxicol.* **2001,** *41*, 535–567.
117. Liu, Y.; She, M.; Wu, Z.; Dai, R. The Inhibition Study of Human UDP-Glucuronosyltransferases with Cytochrome P450 Selective Substrates and Inhibitors. *J. Enzyme Inhibition Med. Chem.* **2011,** *26*, 386–393.
118. Liu, Y. T.; Hao, H. P.; Liu, C. X.; Wang, G. J.; Xie, H. G. Drugs as CYP3A Probes, Inducers, and Inhibitors. *Drug Metab. Rev.* **2007,** *39*, 699–721.
119. Lobo, E. D.; Bergstrom, R. F.; Reddy, S.; Quinlan, T.; Chappell, J.; Hong, Q.; Ring, B.; Knadler, M. P. In Vitro and in Vivo Evaluations of Cytochrome P450 1A2 Interactions with Duloxetine. *Clin. Pharmacokinet.* **2008,** *47*, 191–202.
120. Lynch, T.; Price, A. The Effect of Cytochrome P450 Metabolism on Drug Response, Interactions, and Adverse Effects. *Am. Fam. Phys.* **2007,** *76*, 391–396.
121. Mackenzie, P. I.; Hu, D. G.; Gardner-Stephen, D. A. The Regulation of UDP-Glucuronosyltransferase Genes by Tissue-Specific and Ligand-Activated Transcription Factors. *Drug Metab. Rev.* **2010,** *42*, 99–109.

122. Madan, A.; Graham, R. A.; Carroll, K. M.; Mudra, D. R.; Burton, L. A.; Krueger, L. A.; Downey, A. D.; Czerwinski, M.; Forster, J.; Ribadeneira, M. D.; Gan, L. S.; LeCluyse, E. L.; Zech, K.; Robertson, P. Jr.; Koch, P.; Antonian, L.; Wagner, G.; Yu, L.; Parkinson, A. Effects of Prototypical Microsomal Enzyme Inducers on Cytochrome P450 Expression in Cultured Human Hepatocytes. *Drug Metab. Dispos.* **2003,** *31*, 421–431.
123. Malatkova, P.; Wsol, V. Carbonyl Reduction Pathways in Drug Metabolism. *Drug Metab. Rev.* **2014,** *46*, 96–123.
124. Malhotra, S.; Bailey, D. G.; Paine, M. F.; Watkins, P. B. Seville Orange Juice-Felodipine Interaction: Comparison with Dilute Grapefruit Juice and Involvement of Furocoumarins. *Clin. Pharm. Ther. Clin. Pharmacol. Ther.* **2001,** *69*, 14–23.
125. Martignoni, M.; Groothuis, G. M.; de Kanter, R. Species Differences between Mouse, Rat, Dog, Monkey and Human CYP-Mediated Drug Metabolism, Inhibition and Induction. *Expert Opin. Drug Metab. Toxicol.* **2006,** *2*, 875–894.
126. Mazaleuskaya, L. L.; Theken, K. N.; Gong, L.; Thorn, C. F.; FitzGerald, G. A.; Altman, R. B.; Klein, T. E. PharmGKB Summary: Ibuprofen Pathways. *Pharmacogenet. Genom.* **2015,** *25*, 96–106.
127. McGready, R.; Stepniewska, K.; Seaton, E.; Cho, T.; Cho, D.; Ginsberg, A.; Edstein, M. D.; Ashley, E.; Looareesuwan, S.; White, N. J.; Nosten, F. Pregnancy and Use of Oral Contraceptives Reduces the Biotransformation of Proguanil to Cycloguanil. *Eur. J. Clin. Pharmacol.* **2003,** *59*, 553–557.
128. McLachlan, A. J.; Pont, L. G. Drug Metabolism in Older People—A Key Consideration in Achieving Optimal Outcomes with Medicines. *J. Gerontol. A Biol. Sci. Med. Sci.* **2012,** *67*, 175–180.
129. Meunier, B.; de Visser, S. P.; Shaik, S. Mechanism of Oxidation Reactions Catalyzed by Cytochrome p450 Enzymes. *Chem. Rev.* **2004,** *104*, 3947–3980.
130. Meyer, U. A. Pharmacogenetics and Adverse Drug Reactions. *Lancet* **2000,** *356*, 1667–1671.
131. Miller, G. P.; Jones, D. R.; Sullivan, S. Z.; Mazur, A.; Owen, S. N.; Mitchell, N. C.; Radominska-Pandya, A.; Moran, J. H. Assessing Cytochrome P450 and UDP-Glucuronosyltransferase Contributions to Warfarin Metabolism in Humans. *Chem. Res. Toxicol.* **2009,** *22*, 1239–1245.
132. Ming, D. S.; Pham, S.; Deb, S.; Chin, M. Y.; Kharmate, G.; Adomat, H.; Beheshti, E. H.; Locke, J.; Guns, E. T. Pomegranate Extracts Impact the Androgen Biosynthesis Pathways in Prostate Cancer Models in Vitro and in Vivo. *J. Steroid Biochem. Mol. Biol.* **2014,** *143*, 19–28.
133. Mo, S. L.; Liu, Y. H.; Duan, W.; Wei, M. Q.; Kanwar, J. R.; Zhou, S. F. Substrate Specificity, Regulation, and Polymorphism of Human Cytochrome P450 2B6. *Curr. Drug Metab.* **2009,** *10*, 730–753.
134. Monahan, B. P.; Ferguson, C. L.; Killeavy, E. S.; Lloyd, B. K.; Troy, J.; Cantilena, L. R., Jr. Torsades de Pointes Occurring in Association with Terfenadine Use. *JAMA* **1990,** *264*, 2788–2790.
135. Moore, L. B.; Goodwin, B.; Jones, S. A.; Wisely, G. B.; Serabjit-Singh, C. J.; Willson, T. M.; Collins, J. L.; Kliewer, S. A. St. John's Wort Induces Hepatic Drug Metabolism through Activation of the Pregnane X Receptor. *Proc. Natl. Acad. Sci. USA* **2000,** *97*, 7500–7502.
136. Nebert, D. W.; Gonzalez, F. J. P450 Genes: Structure, Evolution, and Regulation. *Annu. Rev. Biochem.* **1987,** *56*, 945–993.

137. Nebert, D. W.; Wikvall, K.; Miller, W. L. Human Cytochromes P450 in Health and Disease. *Phil. Trans. R. Soc. Lond. B* **2013,** *368*, 20120431.
138. Nelson, D. R.; Kamataki, T.; Waxman, D. J.; Guengerich, F. P.; Estabrook, R. W.; Feyereisen, R.; Gonzalez, F. J.; Coon, M. J.; Gunsalus, I. C.; Gotoh, O. et al. The P450 Superfamily: Update on New Sequences, Gene Mapping, Accession Numbers, Early Trivial Names of Enzymes, and Nomenclature. *DNA Cell Biol.* **1993,** *12*, 1–51.
139. Nelson, D. R.; Koymans, L.; Kamataki, T.; Stegeman, J. J.; Feyereisen, R.; Waxman, D. J.; Waterman, M. R.; Gotoh, O.; Coon, M. J.; Estabrook, R. W.; Gunsalus, I. C.; Nebert, D. W. P450 Superfamily: Update on New Sequences, Gene Mapping, Accession Numbers and Nomenclature. *Pharmacogenetics* **1996,** *6*, 1–42.
140. Niwa, T.; Shiraga, T.; Takagi, A. Effect of Antifungal Drugs on Cytochrome P450 (CYP) 2C9, CYP2C19, and CYP3A4 Activities in Human Liver Microsomes. *Biol. Pharm. Bull.* **2005,** *28*, 1805–1808.
141. Obach, R. S. Pharmacologically Active Drug Metabolites: Impact on Drug Discovery and Pharmacotherapy. *Pharmacol. Rev.* **2013,** *65*, 578–640.
142. Oda, S.; Fukami, T.; Yokoi, T.; Nakajima, M. A Comprehensive Review of UDP-Glucuronosyltransferase and Esterases for Drug Development. *Drug Metab. Pharmacokinet.* **2015,** *30*, 30–51.
143. Oesch, F.; Fabian, E.; Oesch-Bartlomowicz, B.; Werner, C.; Landsiedel, R. Drug-metabolizing enzymes in the skin of man, rat, and pig. *Drug Metab. Rev.* **2007,** *39*, 659–698.
144. Oesterheld, J. R. A Review of Developmental Aspects of Cytochrome P450. *J. Child Adolesc. Psychopharmacol.* **1998,** *8*, 161–174.
145. Ogu, C. C.; Maxa, J. L. Drug Interactions Due to Cytochrome P450. *Proceedings* **2000,** *13*, 421–423.
146. Ohyama, K.; Nakajima, M.; Suzuki, M.; Shimada, N.; Yamazaki, H.; Yokoi, T. Inhibitory Effects of Amiodarone and Its N-deethylated Metabolite on Human Cytochrome P450 Activities: Prediction of in Vivo Drug Interactions. *Br. J. Clin. Pharmacol.* **2000,** *49*, 244–253.
147. Omura, T. Mitochondrial P450s. *Chem. Biol. Interact.* **2006,** *163*, 86–93.
148. Orlando, R.; Piccoli, P.; De Martin, S.; Padrini, R.; Palatini, P. Effect of the CYP3A4 Inhibitor Erythromycin on the Pharmacokinetics of Lignocaine and Its Pharmacologically Active Metabolites in Subjects with Normal and Impaired Liver Function. *Br. J. Clin. Pharmacol.* **2003,** *55*, 86–93.
149. Ortiz de Montellano, P. *Cytochrome P450: Structure, Mechanism, and Biochemistry*; Plenum Press: New York, 1995.
150. Paine, M. F.; Criss, A. B.; Watkins, P. B. Two Major Grapefruit Juice Components Differ in Intestinal CYP3A4 Inhibition Kinetic and Binding Properties. *Drug Metab. Dispos.* **2004,** *32*, 1146–1153.
151. Paine, M. F.; Hart, H. L.; Ludington, S. S.; Haining, R. L.; Rettie, A. E.; Zeldin, D. C. The Human Intestinal Cytochrome P450 "Pie". *Drug Metab. Dispos.* **2006,** *34*, 880–886.
152. Pantuck, E. J.; Hsiao, K. C.; Loub, W. D.; Wattenberg, L. W.; Kuntzman, R.; Conney, A. H. Stimulatory Effect of Vegetables on Intestinal Drug Metabolism in the Rat. *J. Pharmacol. Exp. Ther.* **1976,** *198*, 278–283.
153. Parkinson, A.; Ogilvie, B. W.; Paris, B. L.; Hensley, T. N.; Loewen, G. J. Human Biotransformation. In *Biotransformation and Metabolite Elucidation of Xenobiotics*; Nassar, A. F., Ed; John Wiley & Sons, Inc.: New Jersey, 2010; pp 1–77.

154. Pascussi, J. M.; Drocourt, L.; Gerbal-Chaloin, S.; Fabre, J. M.; Maurel, P.; Vilarem, M. J. Dual Effect of Dexamethasone on CYP3A4 Gene Expression in Human Hepatocytes. Sequential Role of Glucocorticoid Receptor and Pregnane X Receptor. *Eur. J. Biochem./ FEBS* **2001,** *268*, 6346–6358.
155. Pascussi, J. M.; Gerbal-Chaloin, S.; Fabre, J. M.; Maurel, P.; Vilarem, M. J. Dexamethasone Enhances Constitutive Androstane Receptor Expression in Human Hepatocytes: Consequences on Cytochrome P450 Gene Regulation. *Mol. Pharmacol.* **2000,** *58*, 1441–1450.
156. Pelkonen, O.; Turpeinen, M.; Hakkola, J.; Honkakoski, P.; Hukkanen, J.; Raunio, H. Inhibition and Induction of Human Cytochrome P450 Enzymes: Current Status. *Arch. Toxicol.* **2008,** *82*, 667–715.
157. Penner, N.; Woodward, C.; Prakash, C. Drug Metabolizing Enzymes and Biotransformation Reactions. In *ADME-Enabling Technologies in Drug Design and Development*; Zhang, D., Surapaneni, S., Eds; John Wiley & Sons, Inc.: New Jersey, 2012; pp 545–565.
158. Pereira de Sousa, I.; Bernkop-Schnurch, A. Pre-Systemic Metabolism of Orally Administered Drugs and Strategies to Overcome It. *J. Control. Release* **2014,** *192*, 301–309.
159. Pham, S.; Deb, S.; Ming, D. S.; Adomat, H.; Hosseini-Beheshti, E.; Zoubeidi, A.; Gleave, M.; Guns, E. S. Next-Generation Steroidogenesis Inhibitors, Dutasteride and Abiraterone, Attenuate But Still Do Not Eliminate Androgen Biosynthesis in 22RV1 Cells in Vitro. *J. Steroid Biochem. Mol. Biol.* **2014,** *144* (Pt B), 436–444.
160. Polasek, T. M.; Elliot, D. J.; Lewis, B. C.; Miners, J. O. Mechanism-Based Inactivation of Human Cytochrome P4502C8 by Drugs in Vitro. *J. Pharmacol. Exp. Ther.* **2004,** *311*, 996–1007.
161. Projean, D.; Baune, B.; Farinotti, R.; Flinois, J. P.; Beaune, P.; Taburet, A. M.; Ducharme, J. In Vitro Metabolism of Chloroquine: Identification of CYP2C8, CYP3A4, and CYP2D6 as the Main Isoforms Catalyzing N-Desethylchloroquine Formation. *Drug Metab. Dispos.* **2003,** *31*, 748–754.
162. Quintieri, L.; Palatini, P.; Nassi, A.; Ruzza, P.; Floreani, M. Flavonoids Diosmetin and Luteolin Inhibit Midazolam Metabolism by Human Liver Microsomes and Recombinant CYP 3A4 and CYP3A5 Enzymes. *Biochem. Pharmacol.* **2008,** *75*, 1426–1437.
163. Regan, S. L.; Maggs, J. L.; Hammond, T. G.; Lambert, C.; Williams, D. P.; Park, B. K. Acyl Glucuronides: The Good, the Bad and the Ugly. *Biopharm. Drug Dispos.* **2010,** *31*, 367–395.
164. Relling, M. V.; Lin, J. S.; Ayers, G. D.; Evans, W. E. Racial and Gender Differences in N-Acetyltransferase, Xanthine Oxidase, and CYP1A2 Activities. *Clin. Pharm. Ther.* **1992,** *52*, 643–658.
165. Rendic, S.; Di Carlo, F. J. Human Cytochrome P450 Enzymes: A Status Report Summarizing Their Reactions, Substrates, Inducers, and Inhibitors. *Drug Metab. Rev.* **1997,** *29*, 413–580.
166. Richter, T.; Murdter, T. E.; Heinkele, G.; Pleiss, J.; Tatzel, S.; Schwab, M.; Eichelbaum, M.; Zanger, U. M. Potent Mechanism-Based Inhibition of Human CYP2B6 by Clopidogrel and Ticlopidine. *J. Pharmacol. Exp. Ther.* *308*, 189–197.
167. Riddick, D. S.; Lee, C.; Bhathena, A.; Timsit, Y. E. The 2001 Veylien Henderson Award of the Society of Toxicology of Canada. Positive and Negative Transcriptional Regulation of Cytochromes P450 by Polycyclic Aromatic Hydrocarbons. *Can. J. Physiol. Pharmacol.* **2003,** *81*, 59–77.

168. Roberts, M. S.; Magnusson, B. M.; Burczynski, F. J.; Weiss, M. Enterohepatic Circulation: Physiological, Pharmacokinetic and Clinical Implications. *Clin. Pharmacokinet.* **2002,** *41*, 751–790.
169. Rogerson, T. D.; Wilkinson, C. F.; Hetarski, K. Steric Factors in the Inhibitory Interaction of Imidazoles with Microsomal Enzymes. *Biochem. Pharmacol.* **1977,** *26*, 1039–1042.
170. Runge, D.; Kohler, C.; Kostrubsky, V. E.; Jager, D.; Lehmann, T.; Runge, D. M.; May, U.; Stolz, D. B.; Strom, S. C.; Fleig, W. E.; Michalopoulos, G. K. Induction of Cytochrome P450 (CYP)1A1, CYP1A2, and CYP3A4 But Not of CYP2C9, CYP2C19, Multidrug Resistance (MDR-1) and Multidrug Resistance Associated Protein (MRP-1) by Prototypical Inducers in Human Hepatocytes. *Biochem. Biophys. Res. Commun.* **2000,** *273*, 333–341.
171. Rydberg, P.; Jorgensen, M. S.; Jacobsen, T. A.; Jacobsen, A. M.; Madsen, K. G.; Olsen, L. Nitrogen Inversion Barriers Affect the N-oxidation of Tertiary Alkylamines by Cytochromes P450. *Angew. Chem. Int. Ed. Engl.* **2013,** *52*, 993–997.
172. Schmidt, J. V.; Bradfield, C. A. Ah Receptor Signaling Pathways. *Annu. Rev. Cell Dev. Biol.* **1996,** *12*, 55–89.
173. Shenfield, G. M. Genetic Polymorphisms, Drug Metabolism and Drug Concentrations. *Clin. Biochem. Rev./Austr. Assoc. Clin. Biochem.* **2004,** *25*, 203–206.
174. Shimada, T.; Fujii-Kuriyama, Y. Metabolic Activation of Polycyclic Aromatic Hydrocarbons to Carcinogens by Cytochromes P450 1A1 and 1B1. *Cancer Sci.* **2004,** *95*, 1–6.
175. Shimada, T.; Sugie, A.; Shindo, M.; Nakajima, T.; Azuma, E.; Hashimoto, M.; Inoue, K. Tissue-Specific Induction of Cytochromes P450 1A1 and 1B1 by Polycyclic Aromatic Hydrocarbons and Polychlorinated Biphenyls in Engineered C57BL/6J Mice of Arylhydrocarbon Receptor Gene. *Toxicol. Appl. Pharmacol.* **2003,** *187*, 1–10.
176. Shirasaka, Y.; Sager, J. E.; Lutz, J. D.; Davis, C.; Isoherranen, N. Inhibition of CYP2C19 and CYP3A4 by Omeprazole Metabolites and Their Contribution to Drug-Drug Interactions. *Drug Metab. Dispos.* **2013,** *41*, 1414–1424.
177. Sinz, M. W. Drug Metabolism in Preclinical Development. In *Applications of Pharmacokinetic Principles in Drug Development*; Krishna, R., Ed; Springer Science+ Business Media, LLC: New York, 2004; pp 75–132.
178. Siu, E. C.; Tyndale, R. F. Selegiline Is a Mechanism-Based Inactivator of CYP2A6 Inhibiting Nicotine Metabolism in Humans and Mice. *J. Pharmacol. Exp. Ther.* **2008,** *324*, 992–999.
179. Skonberg, C.; Olsen, J.; Madsen, K. G.; Hansen, S. H.; Grillo, M. P. Metabolic Activation of Carboxylic Acids. *Expert Opin. Drug Metab. Toxicol.* **2008,** *4*, 425–438.
180. Smith, D. A.; Allerton, C.; Kalgutkar, A. S.; van de Waterbeemd, H.; Walker, D. K. Metabolic (Hepatic) Clearance. In *Pharmacokinetics and Metabolism in Drug Design*; Wiley-VCH Verlag & Co. KGaA: Weinheim, Germany, 2012; pp 111–158.
181. Smith, D. A.; van de Waterbeemd, H.; Walker, D. K. *Pharmacokinetics, in Pharmacokinetics and Metabolism in Drug Design*; Wiley-VCH Verlag GmbH & Co. KGaA: Weinheim, FRG, 2006.
182. Somers, G. I.; Lindsay, N.; Lowdon, B. M.; Jones, A. E.; Freathy, C.; Ho, S.; Woodrooffe, A. J.; Bayliss, M. K.; Manchee, G. R. A Comparison of the Expression and Metabolizing Activities of Phase I and II Enzymes in Freshly Isolated Human Lung Parenchymal Cells and Cryopreserved Human Hepatocytes. *Drug Metab. Dispos.* **2007,** *35*, 1797–1805.

183. Somogyi, A. A.; Coller, J. K. Drugs against Acute and Chronic Pain. In *Metabolism of Drugs and Other Xenobiotics*; Anzenbacher, P., Zanger, U. M., Eds; Wiley-VCH Verlag GmbH & Co. KGaA: Weinheim, Germany, 2012; pp 403–428.
184. Stachulski, A. V.; Baillie, T. A.; Park, B. K.; Obach, R. S.; Dalvie, D. K.; Williams, D. P.; Srivastava, A.; Regan, S. L.; Antoine, D. J.; Goldring, C. E.; Chia, A. J.; Kitteringham, N. R.; Randle, L. E.; Callan, H.; Castrejon, J. L.; Farrell, J.; Naisbitt, D. J.; Lennard, M. S. The Generation, Detection, and Effects of Reactive Drug Metabolites. *Med. Res. Rev.* **2013,** *33*, 985–1080.
185. Stepan, A. F.; Walker, D. P.; Bauman, J.; Price, D. A.; Baillie, T. A.; Kalgutkar, A. S.; Aleo, M. D. Structural Alert/Reactive Metabolite Concept as Applied in Medicinal Chemistry to Mitigate the Risk of Idiosyncratic Drug Toxicity: A Perspective Based on the Critical examination of Trends in the Top 200 Drugs Marketed in the United States. *Chem. Res. Toxicol.* **2011,** *24*, 1345–1410.
186. Strolin Benedetti, M.; Baltes, E. L. Drug Metabolism and Disposition in Children. *Fund. Clin. Pharmacol.* **2003,** *17*, 281–299.
187. Sugiura, M.; Yamazoe, Y.; Kamataki; Kato, R. Reduction of Epoxy Derivatives of Benzo(a)Pyrene by Microsomal Cytochrome P-450. *Cancer Res.* **1980,** *40*, 2910–2914.
188. Suzuki, H.; Sugiyama, Y. Role of Metabolic Enzymes and Efflux Transporters in the Absorption of Drugs from the Small Intestine. *Eur. J. Pharm. Sci.* **2000,** *12*, 3–12.
189. Svensson, C. K. Biotransformation of Drugs in Human Skin. *Drug Metab. Dispos.* **2009,** *37*, 247–253.
190. Tanaka, E. Gender-Related Differences in Pharmacokinetics and Their Clinical Significance. *J. Clin. Pharm. Ther.* **1999,** *24*, 339–346.
191. Testa, B.; Kramer, S. D. The Biochemistry of Drug Metabolism—An Introduction: Part 3. Reactions of Hydrolysis and Their Enzymes. *Chem. Biodivers.* **2007,** *4*, 2031–2122.
192. Testa, B.; Kramer, S. D. The Biochemistry of Drug Metabolism—An Introduction: Part 4. Reactions of Conjugation and Their Enzymes. *Chem. Biodivers. 5*, 2171–2336.
193. Theobald, D. S.; Maurer, H. H. Identification of Monoamine Oxidase and Cytochrome P450 Isoenzymes Involved in the Deamination of Phenethylamine-Derived Designer Drugs (2C-Series). *Biochem. Pharmacol.* **2007,** *73*, 287–297.
194. Thorn, C. F.; Whirl-Carrillo, M.; Leeder, J. S.; Klein, T. E.; Altman, R. B. PharmGKB Summary: Phenytoin Pathway. *Pharmacogenet. Genom.* **2012,** *22*, 466–470.
195. Tomson, T.; Lindbom, U.; Ekqvist, B.; Sundqvist, A. Epilepsy and Pregnancy: A Prospective Study of Seizure Control in Relation to Free and Total Plasma Concentrations of Carbamazepine and Phenytoin. *Epilepsia.* **1994,** *35*, 122–130.
196. Tracy, T. S.; Venkataramanan, R.; Glover, D. D.; Caritis, S. N.; National Institute for Child H; Human Development Network of Maternal-Fetal-Medicine U. Temporal Changes in Drug Metabolism (CYP1A2, CYP2D6 and CYP3A Activity) during Pregnancy. *Am. J. Obstetr. Gynecol.* **2005,** *192*, 633–639.
197. Treluyer, J. M.; Rey, E.; Sonnier, M.; Pons, G.; Cresteil, T. Evidence of Impaired Cisapride Metabolism in Neonates. *Br. J. Clin. Pharmacol.* **2001,** *52*, 419–425.
198. Tripathi, S. P.; Bhadauriya, A.; Patil, A.; Sangamwar, A. T. Substrate Selectivity of Human Intestinal UDP-Glucuronosyltransferases (UGTs): In Silico and In Vitro Insights. *Drug Metab. Rev. 45*, 231–252.
199. Tsunoda, S. M.; Harris, R. Z.; Christians, U.; Velez, R. L.; Freeman, R. B.; Benet, L. Z.; Warshaw, A. Red Wine Decreases Cyclosporine Bioavailability. *Clin. Pharm. Ther. 70*, 462–467.

200. Ueng, Y. F.; Chen, C. C.; Chung, Y. T.; Liu, T. Y.; Chang, Y. P.; Lo, W. S.; Murayama, N.; Yamazaki, H.; Soucek, P.; Chau, G. Y.; Chi, C. W.; Chen, R. M.; Li, D. T. Mechanism-Based Inhibition of Cytochrome P450 (CYP)2A6 by Chalepensin in Recombinant Systems, in Human Liver Microsomes and in Mice in Vivo. *Br. J. Pharmacol.* **2011,** *163*, 1250–1262.
201. Urquhart, B. L.; Tirona, RG.; Kim, R. B. Nuclear Receptors and the Regulation of Drug-Metabolizing Enzymes and Drug Transporters: Implications for Interindividual Variability in Response to Drugs. *J. Clin. Pharmacol.* **2007,** *47*, 566–578.
202. Varghese, A.; Savai, J.; Pandita, N.; Gaud, R. In Vitro Modulatory Effects of Terminalia Arjuna, Arjunic Acid, Arjunetin and Arjungenin on CYP3A4, CYP2D6 and CYP2C9 Enzyme Activity in Human Liver Microsomes. *Toxicol. Rep.* **2015,** *2*, 806–816.
203. Vose, C. W.; Ings, R. M. J. Drug Metabolism. In *The Handbook of Medicinal Chemistry: Principles and Practice*; Davis, A., Ward, S. E., Eds; The Royal Society of Chemistry: Cambridge, 2015; pp 184–207.
204. Walsh, A. A.; Szklarz, G. D.; Scott, E. E. Human Cytochrome P450 1A1 Structure and Utility in Understanding Drug and Xenobiotic Metabolism. *J. Biol. Chem.* **2013,** *288*, 12932–12943.
205. Wang, B.; Zhou, S. F. Synthetic and Natural Compounds That Interact with Human Cytochrome P450 1A2 and Implications in Drug Development. *Curr. Med. Chem.* **2009,** *16*, 4066–4218.
206. Wang, L. S.; Zhu, B.; Abd El-Aty, A. M.; Zhou, G.; Li, Z.; Wu, J.; Chen, G. L.; Liu, J.; Tang, Z. R.; An, W.; Li, Q.; Wang, D.; Zhou, H. H. The Influence of St John's Wort on CYP2C19 Activity with Respect to Genotype. *J. Clin. Pharmacol.* **2004,** *44*, 577–581.
207. Waxman, D. J.; Dannan, G. A.; Guengerich, F. P. Regulation of Rat Hepatic Cytochrome P-450: Age-Dependent Expression, Hormonal Imprinting; Xenobiotic Inducibility of Sex-Specific Isoenzymes. *Biochemistry* **1985,** *24*, 4409–4417.
208. Wen, B.; Nelson, S. D. Common Biotransformation Reactions. In *Mass Spectrometry in Drug Metabolism and Disposition: Basic Principles and Applications*; Lee, M. S.; Zhu, M., Eds.; John Wiley & Sons, Inc.: New Jersey, 2011; pp 13–41.
209. Wilkinson, G. R. Drug Metabolism and Variability among Patients in Drug Response. *New Engl. J. Med.* **2005,** *352*, 2211–2221.
210. Williams, D. Drug Metabolism. In *Foye's Principles of Medicinal Chemistry*; Williams, D., Ed.; Lippincott Willaims & Wilkins: Baltimore, 2002; pp 174–233.
211. Williams, J. A.; Hyland, R.; Jones, B. C.; Smith, D. A.; Hurst, S.; Goosen, T. C.; Peterkin, V.; Koup, J. R.; Ball, S. E. Drug-Drug Interactions for UDP-Glucuronosyltransferase Substrates: A Pharmacokinetic Explanation for Typically Observed Low Exposure (AUCi/AUC) Ratios. *Drug Metab. Dispos.* *32*, 1201–1208.
212. Wolff, K.; Boys, A.; Rostami-Hodjegan, A.; Hay, A.; Raistrick, D. Changes to Methadone Clearance During Pregnancy. *Eur. J. Clin. Pharmacol.* **2005,** *61*, 763–768.
213. Wu, C. Y.; Benet, L. Z. Predicting Drug Disposition via Application of BCS: Transport/ Absorption/ Elimination Interplay and Development of a Biopharmaceutics Drug Disposition Classification System. *Pharm. Res.* **2005,** *22*, 11–23.
214. Yasuda, K.; Ranade, A.; Venkataramanan, R.; Strom, S.; Chupka, J.; Ekins, S.; Schuetz, E.; Bachmann, K. A Comprehensive in Vitro and in Silico Analysis of Antibiotics That Activate Pregnane X Receptor and Induce CYP3A4 in Liver and Intestine. *Drug Metab. Dispos.* **2008,** *36*, 1689–1697.

215. Yin, O. Q.; Tomlinson, B.; Waye, M. M.; Chow, A. H.; Chow, M. S. Pharmacogenetics and Herb-Drug Interactions: Experience with Ginkgo Biloba and Omeprazole. *Pharmacogenetics* **2004,** *14*, 841–850.
216. Yoshinari, K.; Ueda, R.; Kusano, K.; Yoshimura, T.; Nagata, K.; Yamazoe. Y. Omeprazole Transactivates Human CYP1A1 and CYP1A2 Expression through the Common Regulatory Region Containing Multiple Xenobiotic-Responsive Elements. *Biochem. Pharmacol.* **2008,** *76*, 139–145.
217. Zhang, J.; Ji, L.; Liu, W. In Silico Prediction of Cytochrome P450-Mediated Biotransformations of Xenobiotics: A Case Study of Epoxidation. *Chem. Res. Toxicol.* **2015,** *28*, 1522–1531.
218. Zhang, W.; Kilicarslan, T.; Tyndale, R. F.; Sellers, E. M. Evaluation of Methoxsalen, Tranylcypromine, and Tryptamine as Specific and Selective CYP2A6 Inhibitors in Vitro. *Drug Metab. Dispos.* **2001,** *29*, 897–902.
219. Zhou, Q.; Yao, T. W.; Zeng, S. Effects of Stereochemical Aspects on Drug Interaction in Pharmacokinetics. *Acta Pharmacol. Sinica* **2002,** *23*, 385–392.
220. Zhou, S.; Chan, E.; Lim, L. Y.; Boelsterli, U. A.; Li, S. C.; Wang, J.; Zhang, Q.; Huang, M.; Xu, A. Therapeutic Drugs That Behave as Mechanism-Based Inhibitors of Cytochrome P450 3A4. *Curr. Drug Metab.* **2004,** *5*, 415–442.

CHAPTER 14

Drug Interactions: Highlights of Adverse Effects of Low Molecular Weight Ligands

SANTOSH S. CHHAJED[1*], SANDEEP S. SONANWANE[1], HARSHALA CHAUDHARI[2], and AARTI SONAR[1]

[1]*Department of Pharmaceutical Chemistry, MET Institute of Pharmacy, Bhujbal Knowledge City, Nashik 422003, Maharashtra, India*

[2]*Department of Quality Assurance Techniques, MET Institute of Pharmacy, Bhujbal Knowledge City, Nashik 422003, Maharashtra, India*

[*]*Corresponding author. E-mail: chhajedss@gmail.com*

ABSTRACT

Drug interactions are identified when the effect of one drug is changed by the presence of another drug, herbal medicine, food, drink, or by some environmental chemical agent. The outcome of these drug interactions can be damaging if the interaction enhances the toxicity of the drug. But there are some interactions, which can be advantageous and valuable; for example, coprescription of antihypertensive drugs and diuretics in order to attain the antihypertensive effect is probably not accessible with either drug alone. The mechanisms of both types of interaction (beneficial or adverse) are repeatedly very comparable, but the unfavorable interactions are the focus of this chapter. Irrational polypharmacy is the main cause of drug interactions. The more drugs a patient takes, the greater the likelihood that an adverse reaction will occur. The aim of this study is to bring awareness to all healthcare professionals, with the facts about drug interactions, without them having to do the time-consuming literature searches and full assessment of the papers.

Biologically Active Small Molecules: Modern Applications and Therapeutic Perspectives.
Debarshi Kar Mahapatra, & Sanjay Kumar Bharti (Eds.)

14.1 INTRODUCTION

A drug interaction is an interaction between a drug and some other substance, such as another drug or a certain type of food, which prevents the drug from exhibiting its desired action. An interaction can either amplify or lessen the effectiveness and/or the side effects of a drug, or it can create a new side effect not previously seen before.[1] Different types of drug interactions are reported here.

14.2 DRUG–DRUG INTERACTION

Drug–drug interactions are more known to be common. This is when more than two drugs are concurrently prescribed such that one drug interacts with another to show unusual effects. Here, an amplification or lessening of the side effects of the drug may occur. Further, it has been seen that blood levels and toxicity of few drugs may be known to increase. One such incidence of drug–drug interaction is the concurrent administration of oral hypoglycemic drug gliclazide with other drugs like acarbose or abiraterone. It has been observed that when either acarbose or abiraterone is administered with gliclazide to the diabetic patient, the hypoglycemic activity of gliclazide may be increased. The drug abiraterone decreases the metabolism of gliclazide, whereas acarbose is involved in the inhibition of the enzyme α-glucosidase. This reduces the digestion and absorption of carbohydrates into the small intestine and hence decreases glucose levels in the blood, as a consequence of this glucose-lowering effect of gliclazide become increases. Acetohexamide may increase the hypoglycemic activity of glimepiride.[2]

14.3 DRUG–FOOD INTERACTIONS

The interactions between drugs and food can accidentally increase or reduce the consequence of the drug, leading to therapeutic failure or increased toxicity. Some drugs interact with food or certain types of beverages to show undesired effects. For example, grapefruit juice is known to lower levels of metabolizing enzymes in the liver. When grapefruit juice is taken with drugs, the metabolism of drugs may get affected as metabolizing enzymes are becoming lower in the liver by grapefruit juice. As a consequence, the levels of un-metabolized drug increases in the blood and results in serious toxic effects. The best-known example of this type of interactions is between cholesterol-lowering drug class *statins* and grapefruit juice. Drugs like atorvastatin, lovastatin, or simvastatin when administered alongside grapefruit juice, blood levels of these drugs increases drastically and it leads to muscle pain or on serious end muscle injury may result. Hence, on certain medication container, it is written in a clear manner to avoid grapefruit juice. An alcoholic drink can increase or prolong the effects of insulin or oral diabetic agents (pills) and thus lead to hypoglycemia or low blood sugar. When potassium-sparing diuretics are used, a high intake of potassium-rich foodstuffs such as bananas and spinach may result in hyperkalemia.[3]

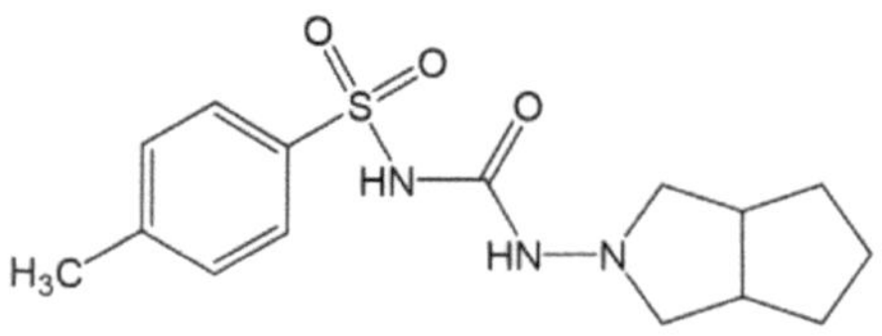

gliclazide

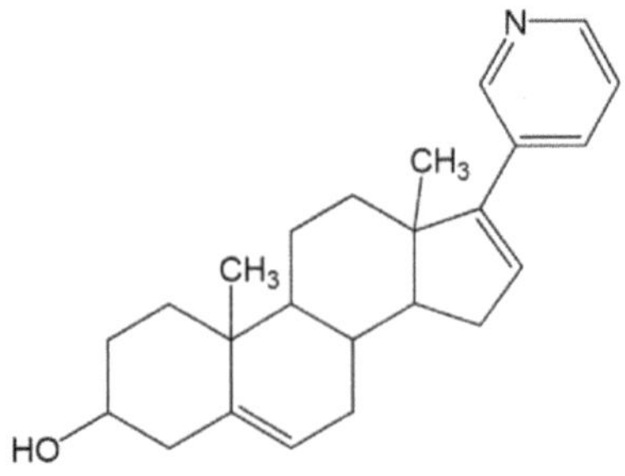

abiraterone

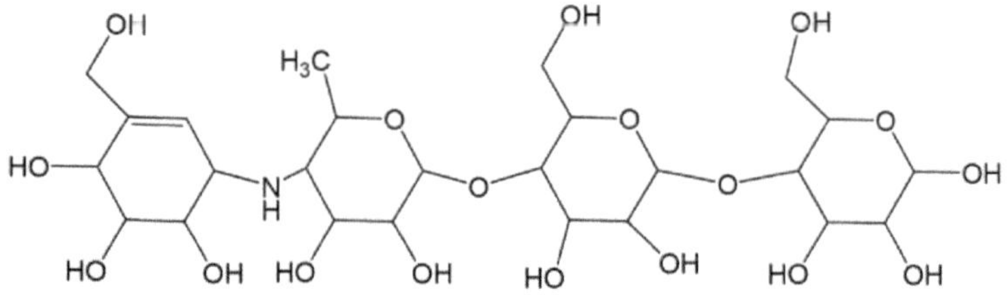

acarbose

14.4 DRUG–DISEASE INTERACTIONS

Drug–disease interaction is altogether a different type of interaction than what was described earlier. Existing disease may affect the way drug produces its action. One of the examples of this type of interaction is when an oral decongestant drug like phenylephrine is dangerous if administered to the patient with elevated blood pressure.

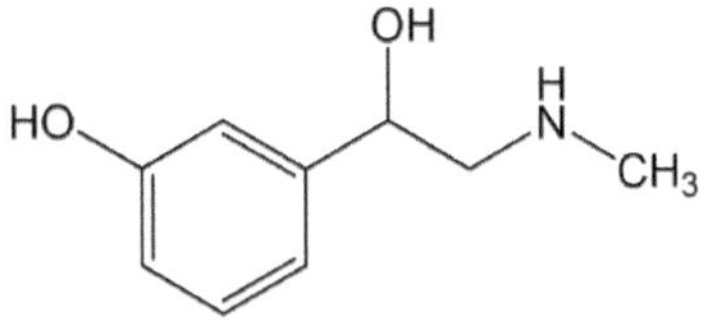

phenylephrine

Drug–drug interaction is one of the most common reasons of a medication error, predominantly in the elderly due to polytherapeutics. The polytherapy augments the intricacy of curative management that leads to drug–drug interaction, which results in the advancement of adverse drug reactions or reduction of clinical efficacy. Drug interactions sometimes occur with concurrent use of two or more drugs. When two drugs are administered simultaneously, they may interact to either enhance or reduce the intended effect of one or both drugs, or they may produce an unintended and potentially harmful reaction, for example, severe hemorrhage may occur if *warfarin* and *salicylates* are combined. Sound knowledge of the pharmacological basis of drug interactions with the physician is essential in prescribing a drug to prevent unwanted interactions.[4]

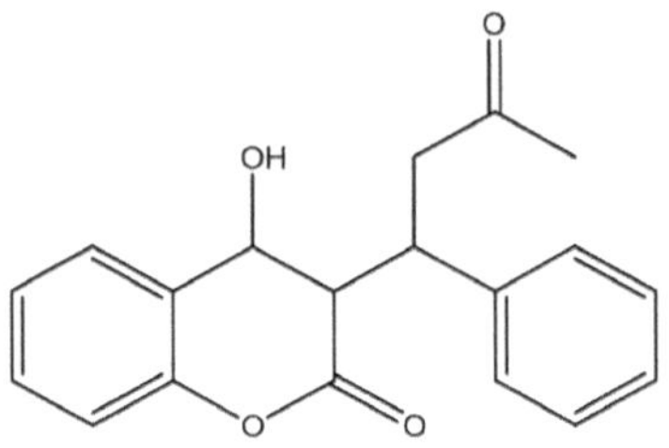

Warfarin

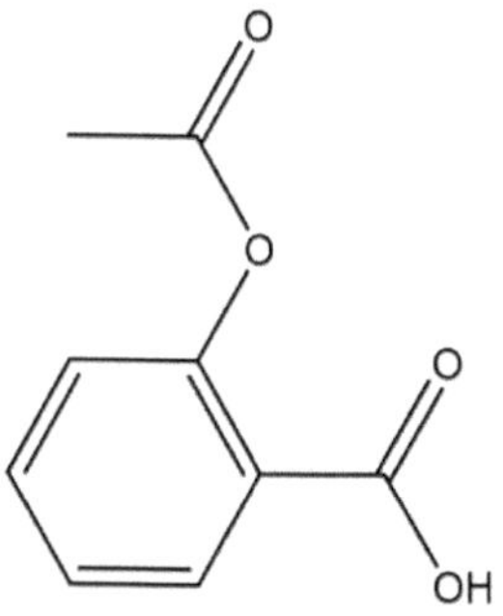

aspirin

Consequences of drug–drug interactions

There are three possible outcomes when drug–drug interactions occur and they are the following:

- One drug may intensify the effects of the other.
- One drug may reduce the effects of the other.
- The combination may produce a new response not seen when either drug is given alone.

The following outline serves as an aid in classifying the drug interactions according to the mechanism involved:

- Outside the body (in vitro)
  1. Physical interactions
  2. Chemical interactions
- Inside the body (in vivo)
  1. Pharmacokinetic interactions
  2. Pharmacodynamic interactions

An incidence of drug interactions has increased mainly due to three factors:

- Drug explosion: extreme use of drugs results in drug interactions.
- Availability of potent drugs: potency is a measure of drug activity expressed in terms of the amount required to produce an effect of a given intensity. A highly potent drug (e.g., *fentanyl, alprazolam, risperidone,* etc.) produces a response at low concentrations. Increase use of such potent drugs leads to drug interactions.
- Irrational polypharmacy: Unnecessary use of multiple drugs also results in drug interactions.

14.5 PHYSICAL INTERACTIONS

Physical interactions occur when the physical state of either drug is altered when the chemicals are mixed. Because of the physical properties, some drugs can directly interact with other drugs or with excipients. This type of interaction usually causes both drugs to be inactive and may cause severe harm to the patient if given together. Never mix two or more drug components in the same container except it has been instituted that a straight interaction will not occur; for example, *amphotericin* precipitates when mixed with normal saline instead of 5% *dextrose.*[5]

14.6 CHEMICAL INTERACTIONS

Some drugs generate actions without altering the cellular role and without binding to a specific receptor. For example, antacids diminish the gastric acidity in human stomach through a very simple chemical reactions where the bases interact chemically with the acidic components to generate salts (neutral nature). The prime action of a bile acid sequestrant, cholestyramine, binds with the bile acids present in the GIT. Chemical interactions occur when the components of a drug mixture interact to form chemically altered products; for example, *Methicillin* and *kanamycin* are chemically incompatible in solution.[5]

14.7 PHARMACOKINETIC INTERACTIONS

Pharmacokinetic interactions are deemed on the foundation of knowledge of each and every drug and are recognized by patient's clinical demonstration controlling as well as the alterations in the drug serum concentrations. Pharmacokinetic interactions affect the processes by which drugs are absorbed, distributed, metabolized, and excreted. These interactions may result in antagonism or synergism.

Synergism: The drug effect is more than the effect produced by a single drug, for example, sulphamethoxazole and trimethoprim.

Antagonism: Antagonism occurs when one drug opposes the action of another drug. It could be a physiological or functional antagonism where two drugs have opposite pharmacological effects, for example, adrenaline and histamine on the bronchial musculature.

14.7.1 INTERACTIONS AT THE SITE OF ABSORPTION

Most of the drugs are orally administered for their absorption through the GIT mucous membranes, and the majority of the drug interactions take place in the gut, which results into reduced bioavailability of drugs.

In the gut, drugs may interfere with each other's absorption by the following mechanisms:

14.7.1.1 EFFECTING GUT MOTILITY

The primary location of drug absorption is the upper section of the small intestine. Drugs that alter the gastric emptying and gastrointestinal motility may have a significant effect on absorption of drugs. For example, when propantheline and paracetamol are administered simultaneously, propantheline delays the gastric-emptying process and reduces paracetamol absorption. Propantheline also increases the absorption of hydrochlorothiazide. Rapid gastrointestinal transit effected by prokinetic agents such as *metoclpramide, domperidone,* and *cisapride* may reduce the degree of absorption of lipophilic drugs or drugs that are usually absorbed from a very limited section of the intestine. Gastric emptying time and gut motility are important factors, which determine the extent and rate of absorption of drugs. For example, drugs like *metoclopramide* cause marked acceleration of gastric emptying and enhance the absorption of drugs.

14.7.1.2 EFFECT OF CHANGES IN GASTROINTESTINAL PH

The rate of drug absorption by passive diffusion is limited by the solubility, or dissolution, of a compound in gastric fluid. Basic drugs are more soluble in acidic fluids, and acidic drugs are more soluble in basic fluids. Drugs are absorbed mainly in nonionized form from the gut. The proportion of ionized to the nonionized drug is dependent on the p*K*a value of the drug and surrounding pH. Antacids increase gastric pH and thus acidic drugs are more ionized, less soluble in lipids, and so absorbed more slowly, for example, oral anticoagulants. Weak acids (p*K*a = 3–8) may have imperfect absorption from alkaline environment and vice versa for weak bases (p*K*a = 5–11). The rise in pH due to proton pump inhibitors, H_2-receptor antagonists, can markedly reduce the absorption of ketoconazole. Many antacids that contain aluminum and magnesium interfere with the absorption of prednisolone, iron, and tetracycline by forming insoluble complexes. *Sucralfate* interferes with the absorption of *phenytoin.*

14.7.1.3 CHANGING GUT FLORA

Antimicrobials can inhibit bacterial synthesis of vitamin K by altering the gut flora and potentiate the action of oral anticoagulants. In addition, absorption of the drug from the injection site may be affected by the simultaneous administration of another drug.[6]

14.7.2 INTERACTION DURING DISTRIBUTION

14.7.2.1 PLASMA PROTEIN BINDING AND DISPLACEMENT

Drugs rapidly distribute around the body by circulation. Drug interactions concerning the distribution process are those that modify the protein binding. Depending on relative plasma concentrations and protein-binding affinities, one drug may displace another with clinically significant results. This interaction is much more likely to occur with drugs that are at least 80–90% bound to plasma proteins, with small changes in protein binding leading to large relative changes in free drug concentration (Table 14.1). When certain drugs are absorbed into the bloodstream, most of their molecules attach themselves to plasma proteins. The portion of the drug that is being transported in the bound form is pharmacologically inactive, and only the free drug molecules exert their effect. The presence of a second drug with higher affinity for protein competes with the first drug for protein binding sites and increases the effect of displaced drug producing serious toxicity. For example, *warfarin* is 98% protein bound, and only 2% is free in the plasma, which is the active fraction. If *phenytoin* is given along with *warfarin, phenytoin* displace the *warfarin* that results in an increase of free drug concentration leading to toxicity. *Tolbutamide* can be displaced by *dicoumarol* resulting in severe hypoglycemia. The competition for plasma protein binding gains clinical significance when the second mechanism operates simultaneously. For example, *sodium valproate* can cause *Phenytoin* toxicity because it not only displaces phenytoin from its binding site on plasma albumin but also inhibits its metabolism. *Sulphonamides, vitamin K*, X-ray contrast media, and *indomethacin* displace *bilirubin* from its binding protein in neonates. The capacity to metabolize bilirubin is not fully developed in the neonate, and this may pose a risk of developing kernicterus.[7]

TABLE 14.1 Displacement of Drugs from Plasma Protein Binding Sites.

Drug displaced	Displacing agent
Acetaminophen	Phenylbutazone, clofibrate, phenytoin, salicylates
Methotrexate	Salicylates, sulphonamides
Pamaquine	Quinacrine
Sulphonamides	Phenylbutazone, oxyphenbutazone, warfarin, salicylates, tolbutamide
Warfarin	Phenylbutazone, oxyphenbutazone, clofibrate, salicylates, phenytoin, sulphinpyrazone

14.7.3 INTERACTIONS DURING METABOLISM

The primary site for the metabolism of drug is the liver, where it converts the lipophilic compounds into ionized metabolites through nonsynthetic (Phase-I reactions like reduction, hydrolysis, and oxidation that takes place at hepatocyte endoplasmic reticula membrane) and synthetic (Phase-II reactions like sulfation, glucuronidation, and so on that takes place at hepatocyte cytosol) phases, that are further eliminated by the kidneys. These reactions are generally mediated by enzymes so that drugs are capable of altering the enzymatic processes involved in drug metabolism can lead to drug interaction. The process of conversion of drugs to their respective metabolites is termed as biotransformation.

14.7.3.1 ENZYME INDUCTION

Metabolism is mediated by enzymes so that drugs capable of altering the enzymatic process involved in metabolism can lead to drug interactions. Some drugs like *barbiturates, hydantoins, griseofulvin, chlorinated hydrocarbon insecticides,* and many others stimulate the liver microsomal enzymes involved in drug metabolism, results in a reduced therapeutic response to those drugs that are metabolized by microsomal enzymes. Enzyme stimulation can be utilized for its beneficial effects also, for example, *phenobarbitone* is used in the treatment of neonatal hyperbilirubinaemia, because it accelerates the metabolism of bilirubin to more readily excreted compounds (Table 14.2).

TABLE 14.2 List of Few Drugs That Induce Metabolism of Other Drugs.

Inducers	Drugs that metabolize
Alcohol	Tolbutamide
Haloperidol	Warfarin
Phenobarbital	Cortisol, cyclophosphamide, digitoxin, griseofulvin, barbiturates
Chlorcyclazine	Steroid hormones

14.7.3.2 ENZYME INHIBITION

Inhibition-based drug–drug interaction constitutes the chief amount of clinically relevant drug–drug interactions. In this process, the enzyme activity is diminished as a result of direct interaction with a drug. The action of the drug may be enhanced by inhibition of its metabolism by other drugs. For example, *Monoamine oxidase inhibitors* (MAOIs) inhibit the normal functioning of the endogenous enzyme monoamine oxidase, elevated levels of biogenic amines, and may produce a hypertensive crisis in the presence of pressor amines (Table 14.3).

TABLE 14.3 List of Few Drugs, Which Inhibit the Metabolism of Other Drugs.

Drug	Metabolic inhibitor
Cyclophosphamide	Chloramphenicol
6-Mercaptopurine	Allopurinol
Nortryptyline	Hydrocortisone
Hexobarbital	Metyrapone, *p*-aminosalicylates, progesterone
Tolbutamide	Alcohol, chloramphenicol, salicylates

14.7.3.3 REVERSIBLE INHIBITION

Reversible inhibition occurs when compounds quickly form weak bonds with CYP isozymes without permanently disabling them. This can occur both competitively (competition for the same binding site between substrate and inhibitor) and noncompetitively (inhibitor binds at a site on the enzyme distinct from the substrate). The magnitude of this type of inhibition depends both on the affinity of inhibitor and substrate for the enzyme and on the inhibitor concentration at the enzyme site.

14.7.3.4 COMPETITIVE INHIBITION

When the substrate and the inhibitor compete for the same binding site of the enzyme, it is known as competitive inhibition. The inhibition mechanism in this type of interaction rapidly reversible and is direct. The CYP enzyme system is located in numerous tissues (liver is the largest source) and comprised of mixed-function mono-oxygenases that catalyzes the majority of oxidative reactions. The ratio of substrate:inhibitor binding constant depends on the relative amount of each components. At present, hepatic CYP isozymes play key role in mediating the pharmacokinetic drug interactions. The drugs metabolize by multiple CYP-dependent cascade to nitroso-derivatives that further gets bound with reduced CYP enzymes. Common CYP3A4 inhibitors include azole antifungals, antihypertensives (*diltiazem*), and HIV protease inhibitors (*nelfinavir mesylate*). Examples of competitive inhibition involve twofold decline in the *metoprolol* oral clearance in the presence of *propafenone* when administered simultaneously.[8]

14.7.4 INTERACTION DURING EXCRETION

The kidney is the main organ that is chiefly accountable for the removal of xenobiotics (drugs) and metabolites. The drugs are largely expelled through:

- Renal tubular excretion (glomerular filtration, active tubular secretion, and tubular reabsorption).
- Biliary excretion.

14.7.4.1 GLOMERULAR FILTRATION

Changes in renal blood flow, protein binding, and cardiac output affect the glomerular filtration rates. With highly protein-bound drugs, when an unbound fraction of the drug increases, drug elimination is also increased by increasing the glomerular filtration. Conversely, with transporter saturation and renal elimination at maximal, elimination rates may decrease significantly with an increased unbound fraction of the drug.

14.7.4.2 TUBULAR SECRETION

Transporter site of tubular secretion is the most common site where the renal drug interactions occur. Many organic anionic and cationic drugs and metabolites compete with each other for secretion at same tubular active transport system. For example, *probenecid* reduces the excretion of *penicillin* and other drugs.

14.7.4.3 TUBULAR REABSORPTION

The drugs reabsorption occurs by passive diffusion and active transport processes. Only nonionized compounds are reabsorbed passively from the renal tubule, and thus the reabsorption of weak organic acids and bases can be altered by manipulating urinary pH. Renal clearance of weak organic bases (p*K*a = 7–10) is enhanced with acidification of urine and reduces with alkalinization of urine. Likewise, renal elimination of weak organic acids (p*K*a = 3–7) is increased with urine alkalinization and decreased with urine acidification. Most important drug interactions occur as a result of either change in urinary pH which alters the fraction of reabsorbed drug, or through competition for the active tubular mechanism.

14.7.4.4 CHANGES IN URINARY pH

Acidification of urine with agents like ammonium chloride results in reduced effectiveness of basic drugs like amphetamine, as they will be largely ionized at acidic pH and readily excreted. Urinary pH also influences the solubility of drug molecules in a tubular fluid, for example, *sulphonamides* are poorly soluble in acidic urine and may precipitate in the renal tubules causing crystalluria and obstruction.

14.7.4.5 COMPETITION FOR TUBULAR MECHANISM

Many drugs and metabolites are actively secreted from renal blood vessels into the tubules by using same tubular transport system. Therefore, the drugs compete with each other and subsequently get eliminated.[9]

14.7.5 BILIARY EXCRETION

14.7.5.1 ENTEROHEPATIC RECIRCULATION

Most of the drugs are excreted in the bile, either in unchanged or conjugated form to make them more water soluble. Some of the conjugates are metabolized to the parent compound by the gut flora and then reabsorbed. This process prolongs the presence of the drug within the body, but if the gut flora is diminished, then the drug is lost more quickly. For example, concurrent use of *penicillins* and *tetracyclines* results into the rare failure of oral contraceptives.[10]

14.8 PHARMACODYNAMIC INTERACTIONS

The pharmacodynamic interactions generally occur at the drug action site by varying the sensitivity of organ resulting in a synergistic or antagonistic effect. Usually, the drug effects result due to the binding of the drug to the receptor site. The drug may act on the same or different receptors or processes and may produce similar biological effects. The magnitude of the effect depends on the free drug concentration at its receptor site, which in turn regulated by

- Drug's physicochemical properties
- Altered gastrointestinal absorption or competition for protein binding sites or receptor sites
- Altered drug metabolism
- Changes in acid–base equilibrium
- Alteration of hemodynamic or renal tubular function influencing renal drug clearance. The interaction between two drugs may produce a desired or beneficial effect or undesired or harmful effect, for example, sulfamethoxazole is combined with Trimethoprim to enhance their antimicrobial spectrum.[11]

Some of the important pharmacodynamic interactions are discussed further (Table 14.4);

Interactions occurring at the receptor site

Drug A + Receptor ⟶ Drug A-receptor complex ⟶ Response

Drug B + Receptor ⟶ Drug B-receptor complex ⟶ No response (blockade)

TABLE 14.4 Pharmacodynamic Interactions.

Receptor	A-type drugs	B-type drugs
Alpha receptors	Noradrenaline	Phentolamine, phenoxybenzamine, imipramine, phenothiazines
Beta receptors	Isoprenaline	Propranolol
Neuromuscular junction	Acetylcholine	*d*-Tubocurarine
Adrenergic neuron	Noradrenaline	Guanethidine, amphetamine, tricyclic antidepressants, phenothiazines
Liver cell	Vitamin K	Salicylates, chloramphenicol, anabolic steroids

14.9 INTERACTIONS AT BODY SITES

The following list gives examples of such interactions:

- *MAOIs* increase the amount of *noradrenaline* stored in adrenergic nerve terminals and thus enhance the action of ephedrine.

- *Tetracycline and Chloramphenicol* prevent bacterial cell division and thus interfere with the antimicrobial action of antibiotics, which kill the bacterial cell when they are dividing (e.g., *penicillin*) when the two are given simultaneously.
- *Thiazide* and other diuretics (except K^+ sparing diuretics) lower plasma potassium concentration and enhance the action of digitalis glycosides and may predispose to digitalis toxicity.
- *Cephalosporins* and *aminoglycosides* may be toxic if given together, especially in elderly patients with renal dysfunction.
- CNS drugs like *benzodiazepines, antiepileptic drugs, antihistamines (H_1 receptor blockers)*, and alcohol interact and enhance the sedative effect.
- *Aminoglycosides* and *loop diuretics* are ototoxic in high dose, and the chances of toxicity are increased if both are administered concurrently.[12]

14.10 MISCELLANEOUS PHARMACODYNAMIC INTERACTIONS

14.10.1 DIRECT PHARMACODYNAMIC INTERACTIONS

It occurs when two drugs either act on the same site or on two different sites with a similar effect.

14.10.2 INDIRECT PHARMACODYNAMIC INTERACTIONS

These types of interactions are unrelated to the effect of the object drug. For example, diuretics like *furosemide* may attenuate the effect of oral hypoglycemic drugs.

The above discussion through light on hazards of drug interactions, so keeping this view in mind, it laments the importance of further reasons to know drug interactions as follows;

1. It perhaps worsens the disease condition which already exists.
2. It affects how the drug works by altering the levels of it in blood.
3. If interactions are happening, there are always chances that it will put the patient at high risk for undesired effects and probably toxicity.

Monitoring for a drug interaction or keeping an eye before it happens may reduce the chances of worsening of disease states. Doctors and pharmacists may have already taken necessary care to avoid these interactions while prescribing, but it is always better to check prescription drugs again by patients and to learn about these types of interactions on their own. Further, if patients take any *"over the counter,"* like vitamins,or any kind of food supplements, *NSAIDS*, they must make sure to check these for interactions with any prescription drugs they are taking. As serious drug interaction may result into hospitalization may increase the cost of health care. Further interactions may be fatal sometimes so extra care is necessary while reviewing it.[13]

10.11 CONCLUSION

This chapter provides all the essential information related to the administration of the combination of drugs to the patient. The pharmacokinetic, physical, chemical, metabolic, drug–drug, drug–food, drug–herb, and pharmacodynamic (direct and indirect) interactions will help medical prescribers in safe recommendation of polypharmacy therapy.

KEYWORDS

- **drug-drug interactions**
- **drug-food interactions**
- **physical interactions**
- **chemical interactions**
- **pharmacodynamics**
- **pharmacokinetics**

REFERENCES

1. Barar, F. S. K. *Essentials of Pharmacotherapeutics*; S. Chand & Company Ltd.; pp 43–48.
2. Satoskar, R. S.; Rege, N. N.; Bhandarkar, S. D. *Pharmacology and Pharmacotherapeutics;* Popular Prakashan, pp 63–67.
3. *Stockley's Drug Interactions*, edited by Karen Baxter, 8th ed.
4. Chemical Interactions by Abimbola Farinde, PhD, PharmD, Professor, Columbia Southern University, Merck manual.
5. *Drug Interactions*—Lehne; Chapter 6.
6. Morgan, T. O. Clinical Use of Potassium Supplements and Potassium Sparing Diuretics. *Drugs* **1973,** *6* (3), 222–229.

7. Schmidt, L. E.; Dalhoff, K. *Food-Drug Interaction*; Department of Clinical Pharmacology, Rigshospitalet, Copenhagen, Denmark.
8. Kashuba, A. D. M.; Bertino, J. S., Jr. *Mechanisms of Drug Interactions I Absorption, Metabolism, and Excretion.*
9. Bogenschutz, S.; Bojrab, A. *Drug Interactions*; Biology of Aging, Fall 2003.
10. Bailie, G. R.; Johnson, C. Af.; Mason, N. A.; St. Peter, W. L *MED Facts Pocket Guide of Drug Interactions*, 2nd ed.
11. *Handbook of Drug Interactions A Clinical and Forensic Guide*, Edited by Mozayani, A, Raymon, L. P.
12. Seth, S. D.; Seth, V. *Textbook of Pharmacology*, 3rd ed.; Elsevier; pp I.54–I.57.
13. Brunton, L. L.; Lazo, J. S.; Parker, K. L. *Goodman & Gilman's, The Pharmacological Basis of Therapeutics*, 11th ed., Mcgraw-Hill, Medical Publishing Division; pp 121–124.

PART V
Evaluation Methods for Small Molecules

CHAPTER 15

Low Molecular Weight Antioxidant Agents and Model for Their Evaluation

BUCHI REDDY PURRA[1], MADHUSUDANA REDDY[1], and ANIMESHCHANDRA G. M. HALDAR[2*]

[1]*Department of Chemistry, REVA University, Bangalore, Karnataka, India*

[2]*Department of Applied Chemistry, Priydarshini Bhagwati College of Engineering, Nagpur, Maharashtra, India*

[*]*Corresponding author. E-mail: animesh2477@gmail.com*

ABSTRACT

Free radicals are chemical components of molecules or atoms that have an unpaired electron on their orbital (outermost). These components are extremely unstable, hence, they swiftly react with other compounds by giving out or accepting a single electron. Antioxidant activity offered by several natural product classes like flavones, flavonones, flavonoids, flavonols, chalcones, proanthocyanidins, anthocyanidins, aurones, and so on and rationally developed synthetic compounds like butyrated hydroxylated molecules are important for the survival of human being in today's life by preventing biomolecular damage (DNA, membrane, etc.) offered by the superoxide anion, alkyl, hydroxyl, nitric oxide, and peroxy radicals. For assessment of antioxidant activity, several in vitro models are available and used followed by animal studies necessary. The chapter focuses exclusively on the screening methods such as 1,1-diphenyl-2-picryl hydrazyl (DPPH) model, *N,N*-dimethyl-*p*-phenylenediamine dihydrochloride model, Ferric reducing ability of plasma model, oxygen radical absorbance capacity model, total radical-trapping model, β-carotene linoleate model, nitric oxide radical

Biologically Active Small Molecules: Modern Applications and Therapeutic Perspectives.
Debarshi Kar Mahapatra, & Sanjay Kumar Bharti (Eds.)

inhibition activity, hydroxyl radical scavenging activity, and superoxide radical scavenging activity.

15.1 INTRODUCTION

Oxygen is an important component of all living organisms; however, the development of different reactive intermediates of molecular oxygen aerobically metabolizes the cellular units, which ultimately results in the progression of "oxidation," which is a destructive process, in which breakdown followed by damages of various molecules results. In one side, while the in vitro oxidation process involves the partaking of molecular oxygen, while on another side, most of the in vivo biological oxidation/metabolism results in the oxygen deficiency conditions. Free radicals are the chemical components of molecules or atoms that have an unpaired electron on their orbital (outermost). These free radicals are highly unstable and can, therefore, react with other molecules by giving out of accepting single electron.[1,2]

The human body though endlessly produces free radicals, it is nullified by different defense systems (Table 15.1), which consist of enzymes and radical scavengers. These are called "first-line antioxidant defense system" but are not completely effective because almost all components of living bodies, tissues, cells, and genes undergo free-radical destruction. The second-line defense systems are constituted of repair systems for biomolecules, which are damaged by the attack of free radicals. Particular enzymes are known to involve in this context and many of them have been identified in prokaryotes and in eukaryotes. The functions of these enzymes involved in repairing directly damaged biomolecules such as lipids, polysaccharides, proteins, nucleic acids, and so on or in the removal of oxidized compounds and are illustrated in Figure 15.1.[3]

TABLE 15.1 Different Types of Radicals and Their Defense System.

Sl. No.	Type of free radical or oxidants	Defense systems
1	Superoxide (O_2)	Superoxide dismutases
2	Hydroxyl radical (OH)	(SOD), Mn-SOD, Cu, Zn-SOD
3	Peroxy radical (ROO)	Tocopherols, Ubiquinone
4	Singlet oxygen (O_2)	Carotenoide
5	Hydrogen peroxide (H_2O_2)	Catalase, Se glutathione peroxide (GPx)
6	Hydroperoxides (ROO)	Glutathione reductase (GR)
7	Transition metals (Fe^{2+}, Cu^{+})	Chelators

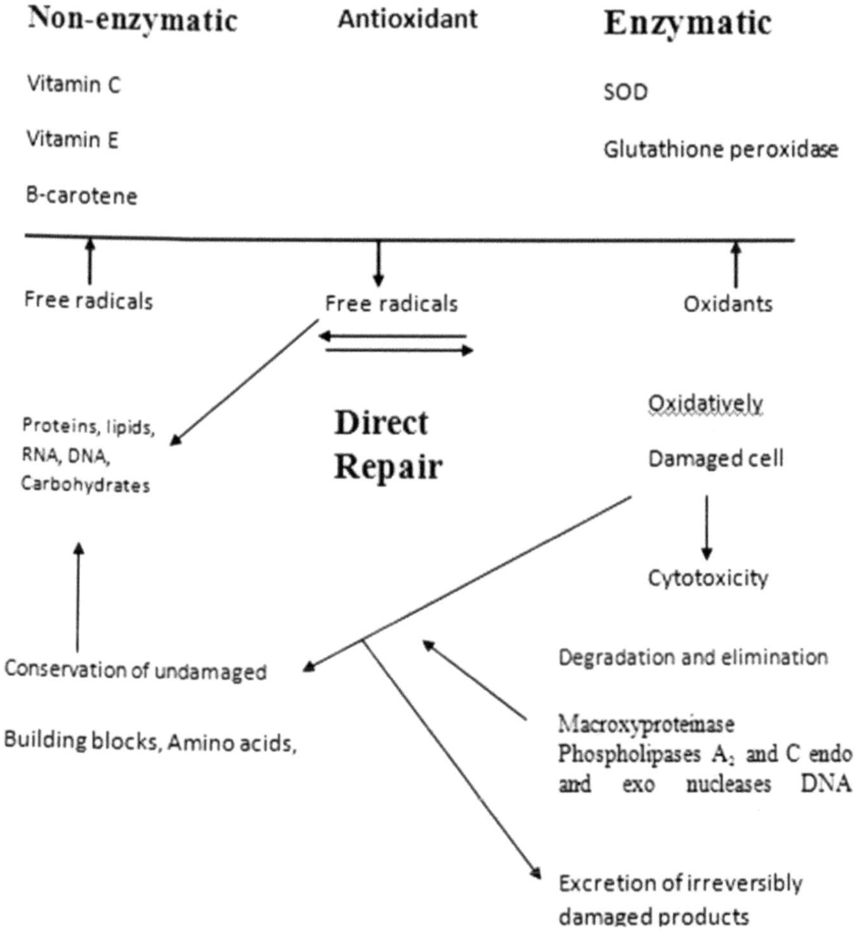

FIGURE 15.1 Free-radicals defense and repair systems for biomolecules.

Hydroperoxides from polyunsaturated fatty acids of membrane phospholipid are reduced by a glutathione peroxidase either directly or after the release of these peroxides by a phospholipase in the cytosol. Thymidine glycol and β-hydroxy-2-deoxyguanosine (HO-2-dG) are some of the metabolites identified while evaluating the outcomes of reactive radicals over the DNA molecules.

According to the theory of aging based on free radical, the action of the antioxidant enzymes is significantly altered in aging cells.[4] The probable participation of the reactive free radical attacks in the process of aging and in some important pathological/degenerative processes leading to an enhancement in aging or disorders like Down's syndrome, Alzheimer's disease, atherosclerosis, Parkinson's disease, hepatic damage, hypertension, diabetes suppressed immune function, and so on have been described (Table 15.2). Mitochondrial defects, glycation, calcium dyshomeostasis are also, moreover, a part of destructive network, as a result of the damage induced by free radicals.[5]

The compounds that inhibit the oxidative process are known as antioxidant agent. They are fairly huge in amount and diverse in nature, which opposes the process of oxidation largely by neutralizing free concentrations have the potential to inhibit the oxidant chain reactions. Antioxidants are also of paramount importance in pharmaceutical formulations because there are innumerable medical agents possessing diverse chemical functions and are known to undergo oxidative decomposition.

The enzymatically potential antioxidants hitherto known are glutathione peroxidase, superoxide dismutase, peroxidases, and catalase.[6] Under nonenzymatic category, some of the documented and well-known antioxidant molecules are vitamin C, vitamin E, vitamin A, β-carotenoids, uric acid, ubriquinone, and synthetic compounds like melatonin, dihydroepiandrosterone (DHEA), etc.

TABLE 15.2 Free-Radical-Mediated Diseases.

Aging	Cardiovascular disorders
Arthritis	Atherosclerosis
Asthma	Reperfusion injury
Bronchopulmonary dysplasia	Drug-induced gastrointestinal tract/liver disorders
Pulmonary fibrosis	Genetic disorders
Autoimmune diseases	Muscular dystrophy
Cancer	Radiation injury
Cataract	Skin disease porphyria
Retinopathy	Senile dementia stroke
Amyotropic lateral Sclerosis	Parkinsons dementia

15.2 DIFFERENT TYPES OF FREE RADICALS AND THEIR ROLE IN CELLULAR DAMAGE

Superoxide anion (O_2) is the most abundantly produced free radical. It is essentially produced enzymatically from NADPH oxidases (during phagocytosis), mitochondrial cytochrome oxidase (cell respiration), liver cytochrome

P450 (oxidative metabolism of xenobiotics), and xanthine-oxidase (ischemia, reperfusion). The biologically generated superoxide anion dismutases into molecular oxygen and hydrogen peroxide in the presence of proton and this reaction are highly favored in the presence of superoxide dismutase.

$$2O_2^- \xrightarrow{2H^+} H_2O_2 + O_2$$

Hydrogen peroxide induces cellular destruction in the presence of ferrous ions by a Fenton reaction resulting in the formation of OH free radicals as follows

$$Fe^{2+} + H_2O_2 \longrightarrow Fe^{3+} + OH^{(-)} + OH$$

Alkoxyl (RO) free radicals and peroxyl free radicals (ROO^-) are also synthesized from polyunsaturated fatty acids by the action of cyclooxygenases and lipoxygenases leading to ROO involving lipid peroxidation. RNA (ribonucleic acid) and DNA (deoxyribonucleic acid) are also targets for favorable attack by free radicals. It is found that the DNA of a cell undergoes about 10,000 free-radical attacks every day.

15.3 SUPEROXIDE DISMUTASES

Superoxide dismutases (SODs) are a group of naturally occurring metalloenzymes, which act as free oxygen radical scavengers. They contain transition metal ion a redox-active like Mn^+, Cu^+, Fe^+, Zn^+,and so forth, as discovered in 1968 and 1969 by McCord and Fridvick.[7]

There are four types of modern-day superoxide families or SODs namely Cu-SOD, Zn-SOD. EC SOD (extracellular SOD) like Cu-SOD, Zn-SOD, Mn-SOD, and Fe-SOD, which is reported in microorganisms.

The function of SOD is primarily to eliminate rapidly the biologically generated free-radical species, the superoxide anion (O_2).

The metabolism of molecular oxygen leads to the production of the superoxide anion radical and this free radical should be eliminated as quickly as possible owing to its toxicity to the aerobic organisms. In the field of therapy and prevention of the majority of diseases associated with inflammation or blood flow disorders, the key role of free radicals and their underlying mechanisms have been envisaged and therefore the developments of antioxidants and antifree radicals are being extensively focused for their use in the biological systems to prevent such toxic, deleterious, and harmful reactions.

Although in almost all the organs of the body, namely, the central nervous system, heart, liver, and pancreas, lungs, gastrointestinal tract, the generations of free radicals and superoxide have been demonstrated in various experimental models, the exact role and mode of action of SOD to protect in particular models are not still very clear and need to be established.

15.4 MEASUREMENT OF ANTIOXIDANT ACTIVITY

For assessment of antioxidant activity, several in-vitro models are available and used followed by animal studies necessary. The chemistry behind the design of antioxidant involves ready donation of electrons to reactive oxygen species (ROS) by antioxidants, which then quenches the ROS the ROS and produces more stable and thus less damaging species.[8] Available in vitro models can be divided into two major categories;

1. Measuring the ability of the antioxidant molecules to donate an electron or hydrogen atom to a specific ROS or to an electron acceptor.
2. Testing ability to remove any source of oxidative initiation, like inhibition of enzymes, chelation of transition of metal ions, and absorption of UV radiation.

Various in vitro models are used to generate the free radicals and the ability of an antioxidant to reduce of disproportionate the radical is measured.

15.4.1 DPPH METHOD (1,1 DIPHENYL-2-PICRYL HYDRAZYL RADICAL)

The DPPH scavenging method is one of the widely used methods for the evaluation of antioxidant activity of the new chemical entity. This assay method is based upon the concept of the reduction of the methanolic solution of 1,1 diphenyl-2-picryl hydrazyl radical by free-radical scavenger observed by a transformation in the solution color. This reduction is monitored by UV absorbance spectroscopy. A reduction in the absorbance of the 1,1 diphenyl-2-picryl hydrazyl radical solution at its concentration of free radical scavenger added to the DPPH reagent solution is indicative of antioxidant activity. The antioxidant activity is largely expressed as EC_{50}.[8]

The percent inhibition activity was calculated according to the formula

$$\% \text{ Inhibition} = \frac{A - A'}{A} \times 100$$

A = Absorbance of Control, A′ = Absorbance of Test

15.4.2 SUPEROXIDE RADICAL SCAVENGING ACTIVITY

In vitro superoxide radical scavenging activity is measured by riboflavin/light/nitro blue tetrazolium (NBT) reduction. Reduction of NBT is the most popular method, which is used for antioxidant activity measurement. The method involves the first generation of superoxide radical by oxidation of riboflavin using of light. The superoxide radical reduces NBT to a blue-colored formazan that can be measured at 560 mm spectrophotometrically.[9] The capacity of solutions of new chemical entities to inhibit the color to 50% is measured in terms of EC_{50}.

15.4.3 HYDROXYL RADICAL SCAVENGING ACTIVITY

This process entails production of hydroxyl radicals, in vitro employing Iron (III)/ascorbate/ethylenediaminetetraacetic acid/hydrogen peroxide system radical in the presence of the antioxidant compound is measured. By the process of oxidation, the OH-radicals are produced, which is further made to facilitate a chemical reaction with dimethyl sulfoxide to yield specific yellow color with the Nash reagent (2M ammonium acetate in distilled water). The color intensity was measured spectrophotometrically at λmax 412 nm against the reagent blank. The antioxidant activity of this assay is articulated as % hydroxyl radical scavenging.[9]

15.4.4 NITRIC OXIDE RADICAL INHIBITION ACTIVITY

The nitric oxide is specifically a free radical that reacts rapidly with other free radicals and different types of protein. This method for screening antioxidant potential arise from the ability of compounds to inhibit the nitric oxide radical generated from sodium nitroprusside and is measured using Griess reagent where the reduced absorbance of the system is evaluated colorimetrically at 546 nm. The antioxidant activity of the compound is expressed as the % nitric oxide reduction.[10]

15.4.5 DMPD (N, N-DIMETHYL-P-PHENYLENEDIAMINE DIHYDROCHLORIDE) MODEL

The assay method is derived from the fact that the buffered solution of colored DMPD gets reduced by ferric chloride in acetate buffer. The reduction in the absorbance of DMPD was measured spectrophotometrically at λ_{max} 505 nm against the reagent blank.[11]

15.4.6 OXYGEN RADICAL ABSORBANCE CAPACITY MODEL

For testing the antioxidant potentials of vitamins, chemical substances, nutritional supplements, and foods, oxygen radical absorbance capacity (ORAC) test tube analysis model is used extensively. The test involves computing the efficacy of a product in potentially scavenging the free radicals. The test is conducted by utilizing a standard water-soluble vitamin E analog, known as Trolox. The antioxidant potential is determined as equivalent to Trolox and stated as ORAC value. A low value indicated a low reducing potential of the compound and vice versa.[12]

15.4.7 β-CAROTENE LINOLEATE MODEL

This is novel method is generally used for testing the antioxidant activity of constituents, which are known to be found in grapes and wines. This is a very fast method, based on the principle that the ROS, produced by the oxygenated water oxidizes the unsaturated fatty acid, Linoleic acid that results in the formation of a discolored product. The extent of decrease in the discoloration under the influence of antioxidants is measured spectrophotometrically at λ_{max} 434 nm.[13]

15.4.8 TOTAL RADICAL-TRAPPING METHOD

This method involves total radical-trapping by using *R*-phycoerythrin, a fluorescent compound. Florescence of *R*-phycoerythrin is quenched by 2,2′-azo-bis(2-amidino-propane)hydrochloride (ABAP). The antioxidant parameter is determined from the degree of quenching in the presence of antioxidants. The free-radical quenching potential is tested by estimating the time required for decoloration.[14]

15.4.9 FERRIC REDUCING ABILITY OF PLASMA METHOD

One of the most rapid antioxidant tests include ferric reducing ability of plasma (FRAP) that is employed in pharmaceutical drug discovery field for the routine analysis. The antioxidative potential of the compound is determined by quantifying the enhancement in the absorbance as a result of ferrous ions formation from the reagent, which contains TPTZ (2,4,6-tri-(pyridyl)-s-triazine) and $FeCl_3.6H_2O$. The absorbance was measured spectrophotometrically at λ_{max} 595 nm against the reagent blank.[14]

15.5 CONCLUSION

Antioxidant activity offered by several natural product classes like flavones, flavanones, flavonoids, flavonols, chalcones, proanthocyanidins, and anthocyanidins, aurones, and rationally developed synthetic compounds like butyrated hydroxylated molecules are important for the survival of human being in today's life by preventing the biomolecular damage (DNA, membrane, etc.) offered by the superoxide anion, alkyl, hydroxyl, nitric oxide, and peroxy radicals. The mentioned screening methods will hold good for the rapid screening of low molecular weight ligands.

KEYWORDS

- **antioxidant agents**
- **free-radical**
- **scavenger**
- **superoxide dismutase**
- **methods**
- **models**

REFERENCES

1. Halliwell, B.; Gutteridge, J. M. C. *Free Radicals in Biology and Medicine*, 2nd ed.; Clarendon Press-Oxford, 1989; p 543.
2. Fullbert, J. C.; Cals, M. J. Free Radicals in Clinical Biology. Origin, Pathogenic Effect and Defence Mechanisms. *Pathol. Biol.* **1991,** *40,* 66–77.
3. Delattre, J.; Bonnefront, R. D. Potential Role of the Interaction between Equine Estrogens, Low-Density Lipoprotein (LDL) and High-Density Lipoprotein (HDL) in the Prevention of Coronary Heart and Neurodegenerative Diseases-in Postmenopausal Women. *Biotech. Lab. Int.* **1992,** *3,* 21–23.

4. Ceballos, P. I.; Trivier, J. M.; Nicole, A.; Sinet, P. M.; Thevenin, M. *Clin. Chem.* **1992,** *38*, 66–70.
5. Ying, W. *Med. Hypoth.* **1997,** *48*, 143–148.
6. Cademas, E. Biochemistry of Oxygen Toxicity. *Annu. Rev. Bio. Chem.* **1989,** *58*, 79–110.
7. McCord, J. M.; Fridovich, I. L. Superoxide Dismutase. An Enzymic Function for Erythrocuprein (Hemocuprein). *J. Biol. Chem.* **1968,** *243*, 5753–5760.
8. Vani, T.; Rajani, M.; Sarkar, S.; Shishoo, C. J. Antioxidant Properties of the Ayurvedic Formulation Triphala and Its Constituents. *Inter. J. Pharmacog.* **1997,** *35*, 313–317.
9. Robak, J.; Gryglewski, R. J. Flavonoids Are Scavengers of Superoxide Anions. *Biochem. Pharmacol.* **1998,** *37*, 837–841.
10. Babu, B. H.; Shylesh, B. S.; Pandikkala, J. Antioxidant and Hepatoprotective Effect of Acanthus Ilicifolius. *Fitoterapia* **2001,** *72*, 272–277.
11. Kanner, J. Natural Antioxidants in Grapes and Wines. *J. Agri. Food Chem.* **1994,** *42*, 64–69.
12. Cao, G.; Alessio, H. M.; Culter, R. G. Oxygen Radical Absorbance Capacity Assay for Antioxidants. *Free Radic. Biol. Med.* **1993,** *14*, 303–311.
13. Ghiselli, A.; Serafini, M.; Maiani, G.; Azzini, E.; Ferro-Luzzi, A. A. fluorescence-Based Method for Measuring Total Plasma Antioxidant Capability. *Free Radic. Biol. Med.* **1995,** *18*, 29–36.
14. Benzie, I. F. F.; Strain, J. J. The Ferric Reducing Ability of Plasma (FRAP) as a Measure of "Antioxidant Power". *Anal. Biochem.* **1996,** *239*, 70–76.

Index

H

L

P

R

S